D0604464

RADIOGRAPHIC ARTIFACTS

RADIOGRAPHIC ARTIFACTS: THEIR CAUSE AND CONTROL

RICHARD J. SWEENEY, R.T., F.A.S.R.T.

Educational Director
School of Radiologic Technology
Albany Medical Center Hospital
Albany, New York

J.B. LIPPINCOTT COMPANY

Philadelphia

London Mexico City New York St. Louis São Paulo Sydney

Acquisitions Editor: Lisa A. Biello
Sponsoring Editor: Sanford J. Robinson
Manuscript Editor: Lee Henderson
Indexer: Julie Schwager
Art Director: Maria S. Karkucinski
Designer: Adrianne Onderdonk Dudden
Production Supervisor: N. Carol Kerr
Production Assistant: S.M. Gassaway
Compositor: York Graphic Services, Inc.
Printer/Binder: Halliday Lithograph

1 3 5 6 4 2

Library of Congress Cataloging in Publication Data

Sweeney, Richard J.
Radiographic artifacts.

Bibliography: p.
Includes index.
1. Diagnosis, Radioscopic—Quality control.
I. Title. [DNLM: 1. Radiography—Atlases. WN 17 S974r]
RC78.S93 1983 616.07'572 82-18637
ISBN 0-397-50554-X

This book is lovingly dedicated to my mother

FOREWORD

As the author points out, using one of Röntgen's earliest radiographs as an illustration, artifactual images have been present since the very first days of radiography; and with the increasing capabilities and complexities of the x-ray machine in the last 80 years, artifacts, instead of disappearing as we all would prefer, still remain. Indeed, they have actually become more numerous, intricate, sometimes difficult to recognize, and hard to understand.

Richard Sweeney has had long and wide experience in both the practice and teaching of radiologic technology, which has afforded him a wealth of material for this volume. Of more importance than time and material is the fact that "quality control" in radiologic technology has always been his chief interest. He is thus "a natural" to present this timely study of image artifacts.

To my knowledge there is in the literature of the radiologic technologist and the radiologist, no work other than this book in which the myriad of artifacts are so well cataloged, illustrated, and explained. It should serve as a marvelous aid to the technologist in improving his technical skills and to the radiologist in understanding and interpreting the images presented to him for his study. Of course, the most important person in all health-care procedures, the patient, should derive significant benefit in the form of diminished exposure to ionizing radiation and more accurate radiologic diagnosis.

It is, therefore, with great pleasure that I heartily recommend this book to the student and graduate technologist as well as the radiology resident and the radiologist himself.

John Faunce Roach, M.D.
Professor of Radiology
The Albany Medical College
of Union University
Albany, New York

PREFACE

Radiographic Artifacts: Their Cause and Control is designed to present an extensive account of the various types of radiographic artifacts that the technologist and radiologist might encounter. Suggestions for detecting their cause and methods of control are included.

Because radiologic technology is a highly visual discipline, the book is arranged in a modified atlas form, based on the premise that a photographic reproduction of the artifact is far superior to a textual description. References are included where the subject is still controversial or where I felt documentation to be necessary.

In some instances, particularly in cases in which the appearance of different artifacts is very similar, the reader is requested to refer to various illustrations for comparative purposes. In this way I have attempted to prepare the reader for more difficult problems on subsequent pages.

Many of the conclusions in *Radiographic Artifacts* are based largely on my personal experience, as well as on the published works of others. Further experience may prove some of these concepts incorrect. All are, therefore, considered subject to future modification or elimination.

The main stimulus for this book was T.E. Keats's *An Atlas of Normal Roentgen Variants That May Simulate Disease* (Chicago, Year Book Medical Publishers, 1973). The illustrations in Keats's book demonstrate that a great deal of knowledge and experience are required in order to interpret the radiographic image. Although this responsibility lies within the domain of the radiologist, I have always felt that the technologist not only should be responsible for the production of the radiograph, he should also be able to make a technical diagnosis of various imaging problems whenever they occur. As a result I started to collect and catalog numerous artifacts, a process that has led to the development of this text. Although this work cannot be compared to that compiled by Keats, it is my hope that *Radiographic Artifacts* will be beneficial to both the radiologist and the technologist who frequently encounters imaging problems.

R.J.S.

ACKNOWLEDGMENTS

The material in *Radiographic Artifacts: Their Cause and Control* represents what I believe to be a fairly comprehensive review of the literature dealing with radiographic artifacts. It would not have been possible without the efforts of a great number of people. I therefore express thanks to all authors, publishers, and equipment manufacturers who not only furnished information but also gave their permission to reproduce many of the illustrations in the text. I am especially indebted to the following individuals, who unselfishly allowed me access to their teaching files: Paul O'Connor, M.D.; Ralph W. Coates, R.T.; William A. Conklin, R.T., F.A.S.R.T.; John C. Cullinan, R.T., F.A.S.R.T.; Lyn Gill, R.T.; and Robert A. Short, R.T.

Sincere thanks to the many commercial firms that furnished technical information and many of the photographs in the book.

A special mention is due to Sharon Ellis for her excellent illustrations and to Gary W. Howe of the Department of Medical Illustrations at Albany Medical College for reproducing the photographs. Gary not only did a splendid job in this respect but also provided useful comments that helped to provide a framework for the book.

I wish to acknowledge the editorial and production assistance of the J.B. Lippincott Company. A particular note of gratitude goes to Lisa Biello and Lee Henderson, whose advice and encouragement contributed significantly to the completion of the manuscript.

To my children, Marybeth, Colleen, and Kerry Ann, for their understanding while their father was involved in the preparation of this text. I hope they will be proud of my efforts.

And finally, to a very special person, Gunta Av[illegible]. I am most appreciative of her faith in my ability and her continuous encouragement, which was responsible for helping me complete this task.

CONTENTS

RADIOGRAPHIC ARTIFACTS

1 INTRODUCTION TO THE DETECTION OF ARTIFACTS

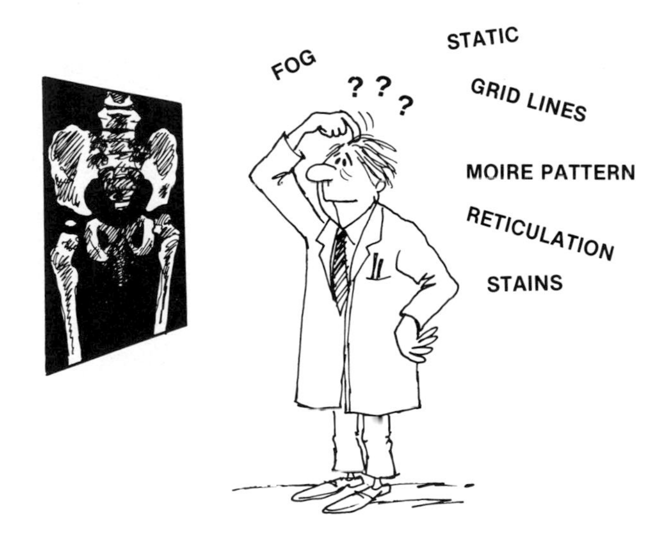

Since the early days of radiography, every effort has been directed toward improving the quality of the radiographic image. Standardization of the technical procedure and refinements in equipment and accessories have done much to diminish the problems associated with the production of high-quality radiographs. These accomplishments did not come easily to our profession. Radiographic equipment, like radiographic technique, has evolved from its beginning up to the present time—and the end is not in sight.

In order to appreciate fully and utilize the many new and unique methods of enhancing the quality of the radiographic image, the technologist of today must possess a greater knowledge of radiographic technique than ever before. This advancement has been accomplished through the steady improvement in the general curriculum for radiologic technologists, as well as through expansion of the curriculum to include such areas as computed tomography, ultrasound, and vascular imaging.

If you consider the progress made in our profession over the years, however, you have to question why retake rates of 10% or more are being reported in this country. The problem will never be totally eliminated until we as technologists are willing to accept the responsibility for correcting these problems in our own radiology departments.

The first step in reducing film waste is to develop a quality-control program in order to provide a method whereby the various conditions responsible for retakes can be identified and corrected. This implies that quality control involves a great deal more than the daily sensitometric evaluation of the film processor and its chemistry.

I have seen a number of logbooks in radiology departments that contain *daily* sensitometric graphs of the evaluation of the processor. But the yearly film-waste analysis in these departments is well above the national average. Although the importance of monitoring the processor on a daily basis cannot be overemphasized, the individual responsible for quality control must evaluate all of the aspects of producing a radiograph in order to minimize the number of retakes. Moreover, he must keep the staff well informed, not only of the problems that he encounters, but also of the progress that is being made in correcting them. This can be accomplished by conducting informal sessions or in-service educational meetings.

Classification and Prevention of Artifacts

Regardless of the attention given to reducing the incidence of radiographic artifacts, we may suddenly encounter an artifact for which identification is most difficult, if not impossible. The recognition and classification of artifacts is not an easy task. Many of the artifacts depicted in this text could have been the result of any of a number of conditions. Consider, for example, Figures 1-1 through 1-4. In the first two cases the radiographs have been cropped in order to demonstrate that even the most obvious artifacts may be nearly impossible to identify under some circumstances, particularly if only a segment of the artifact is visible. Figure 1-3 shows an ordinary object that becomes almost invisible in a tangle of electrical leads. In Figure 1-4 a common object seen at different angles still managed to stump nearly half of the technologists who studied the radiographs.

The problem of identification is further compounded by the fact that the artifact may be brought to our attention long after the radiographic examination has been completed. Because clues are needed in order to determine the probable cause of various artifacts, it is best to investigate each artifact immediately in a systematic manner. In this way the cause can be determined by simple observation of the

Classification of Artifacts by Appearance

Spot	✓				
Streaks					
Clear					
Dark	✓				
Cloudy					
Patchy					

radiographic conditions present at the time of the exposure. This is where established quality-control procedures can be of invaluable assistance by providing a means of detecting and eliminating the conditions responsible for such artifacts.

A useful means of classifying artifacts is according to their appearance on the radiograph—that is, whether there are spots or streaks and whether the image is clear, dark, cloudy, or patchy. In addition, by taking note of the character of the film surface and the degree of image sharpness, most faulty marks can be recognized and correctly interpreted.

Since radiographic artifacts are the cause of needless repeat examinations, it is the responsibility of the technologist to develop a system whereby a greater concern is focused on the prevention of artifacts than on their recognition. This can easily be accomplished if the technologist makes a conscientious effort to keep the imaging and processing equipment in excellent condition.

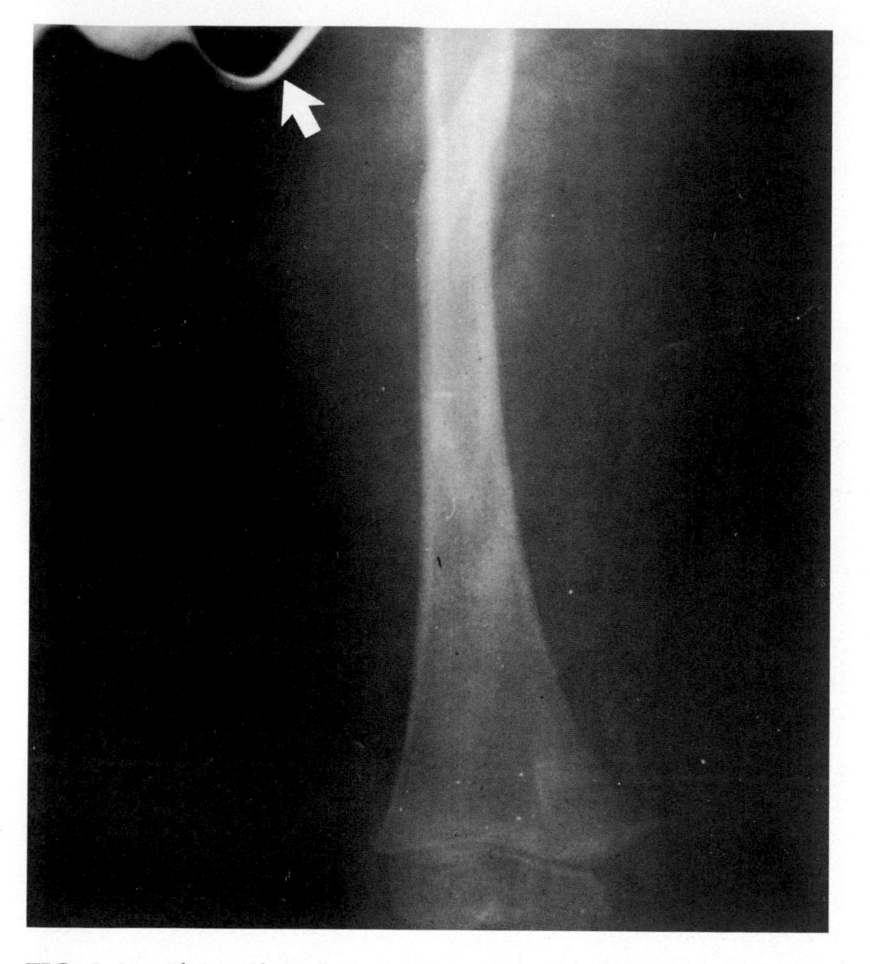

FIG. 1-1. *The artifactual image* (arrow) *in this femoral radiograph of a child is a segment of the total artifact. Can you identify the object responsible for the appearance of the artifact?* *(Answer on p 22)*

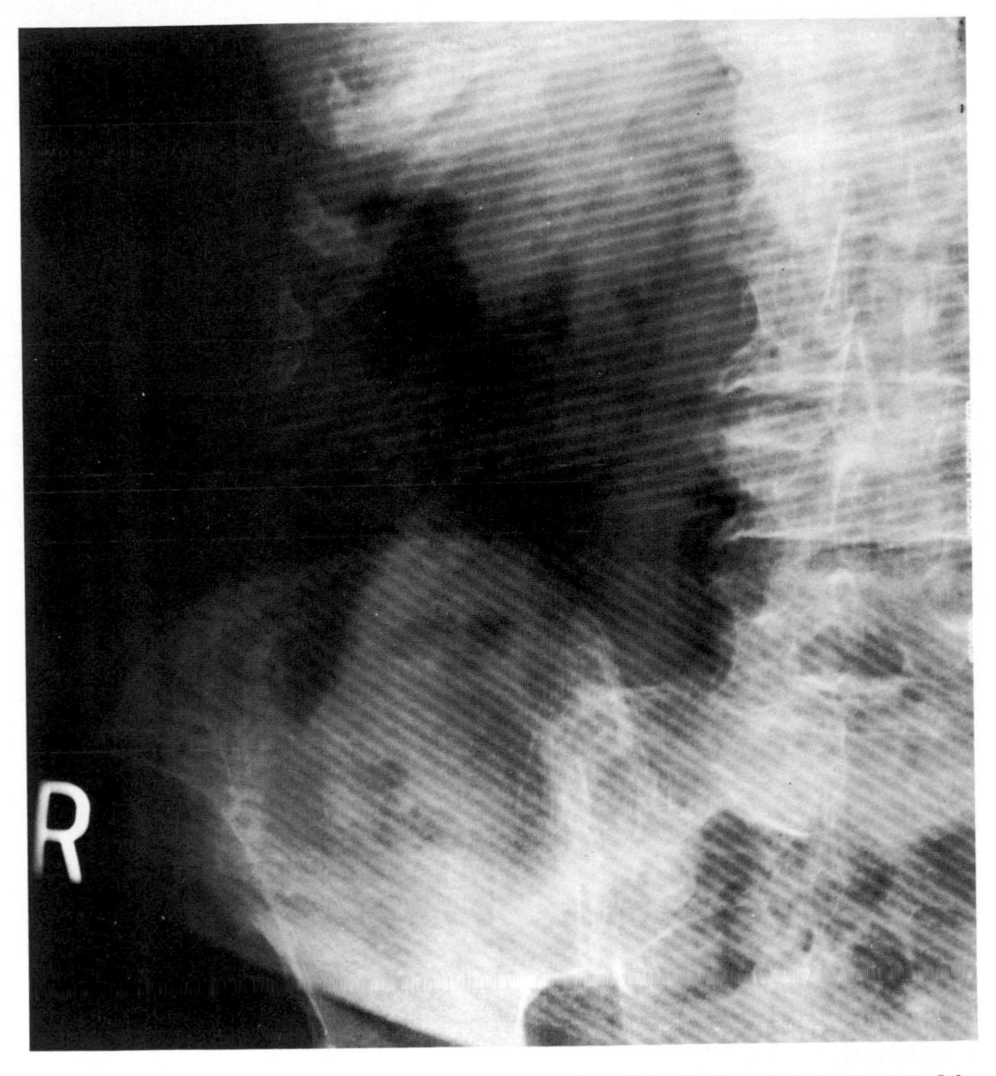

FIG. 1-2. *The wavy pattern in this partial view of an abdominal study is a segment of the total artifact. Identify the artifact. (Answer on p 23)*

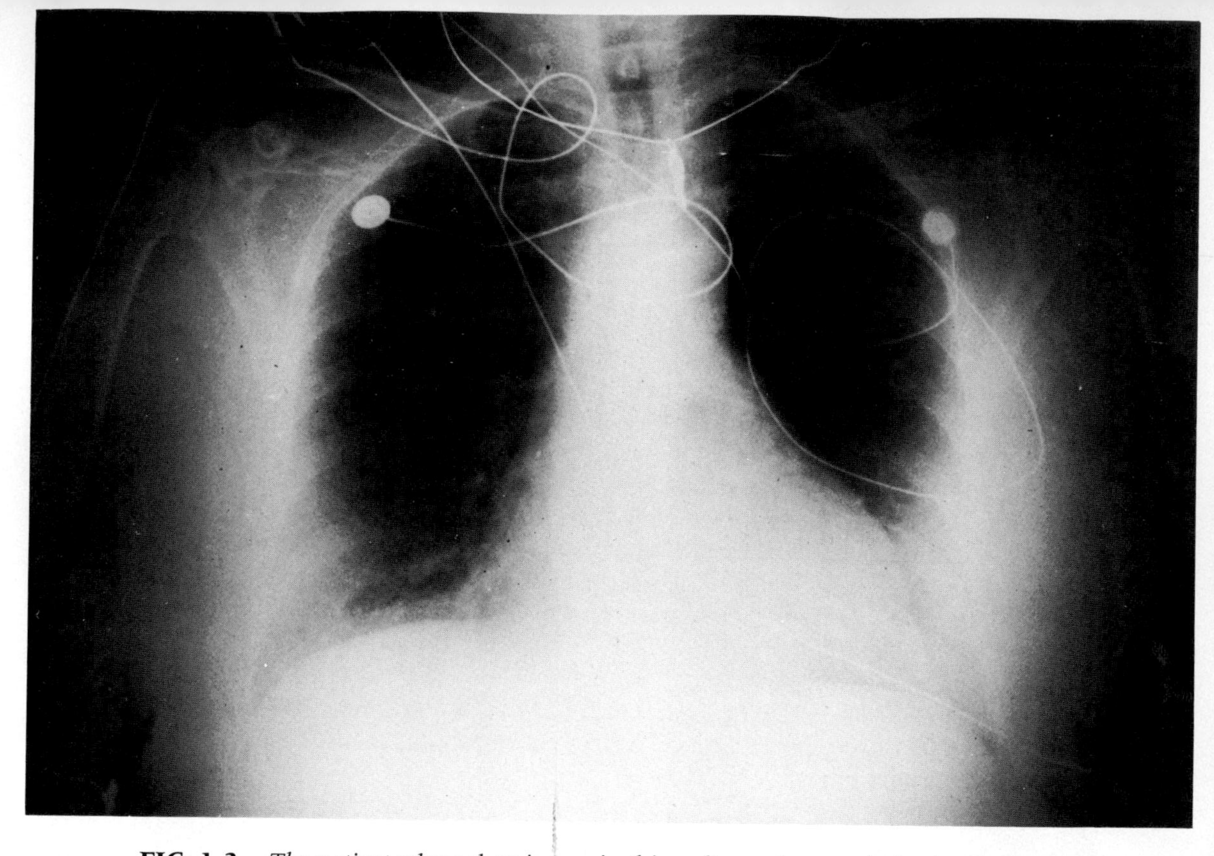

FIG. 1-3. *The patient whose chest is seen in this radiograph was admitted to the hospital following a heart attack. The presence of the wire leads indicates that he is attached to a cardiac monitor. Although the quality of the study is below optimal, an artifact is visible. Identify the artifact.* *(Answer on p 24)*

FIG. 1-4. *Even a common object may be difficult to identify if it happens to be projected axially. Identify the artifact.* *(Answer on p 25)*

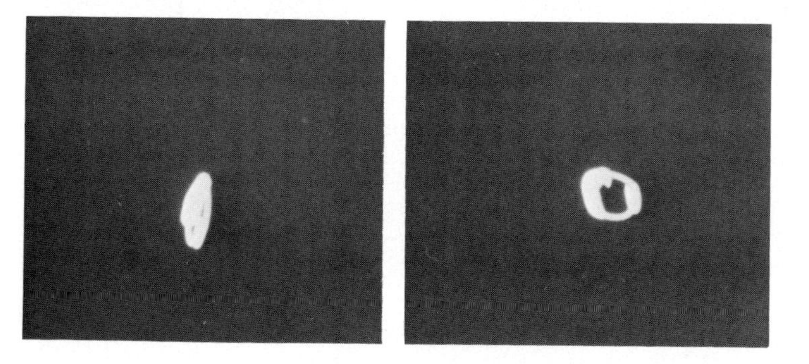

"Film Artifacts"

Unfortunately, the term *film artifact* is used in an all-encompassing manner to describe what is actually a radiographic artifact. Although a radiograph is often referred to as an x-ray, a film, or the like, to many people the term *film artifact* implies that the artifact is within the film itself and occurred during the manufacturing process. Moreover, when an artifact of undetermined origin is encountered, it is often blamed on the manufacturing process. However, one need only tour a film-manufacturing facility to see that such conditions are unlikely to occur, owing to the high level of quality control employed during manufacturing. Following manufacturing, however, the film is shipped, stored, and eventually distributed to the consumer. The film may be damaged in any number of ways at these stages. Consequently, it may be impossible to determine the cause of some film-related artifacts.

Consider, for example, Figure 1-5. The large circular density seen in both radiographs appeared in the same location on at least 30 films from a box containing 100 sheets. The smaller density, which appears only on the left-hand radiograph, resulted from a static discharge. The density of the larger artifact varied in each film and finally disappeared after the technologist ran at least half of the films in the box through the processor. Notice that before the imaging problem was discovered, the large circular density appeared in another radiographic study made with a sheet of film from the same box (Fig. 1-6).

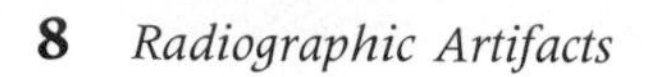

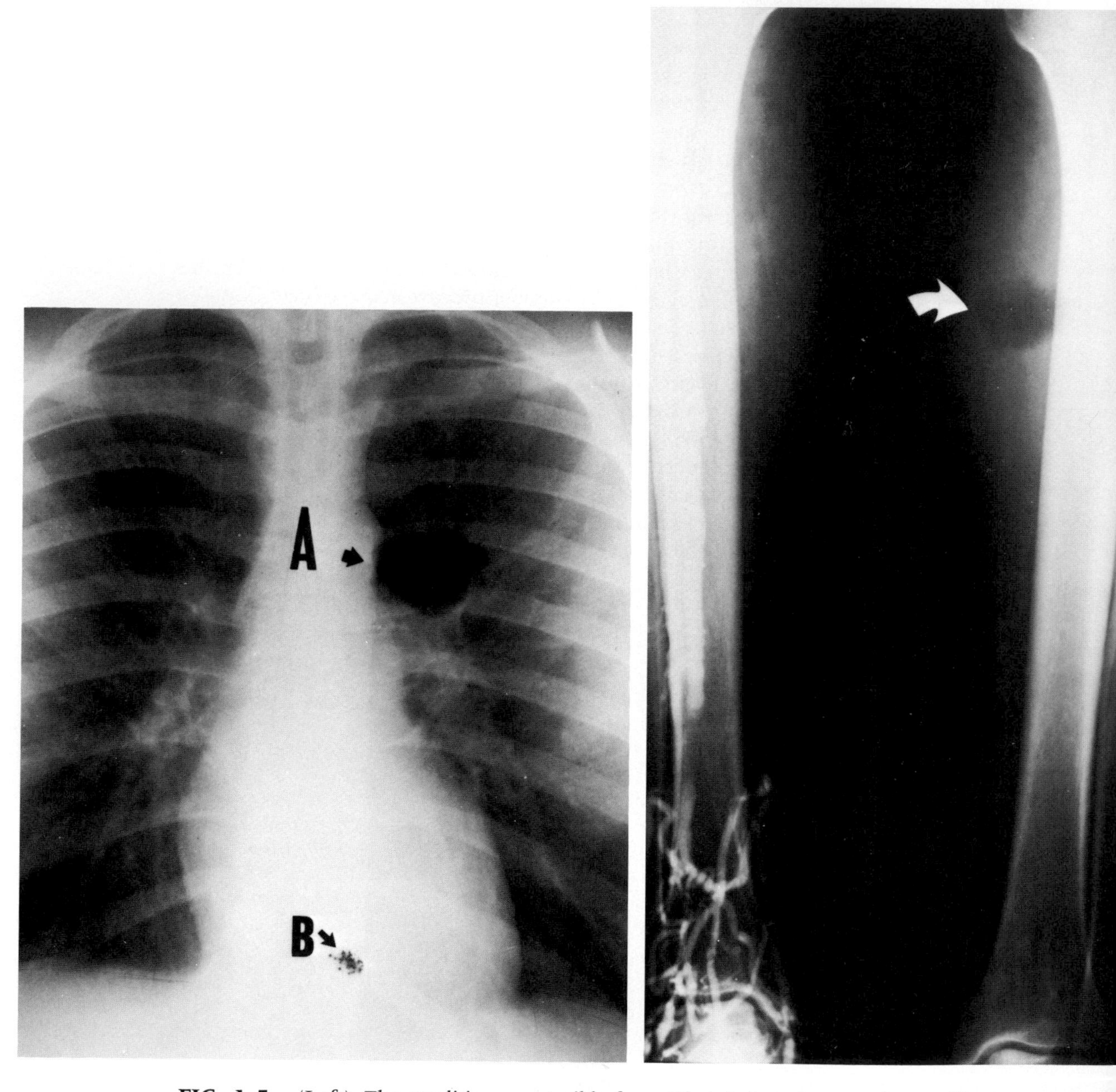

FIG. 1-5. (Left) *The condition responsible for artifact* A *is unknown, but artifact* B *resulted from a static discharge.* (Right) *The mysterious artifact* (arrow) *appears in the same location on a second radiograph. Although the configuration of the artifact is somewhat similar in the second example, it is not of the same density.*

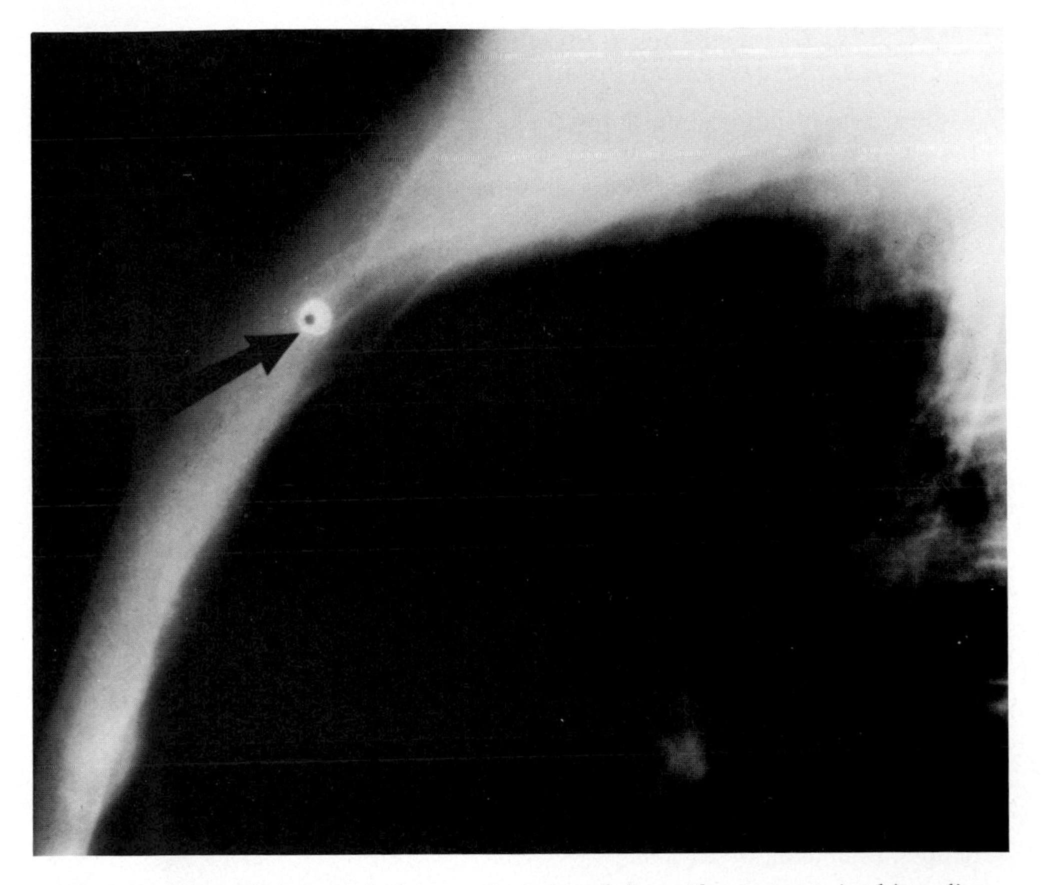

FIG. 1-6. *The well-defined dark central portion of the artifact* (arrow) *in this radiograph indicates that it might have been caused by something attached to the patient's clothing. However, in this instance the patient was properly attired in an examination gown. The condition responsible for the appearance of the artifact was never determined, and the same image appeared in the same location on at least 30 sheets of film from a single box containing 100 sheets of film.*

The Terminology of Image Formation in Radiography and Photography

A great deal of practical and useful information can be obtained by comparing radiography with photography, but the similarity ends when the mechanism responsible in each instance for producing the image is examined. In photography the image is recorded on the film by varying amounts of light reflected from the subject. In radiography the image is produced when varying amounts of radiation are transmitted through an object and onto the film. In each instance, however, the spectral response of the light-sensitive coating on the film is responsible for the resultant density following exposure and processing. This similarity often results in some confusion when the terms *negative* and *positive* are used to describe the images in radiographs and photographs. Figure 1-7, *A*, is a positive image of the negative in Figure 1-7, *B*. In radiography, if the image from the surface of the intensifying screen could be seen during the exposure, it would appear in a positive form, as seen in Figure 1-7, *C*, rather than in the negative form we are accustomed to seeing (Fig. 1-7, *D*).

As we shall see, the negative or positive appearance of a particular artifact may result from actual differences in the absorption of x-rays. Some artifacts, however, are not caused by differential x-ray absorption but by other imaging conditions. Consequently, when an artifact is described as being of *minus density,* the term may not imply that the artifact was produced by a greater absorption of radiation in the object, since a minus density could be the result of dust or dirt particles on the surface of the intensifying screen. Conversely, although the term *plus density* can indicate that the image is dark owing to a greater transmission of x-rays through low-density structures, a plus-density artifact may be the result of a static discharge or rough handling of the film.

FIG. 1-7. *In photography the positive image* (A) *that we are accustomed to seeing is made from a photographic negative* (B). *In radiography the positive image* (C) *that appears on the intensifying screen results in a negative print* (D). ▷

FIG. 1-7. *continues*

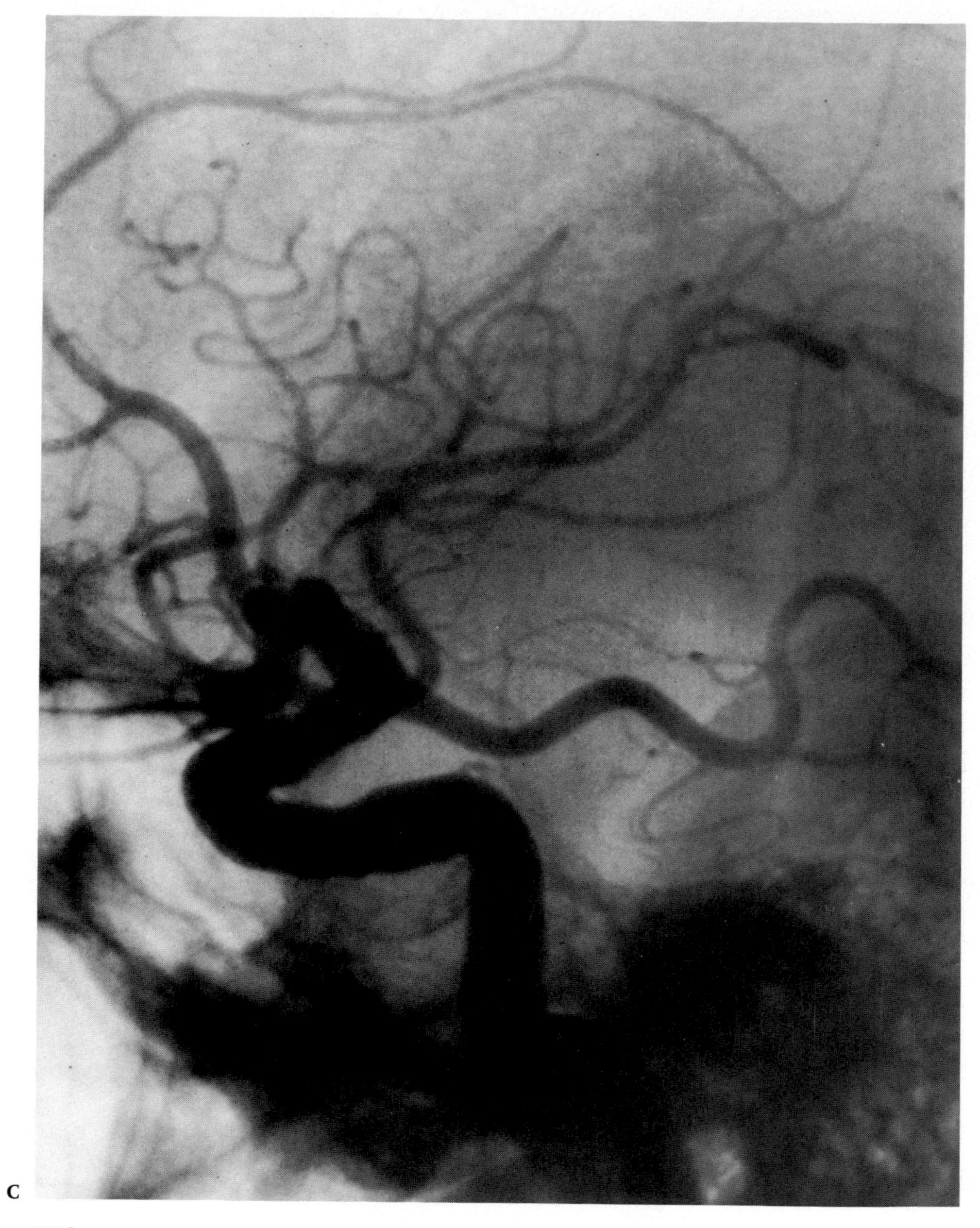

C

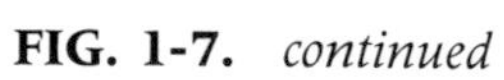

FIG. 1-7. *continued*

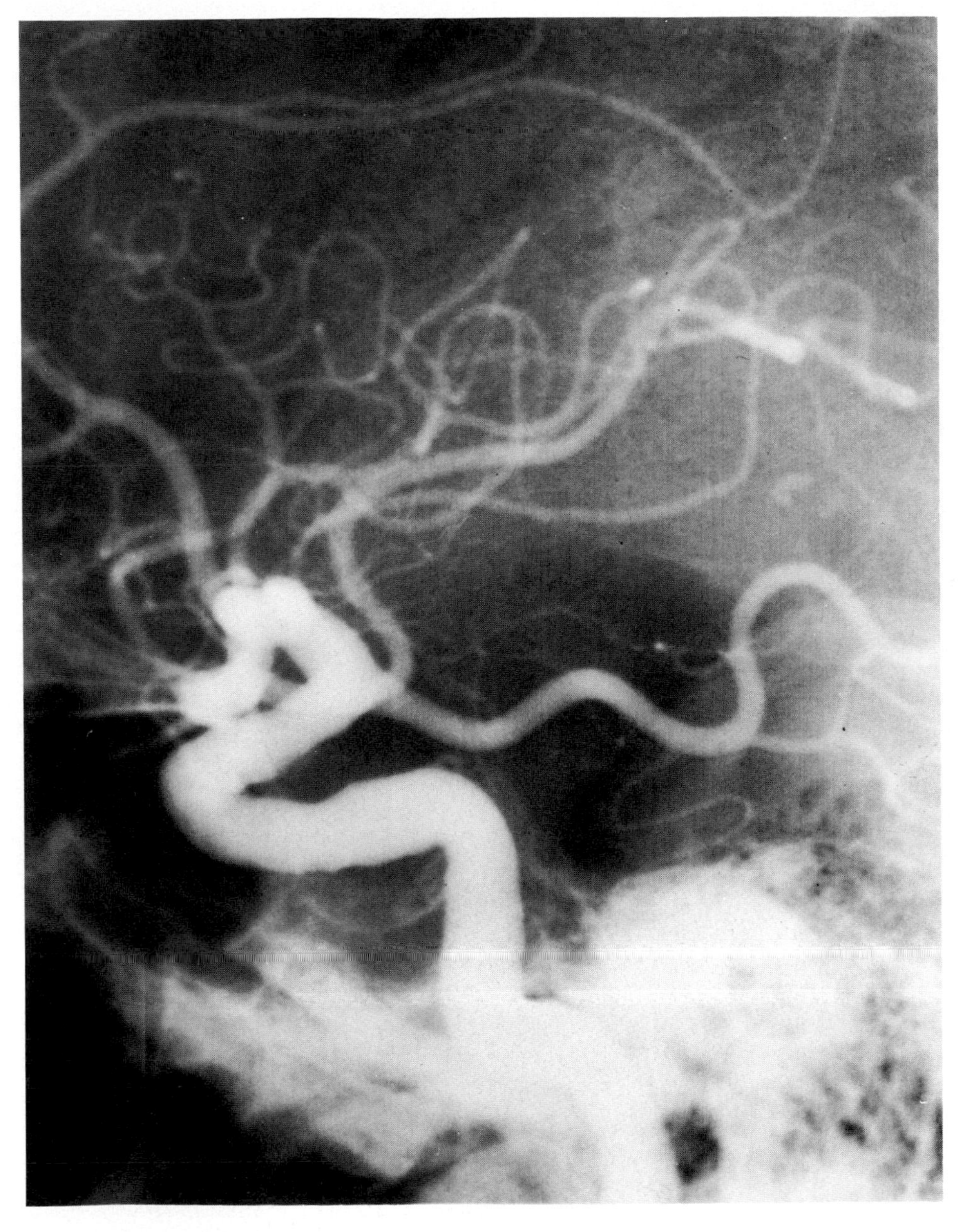

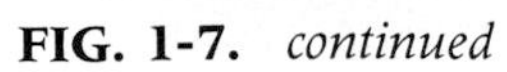

FIG. 1-7. *continued*

Image Analysis

ENHANCING IMAGE QUALITY

By carefully positioning his subject a photographer can create an illusion that may be further enhanced by varying the manner in which the negative is printed and processed. For example, it may be difficult in a photograph to identify a human form posed among the roots of an uprooted tree because of the difference in the density of the illustration (Fig. 1-8). Essentially the same problem can occur in radiography if the exposure is incorrect or if the anatomic region of interest is not parallel to the plane of the film. Notice that in Figure 1-9, *A*, the image of the gallbladder is not well defined owing to its relationship to the film. With the patient in a 45° oblique position, however, the image of the gallbladder is improved, and three large, radiolucent stones are now visible (Fig. 1-9, *B*).

On occasion, the image of a particular artifact can be enhanced by duplicating the radiograph. Using such a technique, you can alter the density and contrast of the copy in order to visualize certain details not seen in the original. In addition, the appearance of an artifact can also be improved by simply reversing the tone of the image—that is, by making a positive print rather than a negative one (Fig. 1-10).

FIG. 1-8. *A photographic negative can be manipulated in processing and printing in order to achieve a desired effect. The roots of an uprooted tree in an underexposed photograph* (A) *suddenly assume human form in a properly exposed photograph* (B). ▷

A

B

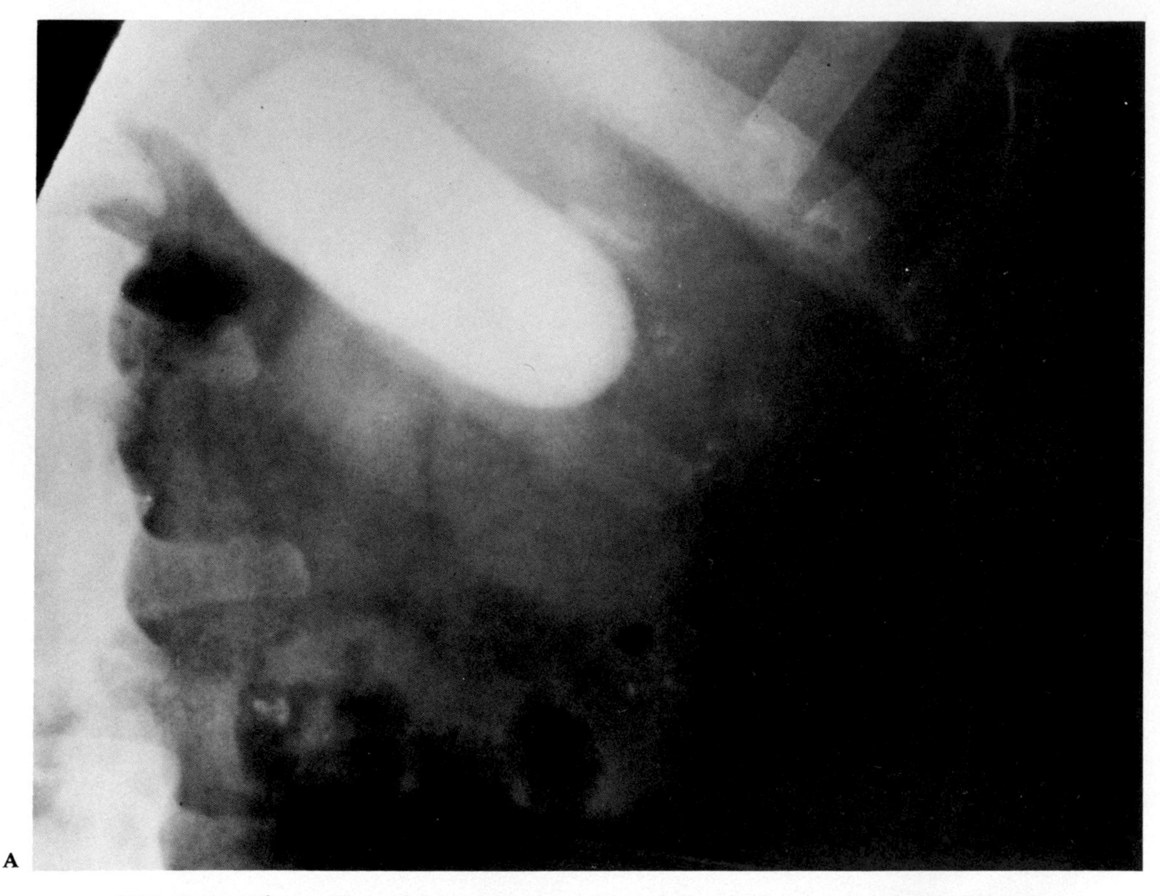

FIG. 1-9. *The position of the anatomic part in relation to the x-ray source and the plane of the film can be responsible for producing a poorly defined image. The lack of a parallel relationship of the gallbladder to the film in* A *prevented the radiologist from seeing three large stones, visible in* B.

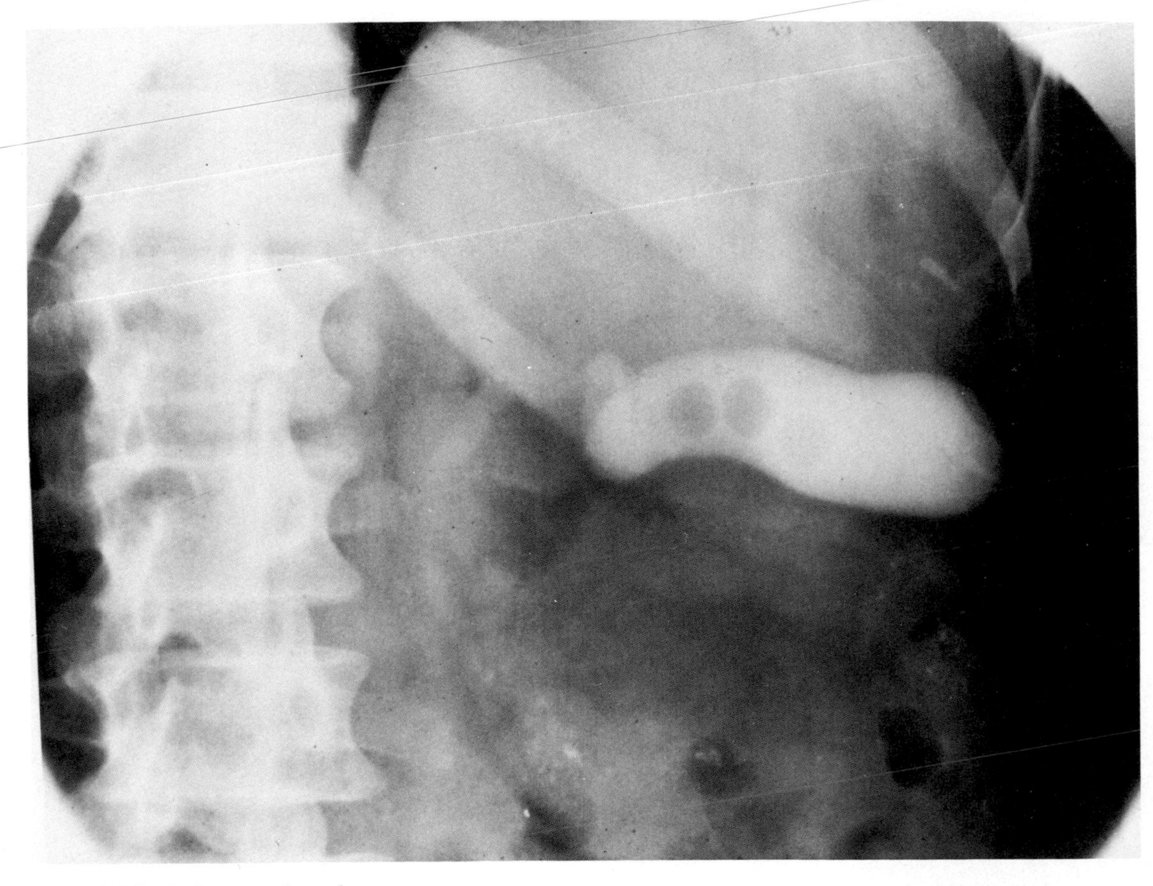

FIG. 1-9. *continued*

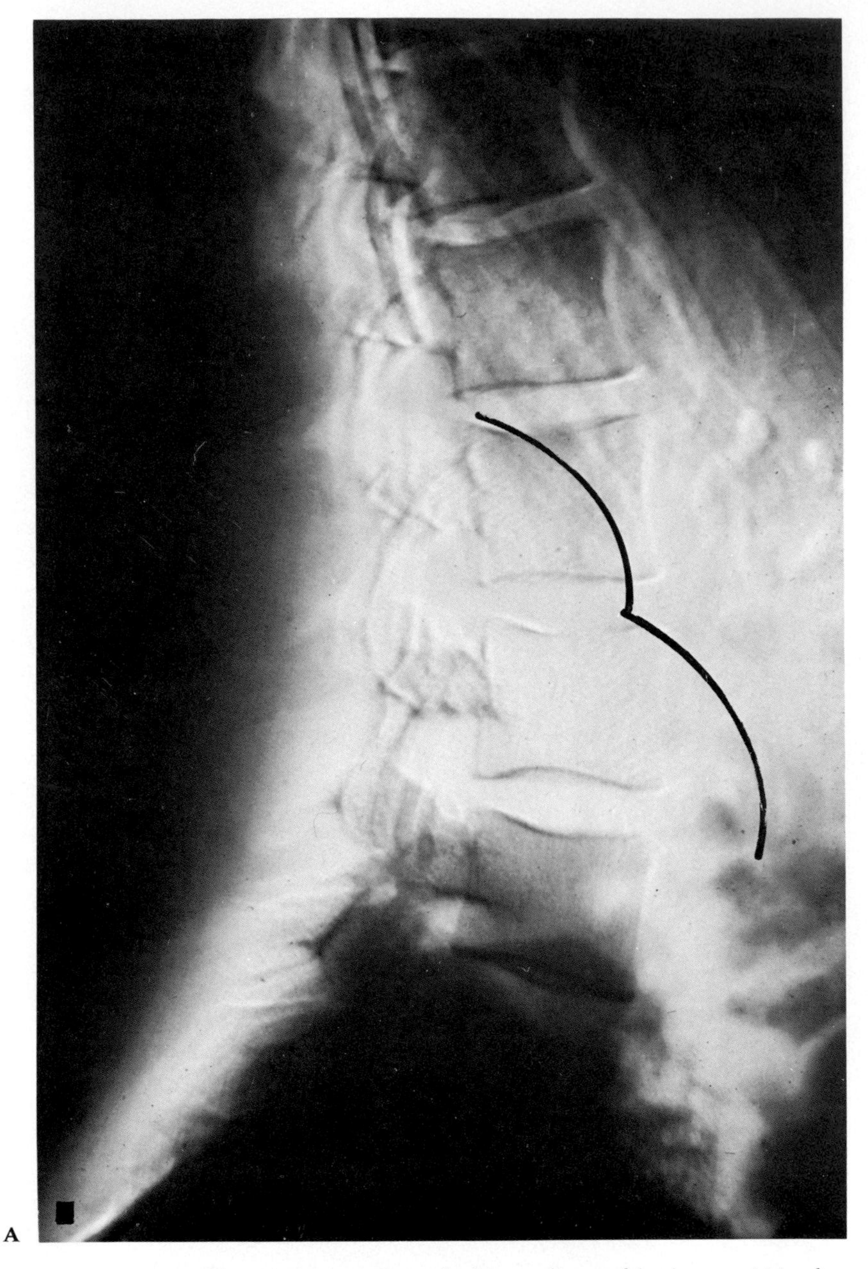

FIG. 1-10. *The positive print of this radiographic image* (A) *demonstrates the outline of the teeth of a large plastic comb* (beneath line) *located in the pocket of the patient's pajamas.* (B) *The image of the comb is barely visible in the original radiograph.*

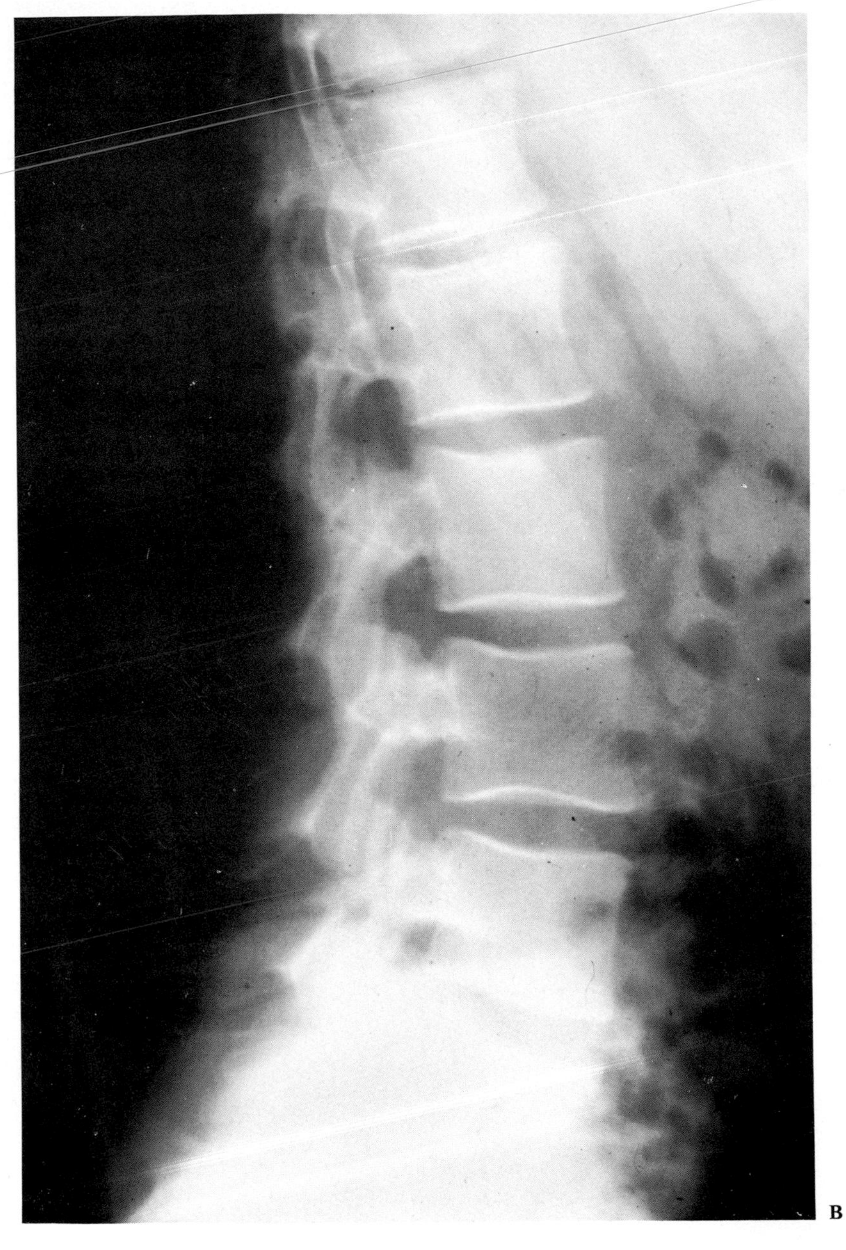

FIG. 1-10. *continued*

Identification of the Radiograph

Although a radiograph can never be over-identified, the improper use of identification can result in various problems if not immediately recognized and corrected. Every radiograph should exhibit the following information: the patient's name, the patient's age, and the hospital or clinic where the radiograph was made. This identification can be accomplished by typing the information on an identification card used in conjunction with a photographic printer. Considering the possibility of human error, the technologist should always check the information on the card for accuracy prior to identifying the films. Although isolated instances involving the improper use of the identification card do occur, most of the major complaints arise from the inaccurate placement of the lead identification marker on the cassette. These errors include identifying the extremity incorrectly or positioning the lead marker in a manner that interferes with the visualization of the anatomic region of interest. In order to eliminate such problems the technologist should follow these recommendations:

Use as much identification as possible but be sure that it is accurate.

Place markers so that they do not obscure the most important area in the radiograph.

Place the marker to correspond to the patient's position.

When doing more than one projection, try to place the marker in the same location on each film.

When doing a series of radiographs in which the time of exposure is important, mark each film accordingly.

When marking a film, do it neatly and accurately, work out your own system, and be consistent.

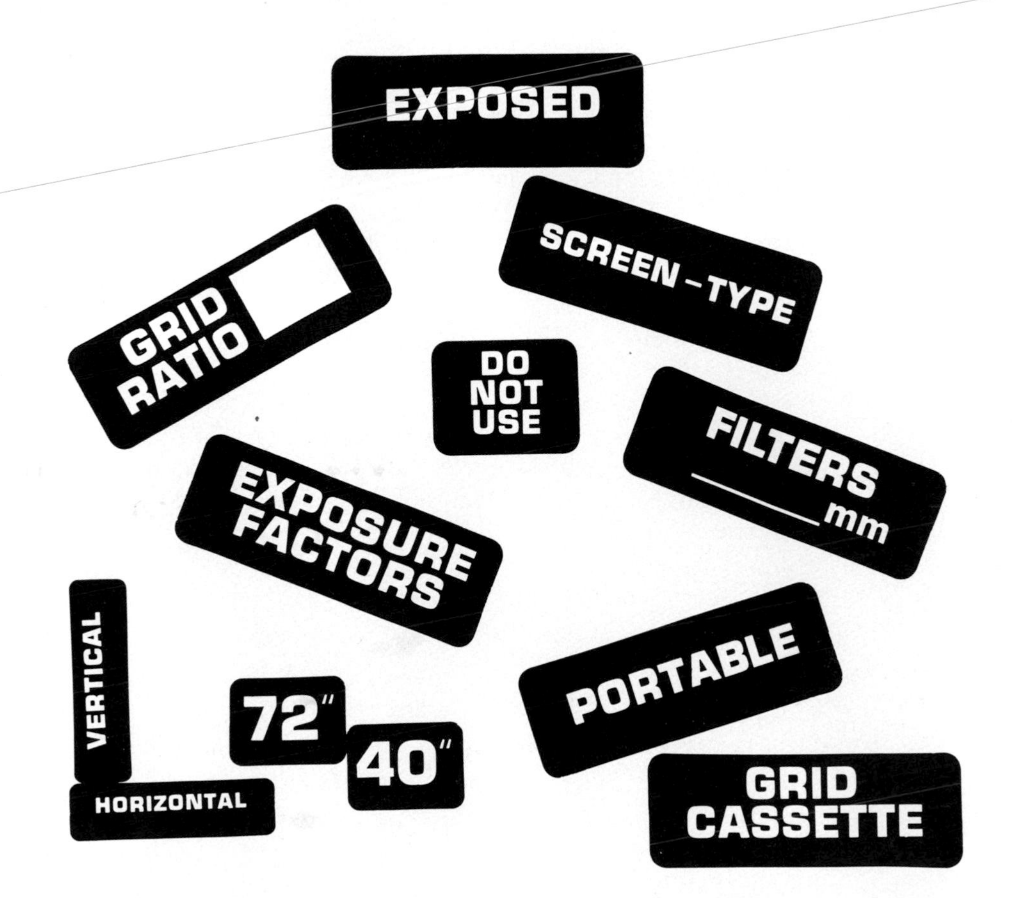
EXPOSED
GRID RATIO
SCREEN-TYPE
DO NOT USE
FILTERS
mm
EXPOSURE FACTORS
VERTICAL
HORIZONTAL
72"
40"
PORTABLE
GRID CASSETTE

Identify the Artifact: Solutions

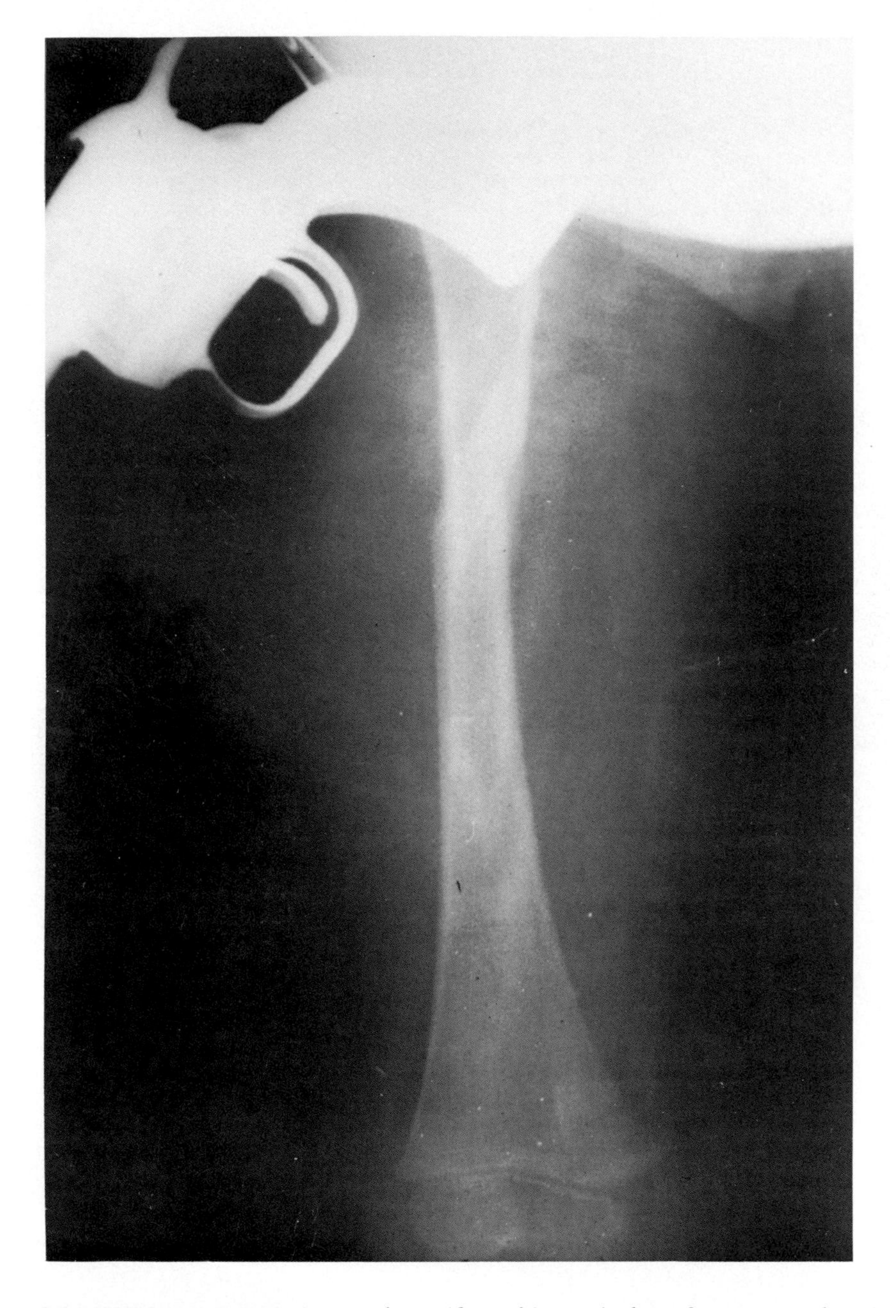

SOLUTION TO FIG. 1-1. *The artifactual image is that of a toy gun, the trigger guard of which is seen in Figure 1-1. The youngster refused to part with the gun each time a bedside radiograph was made in order to evaluate his fractured femur.*

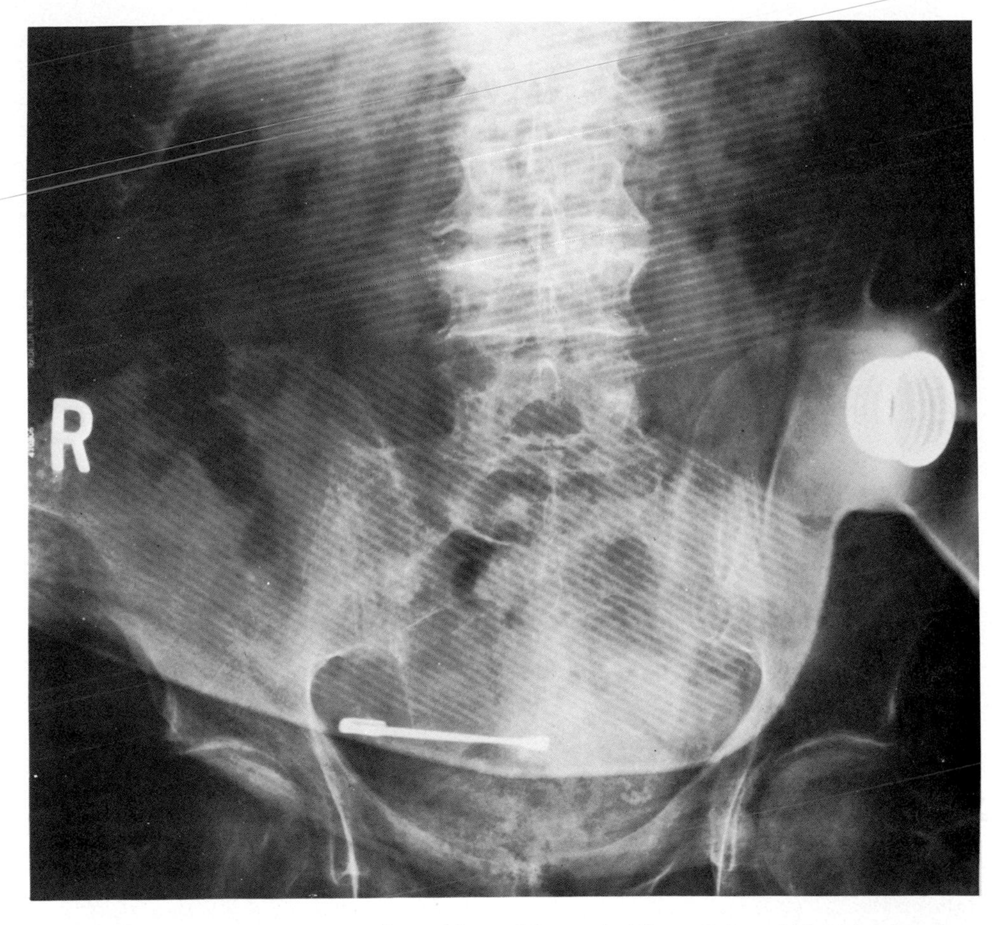

SOLUTION TO FIG. 1-2. *The artifactual image in Figure 1-2 could be mistaken for a grid-related problem until one notices that the lines are running in different directions. If you guessed that this was something on or under the patient, you are correct. The artifact is an image of a hot water bag located on the patient's abdomen. (Courtesy of Ralph Coates, R.T. [R])*

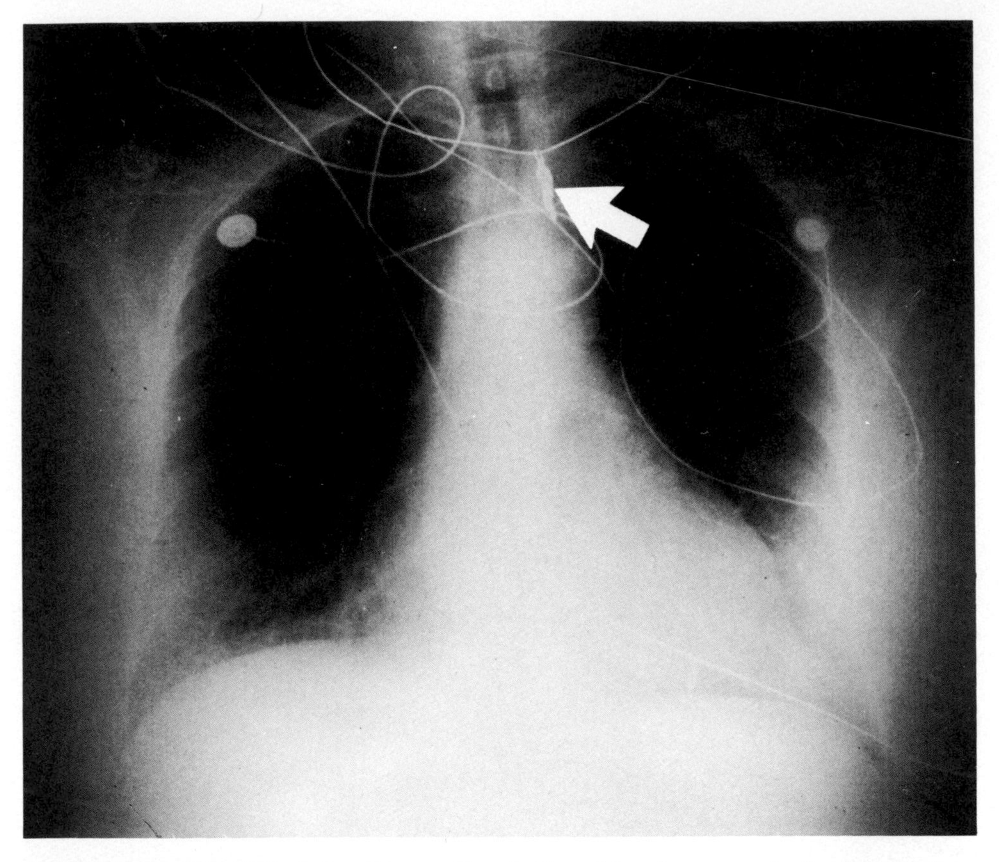

SOLUTION TO FIG. 1-3. *The artifact indicated by the* arrow *is the image of an Italian horn, a type of good-luck charm. The charm is attached to a necklace that appears to be one of the leads of the cardiac monitor.*

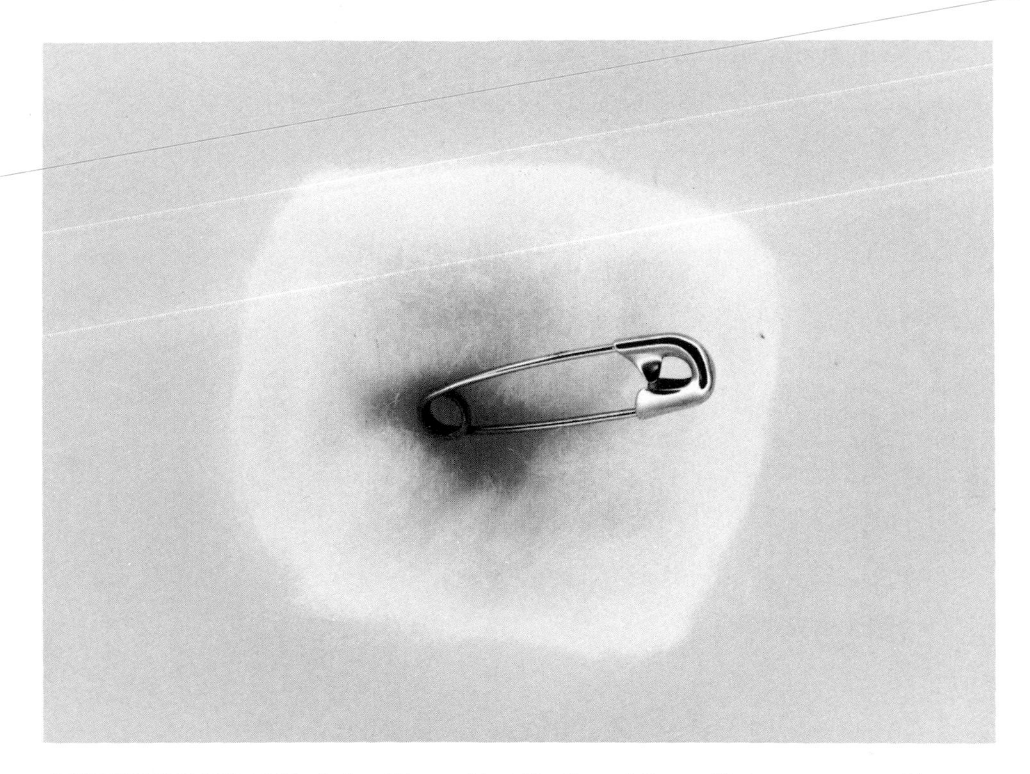

SOLUTION TO FIG. 1-4. *If your identification of the artifact was correct, either you have excellent perception or you have looked at this page before being instructed to do so. The radiographs in Figure 1-4 have been shown to a number of your colleagues, and fewer than half of them identified the images to be those of a safety pin.*

Suggested Reading

Fischer HW: Radiology Departments: Planning, Operation and Management, pp 243–250. Ann Arbor, Edwards Brothers, 1982

2 DETECTION OF ARTIFACTS BASED ON THE GEOMETRY OF THE IMAGE

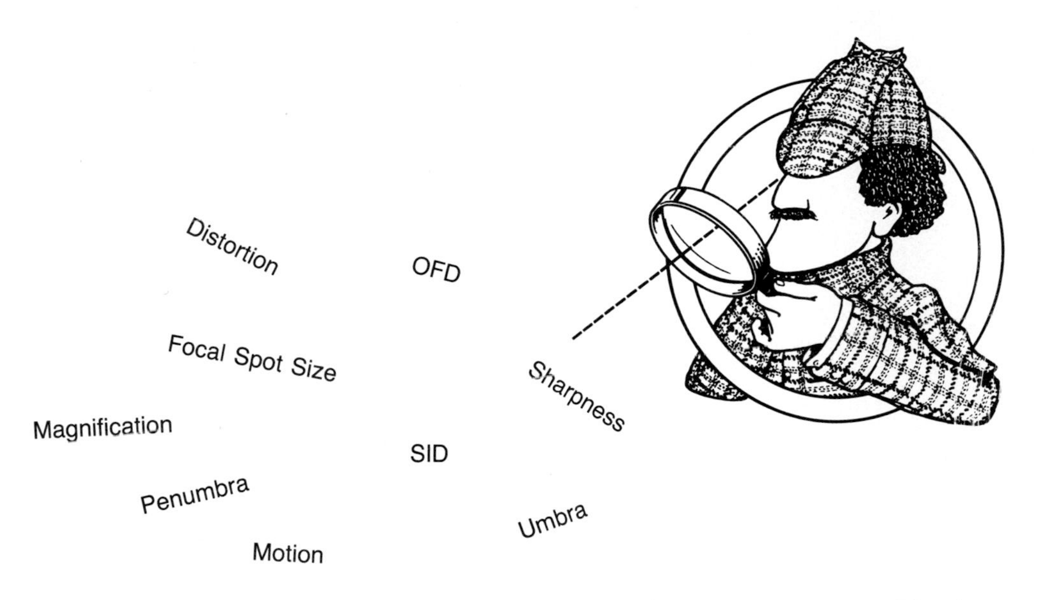

The degree of structural distinctness of the radiographic image is governed by the geometric factors of focal-spot size, object–film distance, and source–image distance, and by motion, intensifying screens, and film. The effect of each of these factors on image sharpness can readily be seen when each factor is manipulated and the results are compared to a standard. As will be demonstrated in the figures that follow, source–image distance, object–film distance, and motion are the primary factors that contribute to the problems associated with the detection of various artifacts. If not controlled, voluntary and involuntary motion produces a blurred image that obscures the visualization of the artifact. Variations in source–image distance and object–film distance influence the amount of image magnification and are responsible for producing a geometrically unsharp image of the artifact in the field of exposure.

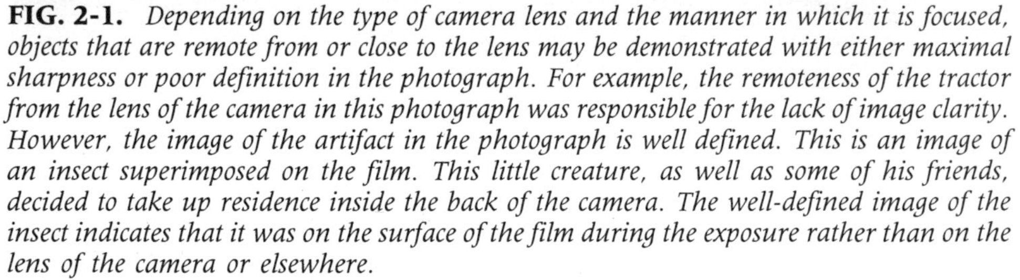

FIG. 2-1. *Depending on the type of camera lens and the manner in which it is focused, objects that are remote from or close to the lens may be demonstrated with either maximal sharpness or poor definition in the photograph. For example, the remoteness of the tractor from the lens of the camera in this photograph was responsible for the lack of image clarity. However, the image of the artifact in the photograph is well defined. This is an image of an insect superimposed on the film. This little creature, as well as some of his friends, decided to take up residence inside the back of the camera. The well-defined image of the insect indicates that it was on the surface of the film during the exposure rather than on the lens of the camera or elsewhere.*

FIG. 2-2. *A rather unsharp, chainlike configuration is superimposed over the distal fibula in this radiograph. The unsharpness of the image suggests that it was remote from the film during the exposure. This is an image of the metal sheath surrounding the high-voltage cable that was under the edge of the collimator. Today, newer designs and methods used to attach the cables to the x-ray tube and column have eliminated such problems. However, with some of the older units, the technologist should periodically evaluate the length of the cable and the manner in which it is attached to the tube column. This is especially important when using long source-image distances or tube-angle techniques, which are responsible for the variation in the cable's position and length during the exposure.*

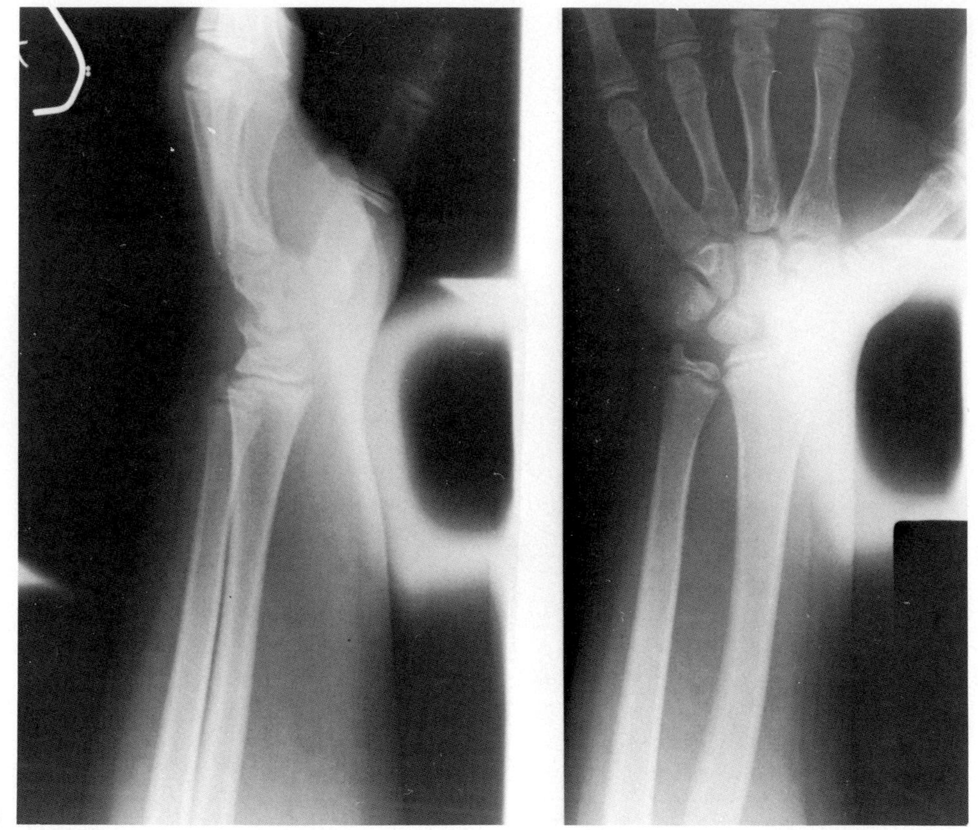

A B

FIG. 2-3. *The unsharp appearance of the artifactual image in* A *and* B *indicates that the source of the image is remote from the film. The artifact is an image of a bolt that became detached from a multiple filtration device and was imaged in the radiograph while located on the plastic window of the collimator. (See Radding MB: A loose screw. Med Radiogr Photogr 32, No. 2:72, 1956)*

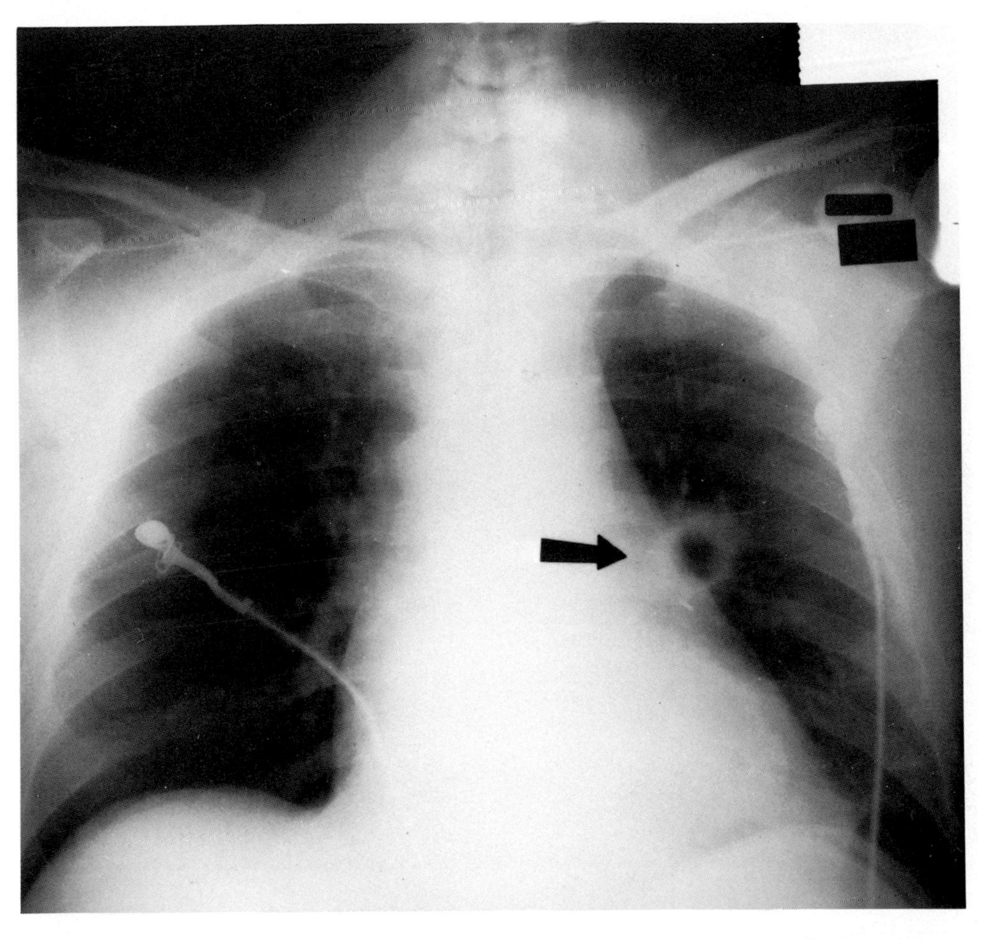

FIG. 2-4. *The unsharp artifactual image* (arrow) *in this chest radiograph is that of a small nut located inside the housing of the x-ray collimator. The nut became detached from the lead shutter mechanism and fell onto the plastic window of the collimator. (Courtesy of Ralph Coates, R.T. [R])*

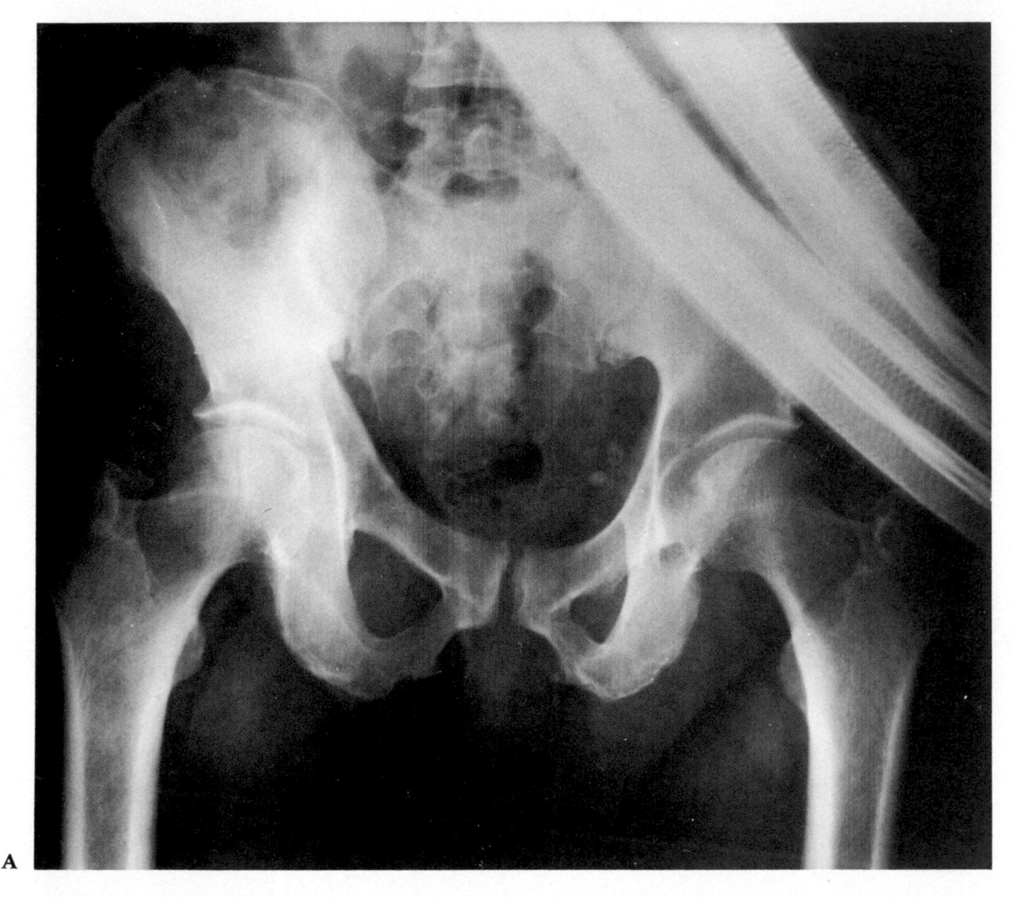

FIG. 2-5. (A) *The snakelike artifact in this pelvic radiograph is the magnified image of two high-voltage cables located on the underside of the collimator during the exposure.* (B) *In this radiograph of a portion of a high-voltage cable, notice the appearance of the protective metal sheath surrounding the cable. The same snakeskin pattern is visible in the cables in both* A *and* B. *(*A, *courtesy of Ralph Coates, R.T. [R])*

FIG. 2-5B.

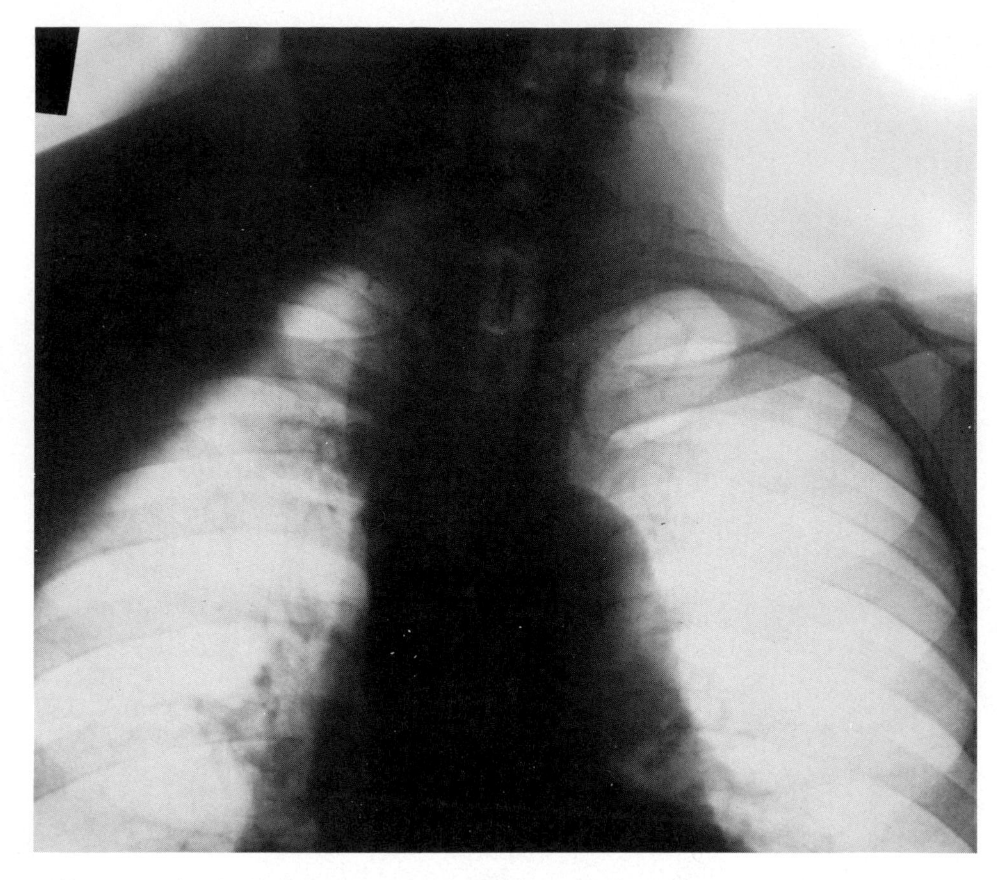

FIG. 2-6. *The shadow obscuring the shoulder and lung in this radiograph is the image of a segment of a high-voltage cable draped over the edge of the collimator. The structure of the cable is not seen because of its remoteness from the film. (Courtesy of Ralph Coates, R.T. [R])*

FIG. 2-7. *The unsharp appearance of the coil-like configuration in this abdominal radiograph was due to its remote location from the film. This is the image of an anode stator wire that became detached from a clip on the high-voltage cable and was imaged in the radiograph while underneath the collimator.* ▷

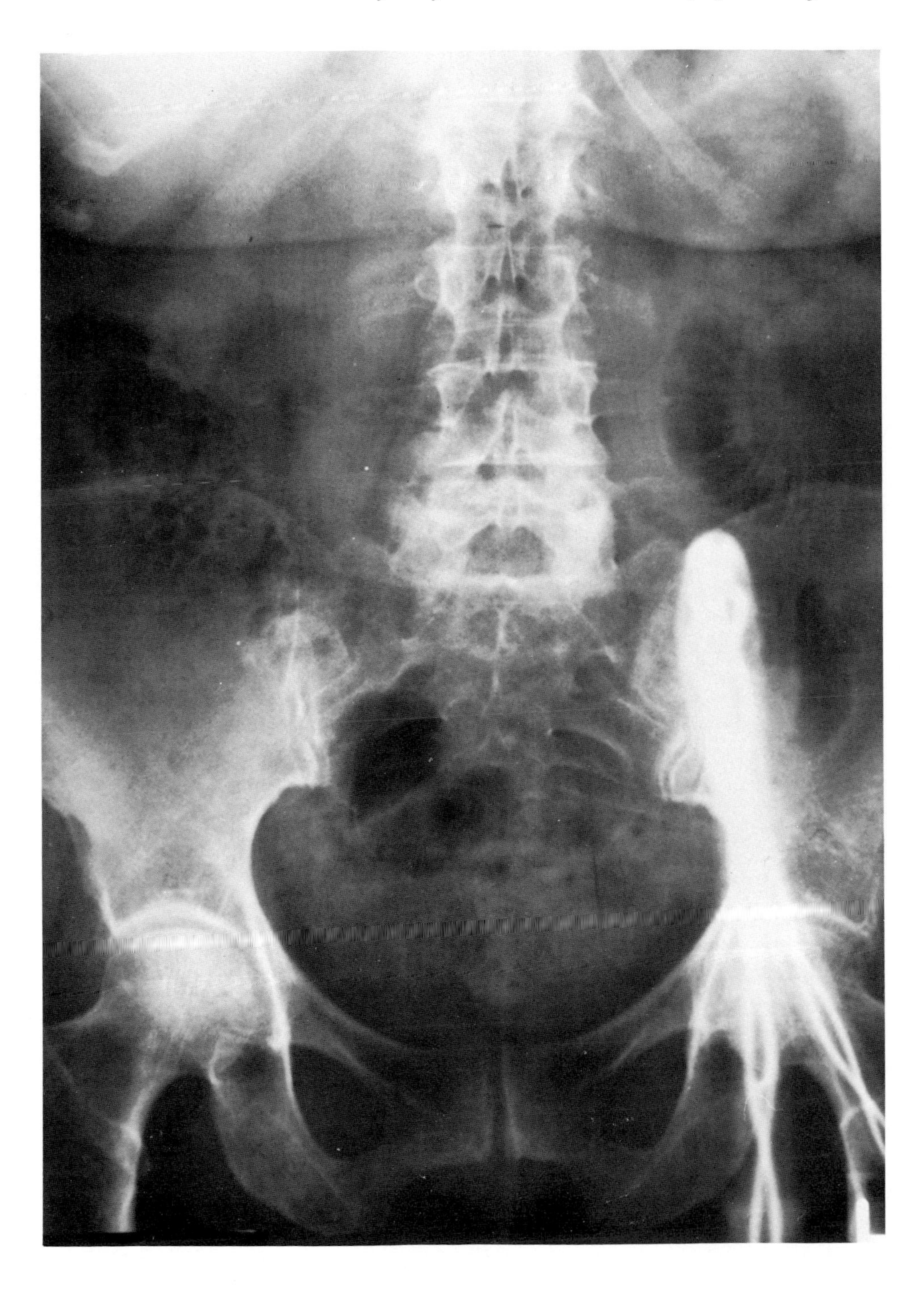

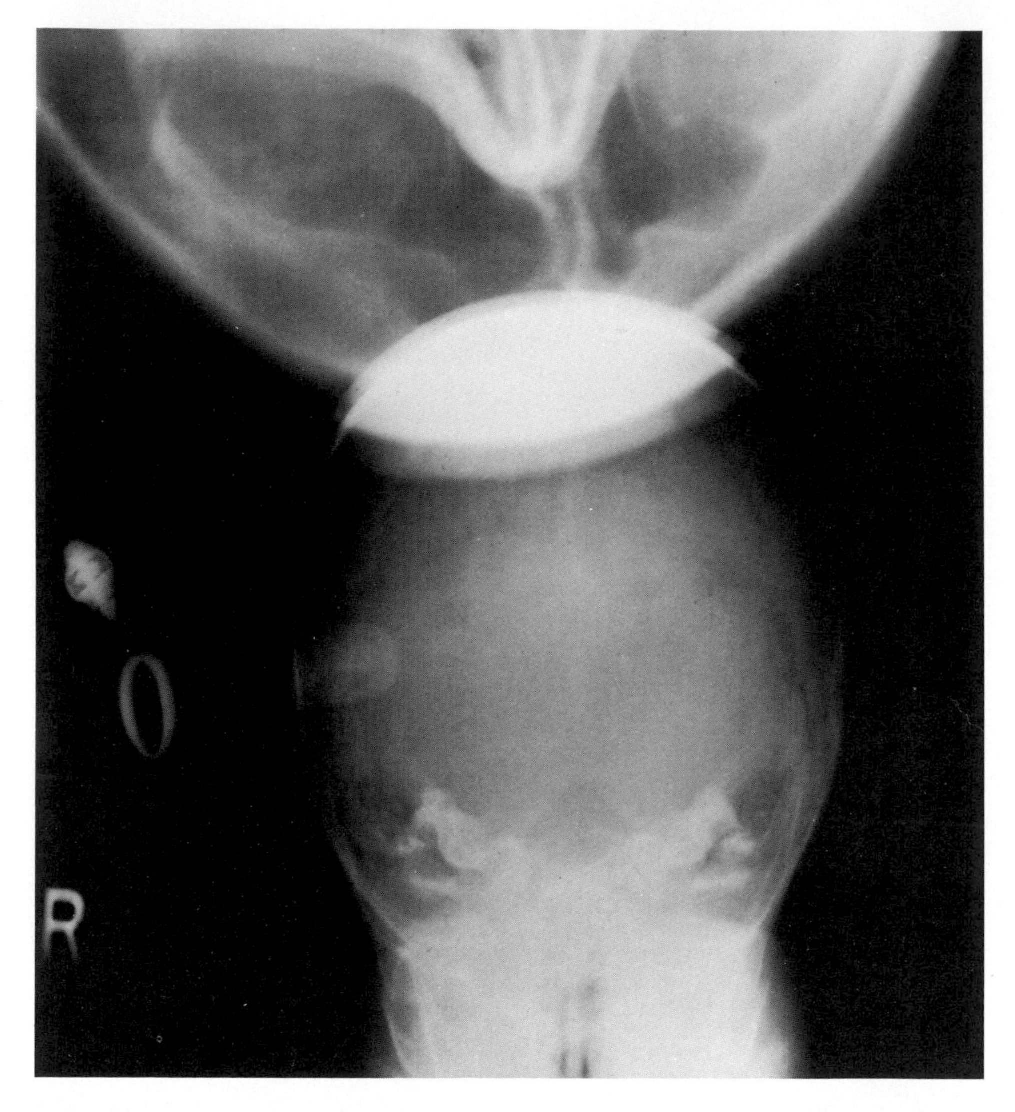

FIG. 2-8. *When mechanical restraints, sandbags, and the like cannot effectively restrain a child, it may be necessary to have one of the parents assist in the procedure. In such cases the parent should be provided with protective lead gloves and an apron. The image of the finger and rings in this radiograph indicates that lead gloves were not worn. The skull of the parent is imaged in the radiograph as well. (Courtesy of Ralph Coates, R.T. [R])*

Unusual Artifacts Associated with Tomographic Procedures

Tomography has long been recognized as a valuable radiographic procedure for investigating certain anatomic structures. Although the varied movement of the x-ray tube and film employed in tomography occasionally results in some bizarre streaking patterns in the radiograph, such imaging problems can usually be overcome by selecting a different tomographic movement or by changing the position of the patient for the radiographic examination. The purpose of this section is not to explore the reasons for such bizarre imaging patterns, but to demonstrate how the location of a particular artifact, as well as the blurring movement of the x-ray tube during tomography, may be responsible for obscuring the cause of a particular imaging problem.

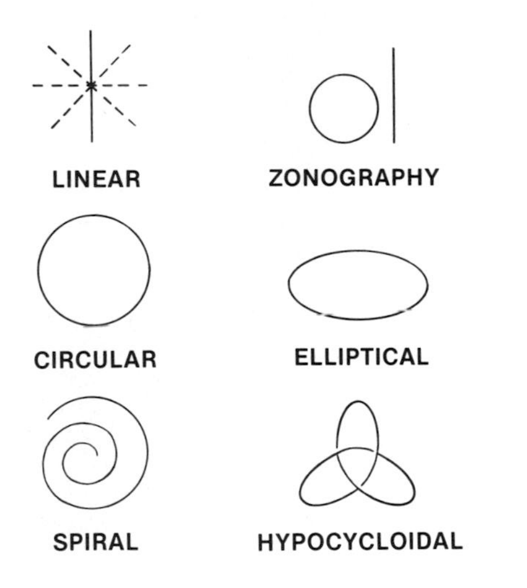

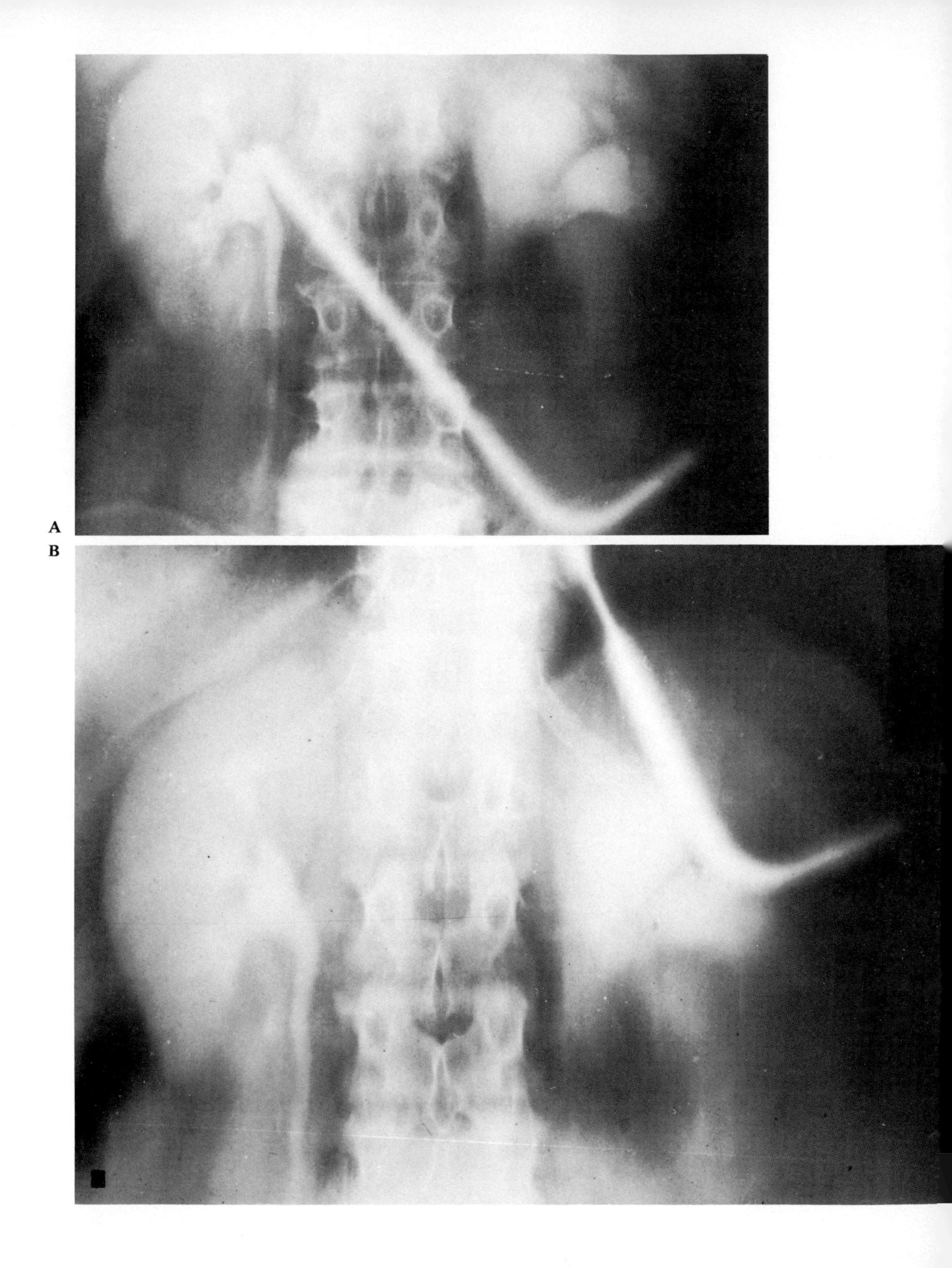
A
B

◁ **FIG. 2-9.** (A) *This tomogram of the kidneys shows an image of a small allen wrench located inside the housing of the collimator. The wrench was left in the collimator during a repair visit by x-ray service personnel.* B *demonstrates the wrench in a different location owing to the movement of the tomographic device. (From Sweeney RJ: On the Technical Side. Radiol Technol 50, No. 4, 1979)*

FIG. 2-10. *This radiograph demonstrates an unsharp image in the central portion of the lung. The artifact would appear on occasion and not reappear for weeks at a time, a pattern that baffled both the staff and the x-ray repairman. The condition responsible for the artifact was discovered following an extensive disassembly and inspection of the collimating device. When the repairman activated the tomographic centering switch, he noticed something in the window of the port of the x-ray tube. This was a tiny speck of solder suspended in the oil surrounding the envelope of the x-ray tube. The movement of the tomographic device and the location of the solder in the oil explained why it would appear and disappear in such an erratic manner.*

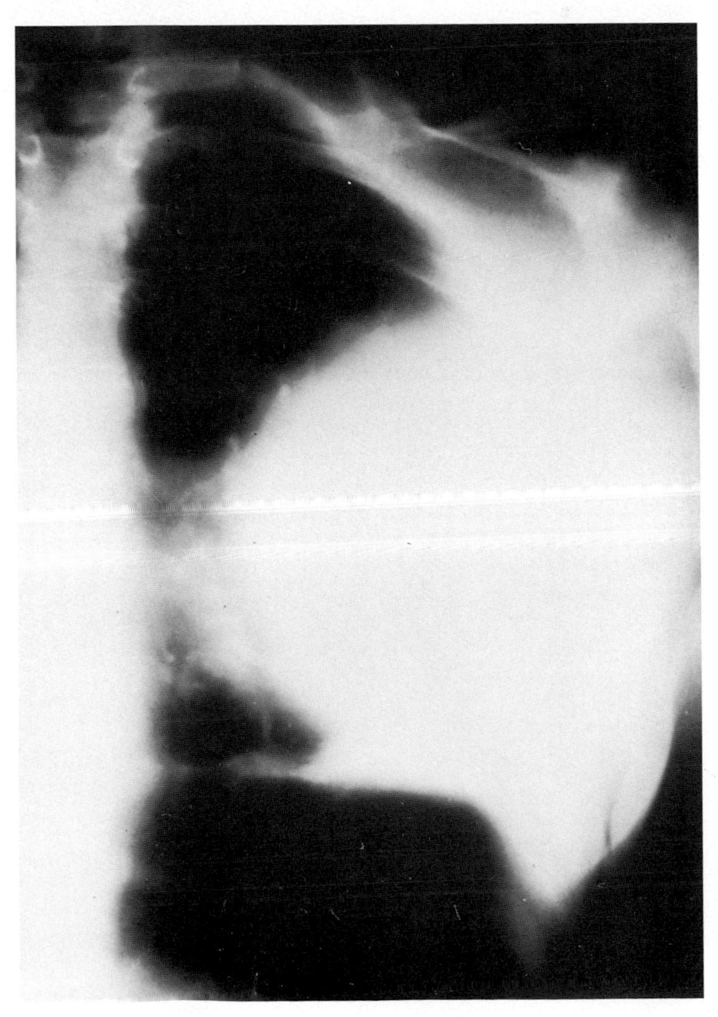

A

FIG. 2-11. *The mushroomlike configuration of the image in the radiograph* (B) *appears as if an atomic bomb were being detonated* (A) *inside the patient. The elongated image is the barium-filled stomach, which was distorted when the technologist accidentally engaged the linear tomographic mode during the exposure. (*A, *courtesy of the Nuclear Regulatory Commission;* B, *courtesy of William A. Conklin, R.T., F.A.S.R.T.)*

B

FIG. 2-11. *continued*

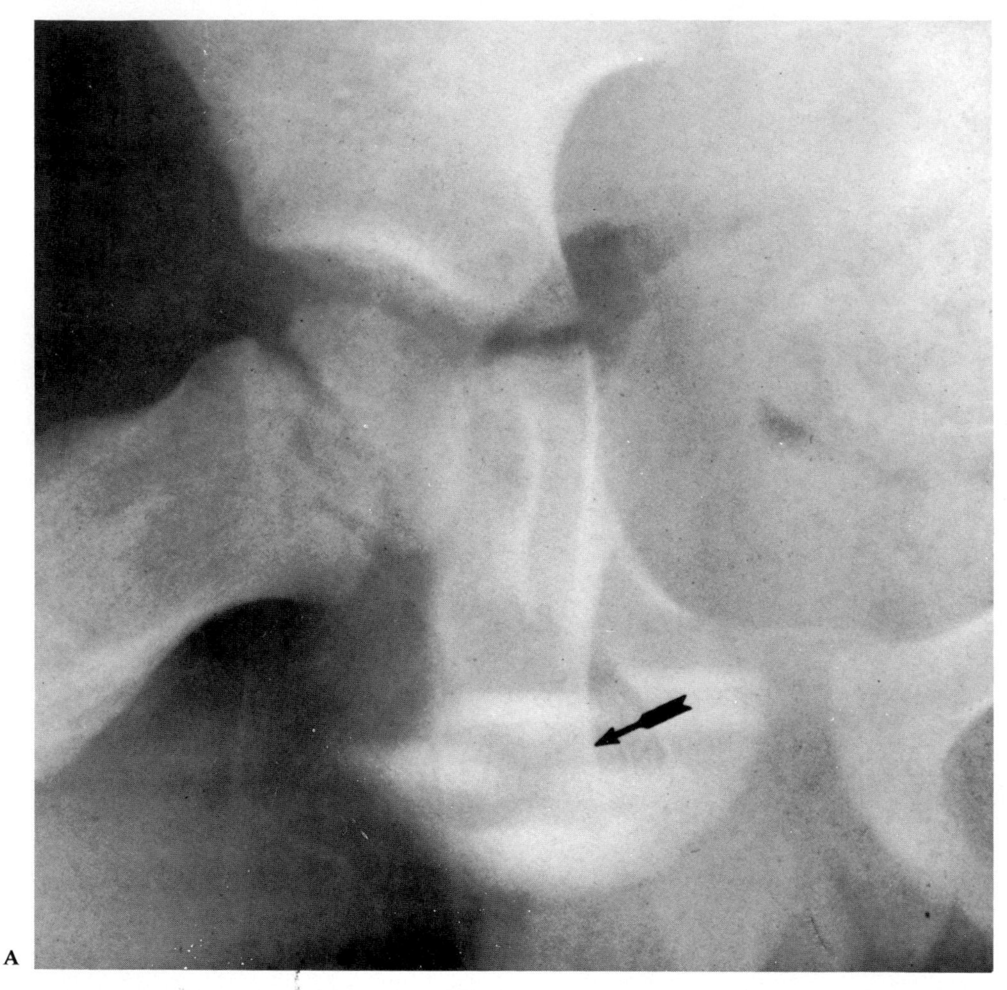

FIG. 2-12. (A) *The artifact seen in this radiograph* (arrow) *is blurred, owing to its movement during the exposure. It is the image of a lead identification marker that was dragged by the undersurface of a moving grid. Because there was too much tape wrapped around the lead marker, its thickness exceeded the tolerance between the grid and cassette. This same condition was responsible for the appearance of the artifact seen in* B (arrow). *(B, courtesy of William A. Conklin, R.T., F.A.S.R.T.)*

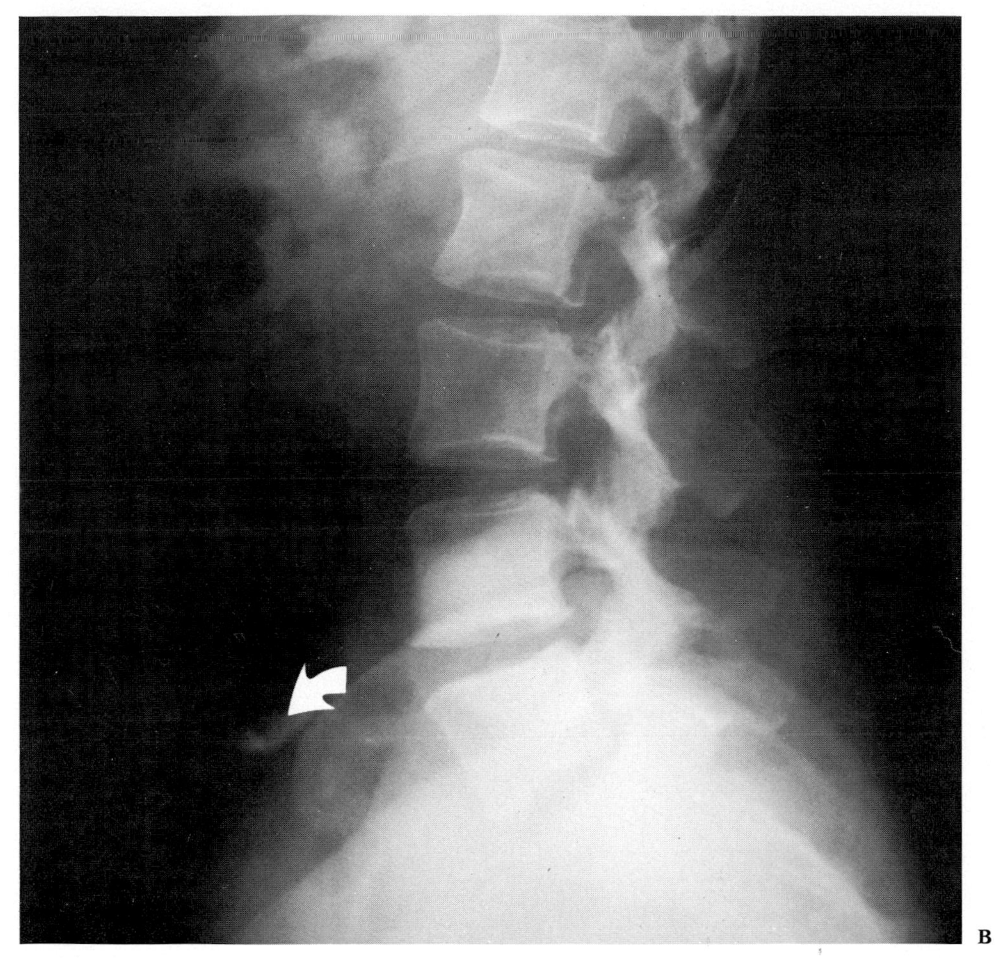

FIG. 1-12. *continued*

Suggested Reading

Cullinan JE, Cullinan AM: Illustrated Guide to X-ray Technics, 2nd ed, pp 171–193. Philadelphia, JB Lippincott, 1980

Durizch ML: Technical Aspects of Tomography. Baltimore, Williams & Wilkins, 1978

Poznanski AJ: Focal-spot artifacts on breast radiographs. Radiology 92, No. 3:644, 1969

Selman J: The Fundamentals of X-ray and Radium Physics, 6th ed, pp 417–438. Springfield, IL, Charles C Thomas, 1978

3 ARTIFACTS CAUSED BY IMPROPER FILM HANDLING

Some of the most persistent and baffling problems related to the appearance of radiographic artifacts are associated with the improper handling of film.

The most common artifact is the so-called kink, crinkle, or half-moon mark (Fig. 3-1). It can occur if a sheet of film is allowed to bend in the manner shown in Figure 3-1, *B*. The excess pressure exerted at the point where the film buckles results in the crinkle mark.

Perhaps one of the most often discussed questions concerning this artifact is whether it should appear black or white in the radiograph. Although the point may seem academic, many feel the crinkle mark will be white if the film is bent prior to exposure and black if the film is bent after exposure and prior to processing. Unfortunately, this is not always the case. For example, the black crinkle mark seen in Figure 3-2 was made on a sheet of film that had never been exposed to x-ray. Actually, the most important factor in determining whether the artifact will be black or white is the amount of pressure exerted on the film.

Although crinkle-type artifacts are normally easy to identify, there may be an occasion when such an artifact could lead to very deceptive results. For example, the radiograph in Figure 3-3, *A*, was interpreted by an emergency-room physician as exhibiting a fracture of the skull, and the patient was admitted to the hospital for observation. A repeat study the following day, which was requested by the radiologist, revealed the pseudofracture to be the image of a pressure mark caused by mishandling the film (Fig. 3-3, *B*).

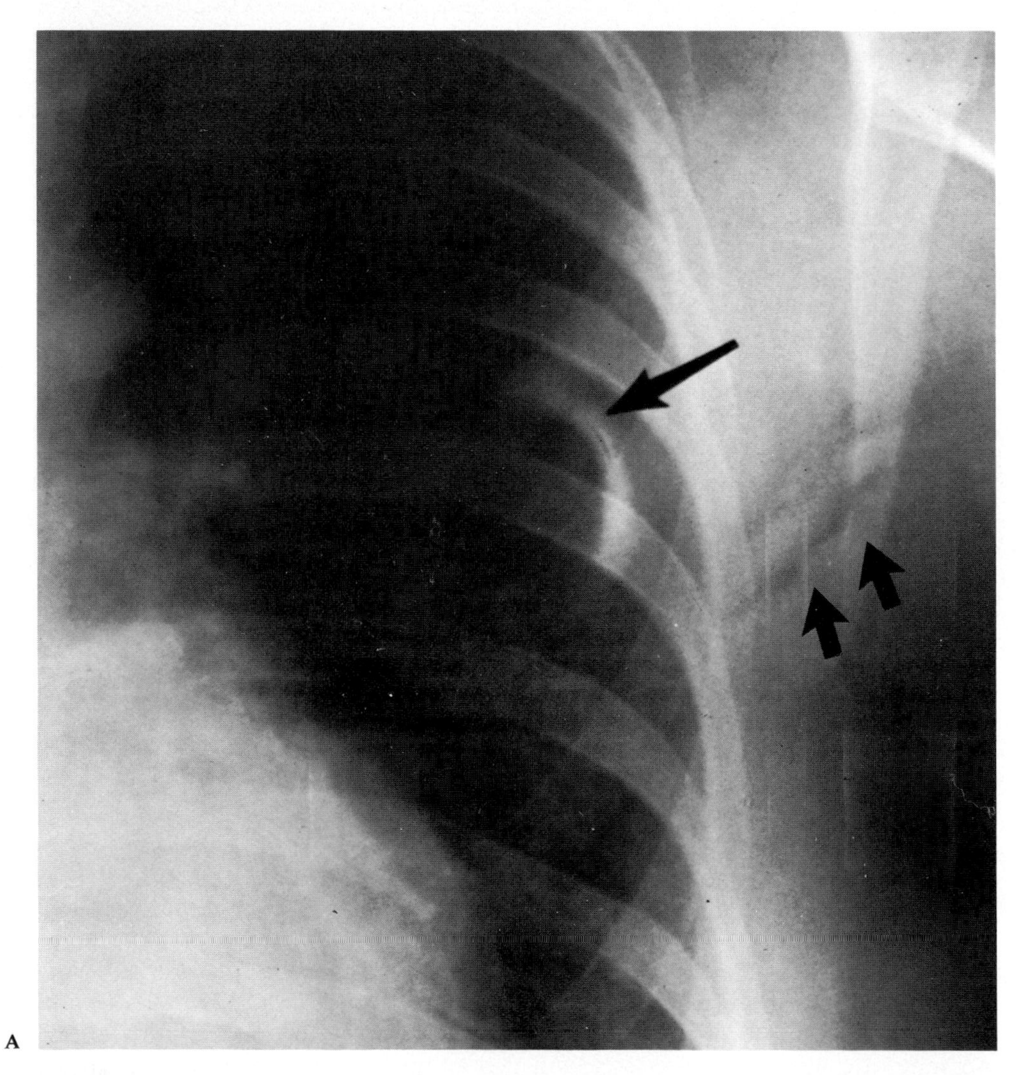

FIG. 3-1. *continued*

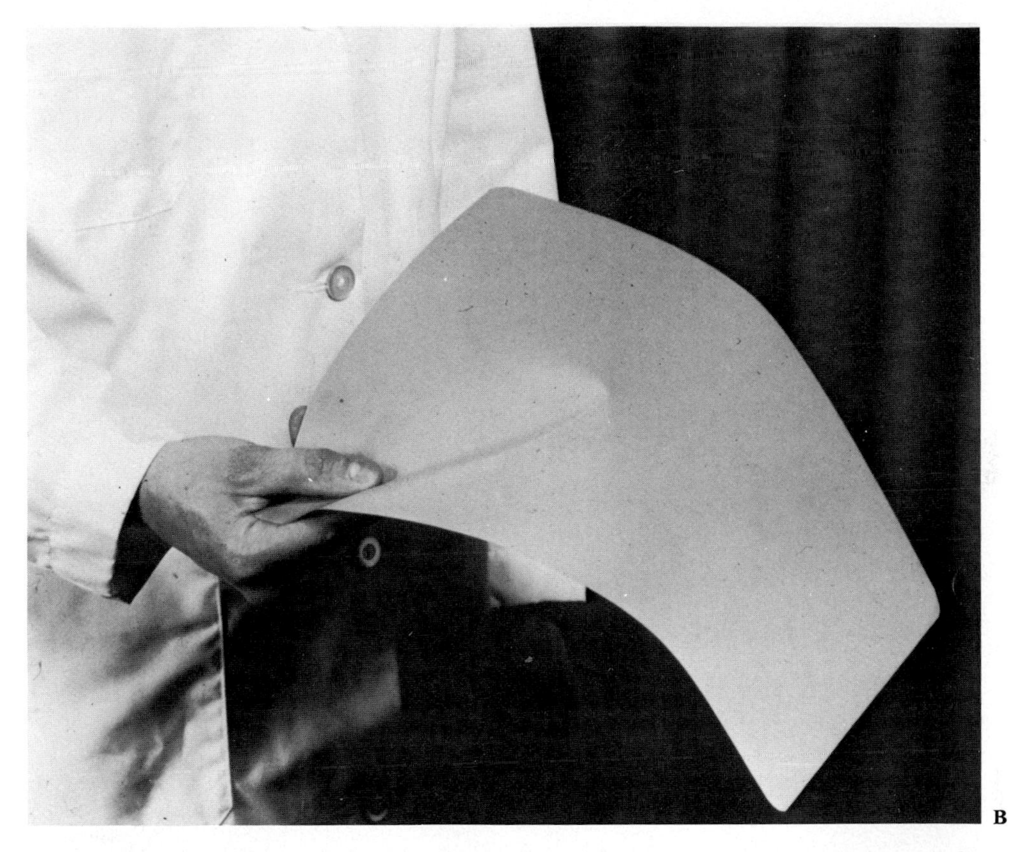

FIG. 3-1. (A) *Both of the images in this radiograph* (arrows) *were caused by mishandling the film. The* single arrow *demonstrates a crinkle mark, which can occur if the film is bent in the manner shown in* B. *The* double arrows *show an artifact caused by exterting pressure on the surface of the film. Compare the appearance of these images to those in Figure 3-8.*

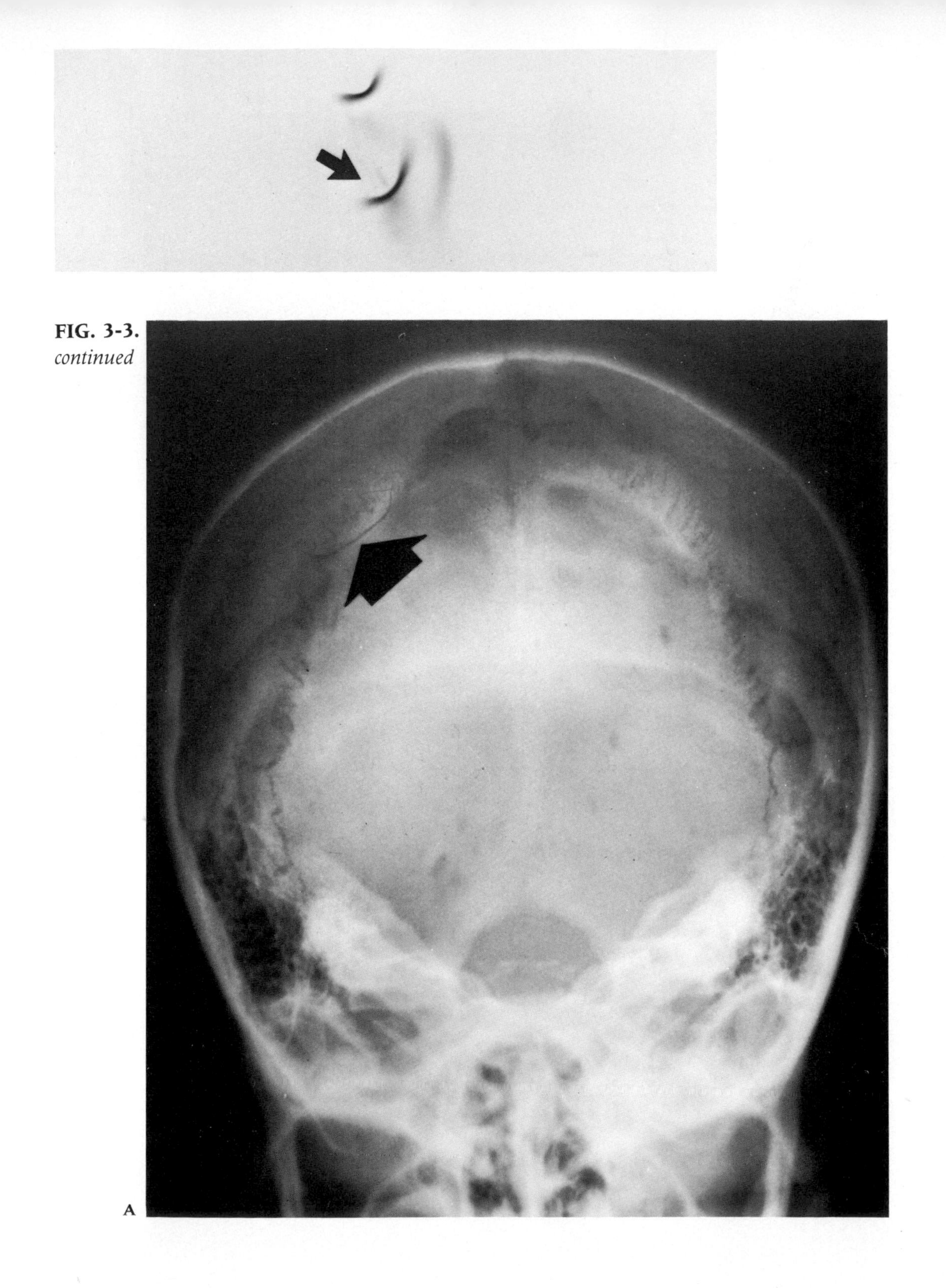

FIG. 3-3. *continued*

A

◁ **FIG 3-2.** *The artifact seen in this radiograph* (arrow) *is called a kink, crinkle, or half-moon mark. In this case the black mark appears on a film that had not been exposed to x-ray.*

FIG. 3-3. (A) *The crinkle-type artifactual image* (arrow) *in this radiograph of the skull was mistaken by an emergency-room physician for a fracture.* (B) *A repeat examination the following day revealed the pseudofracture to be artifactual.*

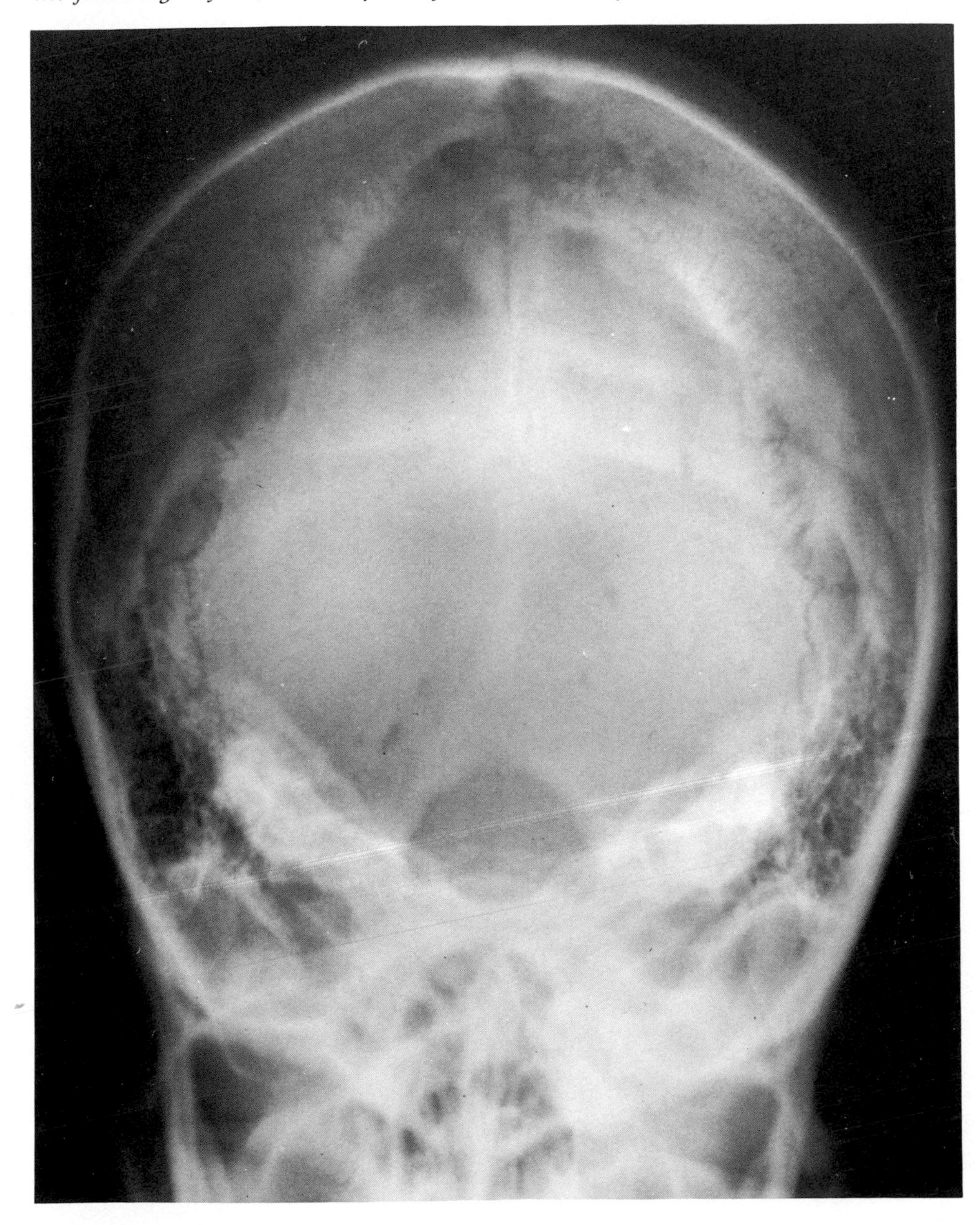

B

Storing Film to Prevent Pressure-Related Artifacts

Radiographic film is packaged in various quantities, up to 125 sheets per box. Although the types of packaging materials, and the methods of packaging the film are carefully selected to provide the best possible means of safely transporting the film, damage due to mishandling and improper storage can occur at any stage after the film leaves the factory. Upon receipt, the heavy film cartons should be stored and transported to the various areas in the hospital on edge in order to prevent pressure marks (Fig. 3-4).[1,2]

FIG. 3-4. *Pressure-related artifacts can often be avoided if the film is properly transported and stored.* ▷

NIF
RX
100
EXP. DATE
EM. NO.
1983 7 69505
1983 7 69673
24×30cm
WRONG

NIF
RX
100
24×30cm
EXP. DATE
EM. NO.
1983 7 69505
1983 7 69673
RIGHT

FIG. 3-5. *The ripple or shock artifact on this film resulted from dropping the box containing the film onto the floor.*

FIG. 3-6. *The multiple pressure marks on this film resulted from deliberate abuse to demonstrate the effect of improper film handling.*

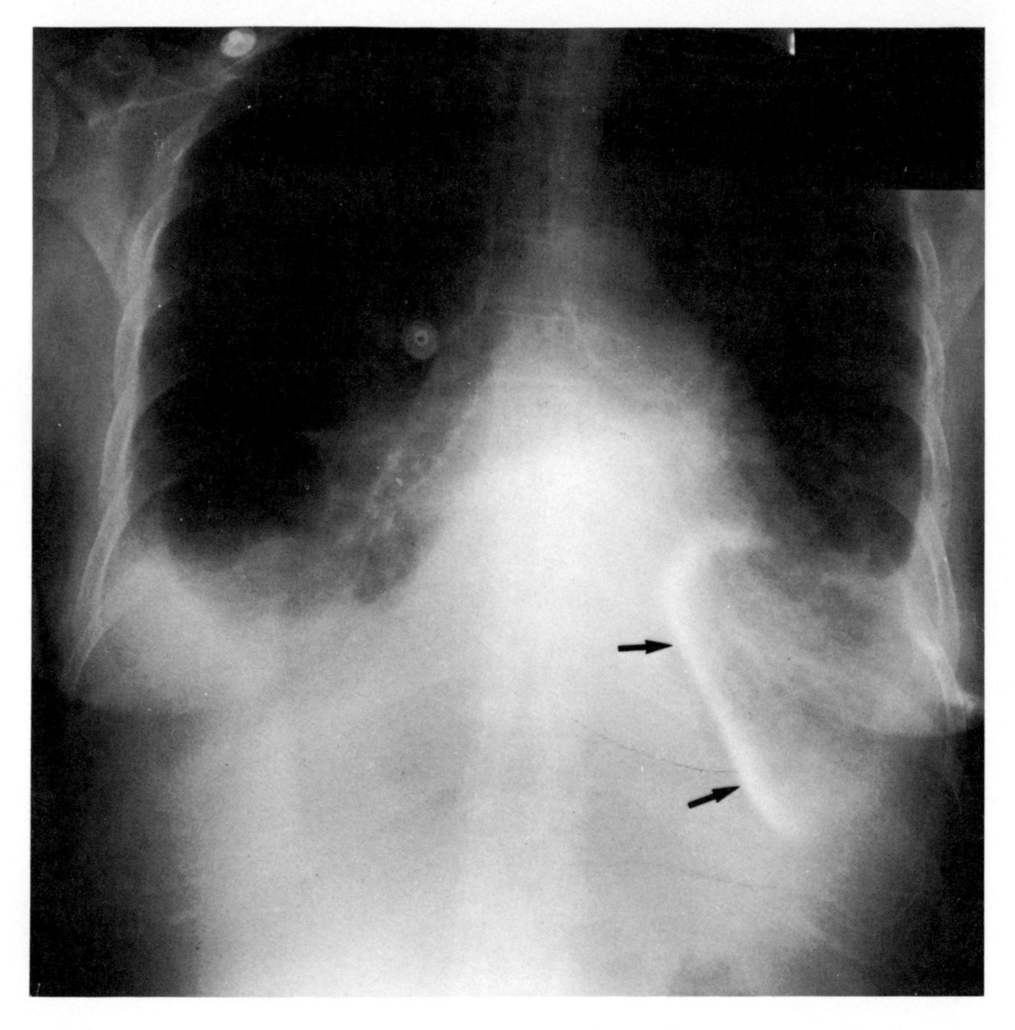

FIG. 3-7. *The hooklike image* (arrows) *in this radiograph is a pressure-related artifact caused by mishandling the film during the loading procedure.*

FIG. 3-8. (A) *The wavy image* (arrows) *is a pressure-related artifact caused by mishandling the film. Compare the appearance of this artifact with the one in* B (bottom arrow), *which resulted from deliberate abuse of the film during handling. In the process of simulating this condition, a few crinkle marks* (B, top arrow) *appeared as well.* ▷

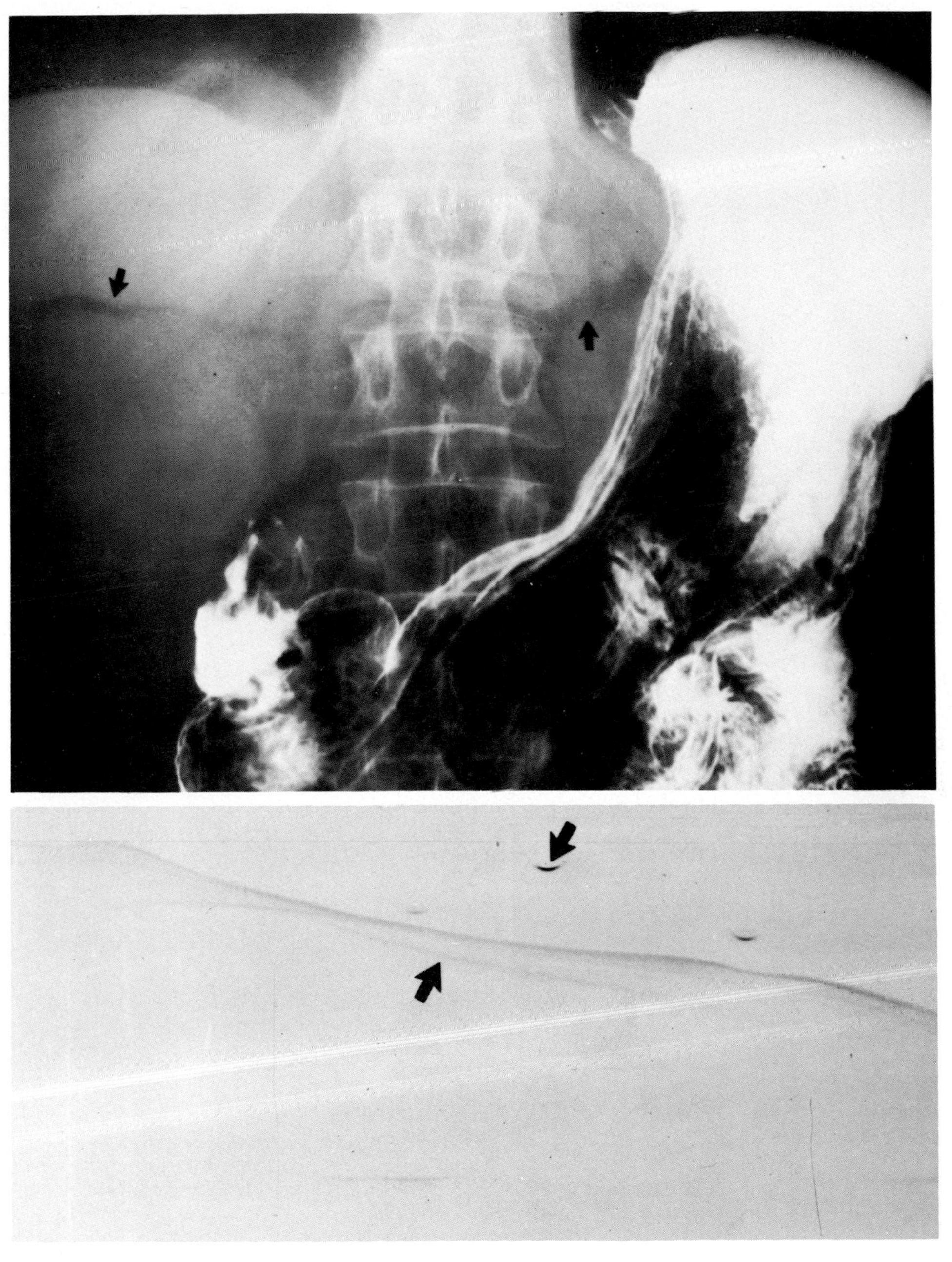
A
B

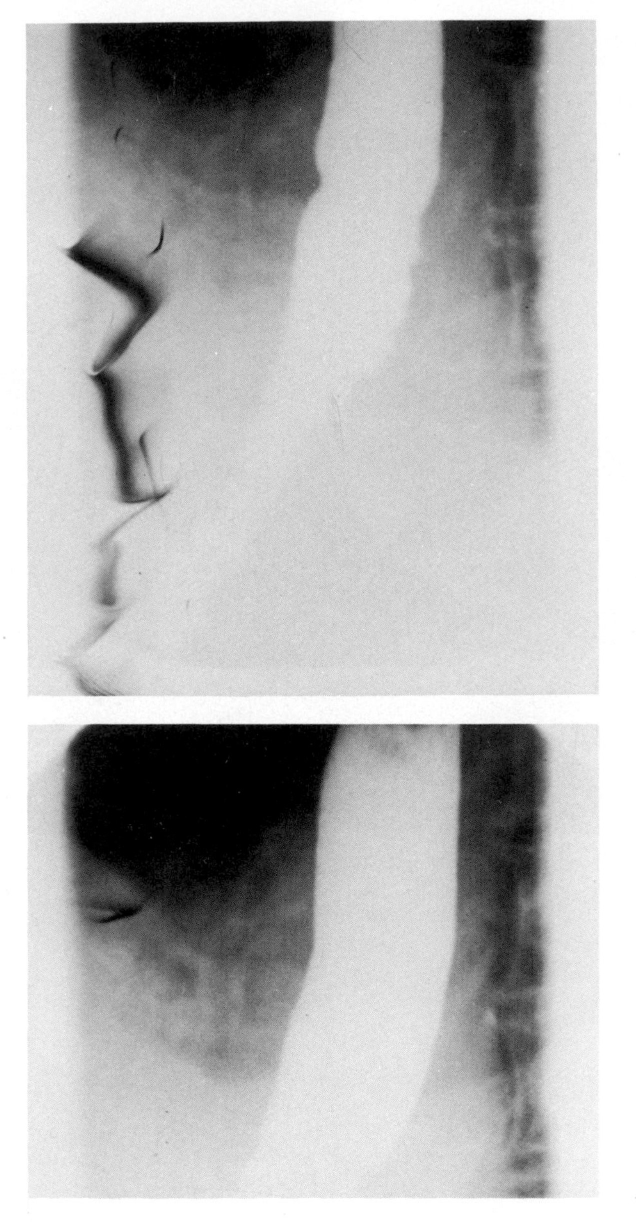

FIG. 3-9. *The pressure and crinkle marks on this segment of 105-mm Fluorospot film were caused by mishandling the film prior to processing.*

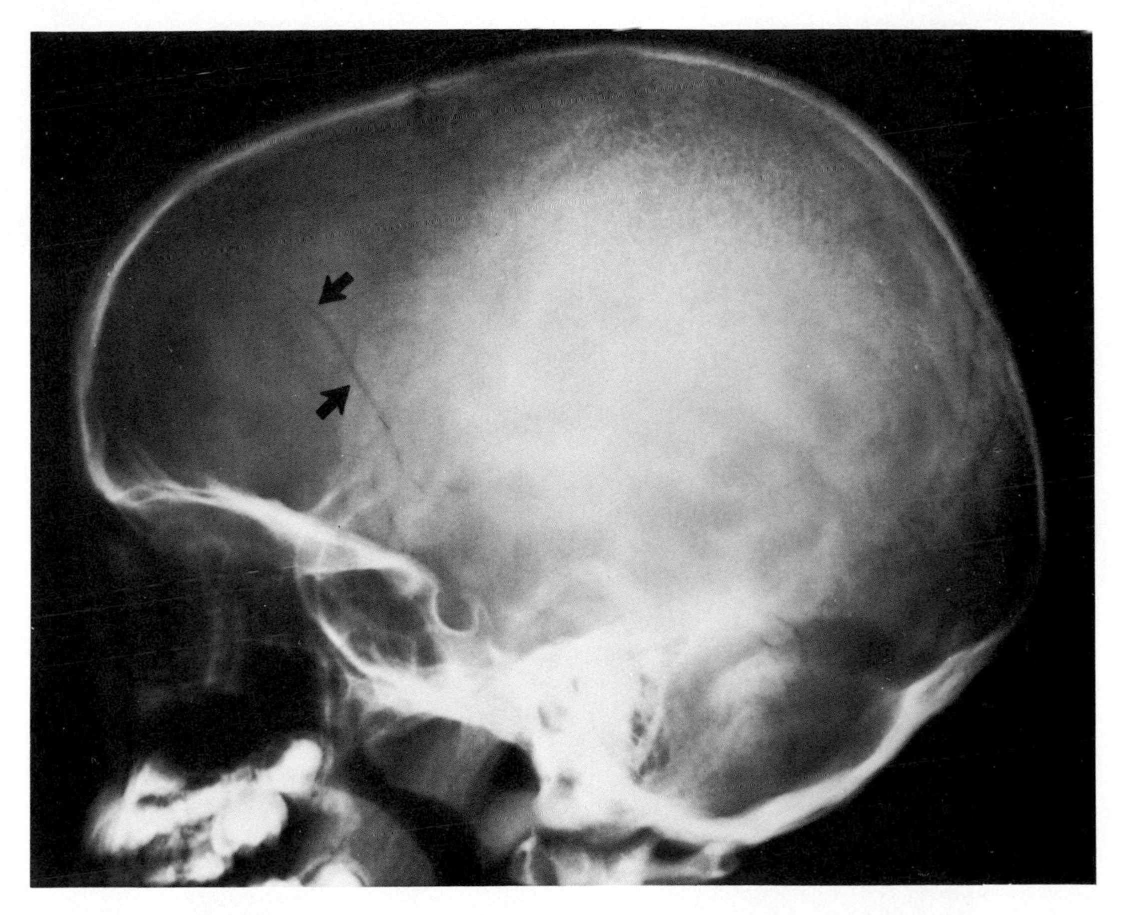

FIG. 3-10. *This radiograph is one example in which being able to examine the surface of the film would be most helpful in identifying the problem. The artifact* (arrows) *is an abrasion mark caused by mishandling the film.*

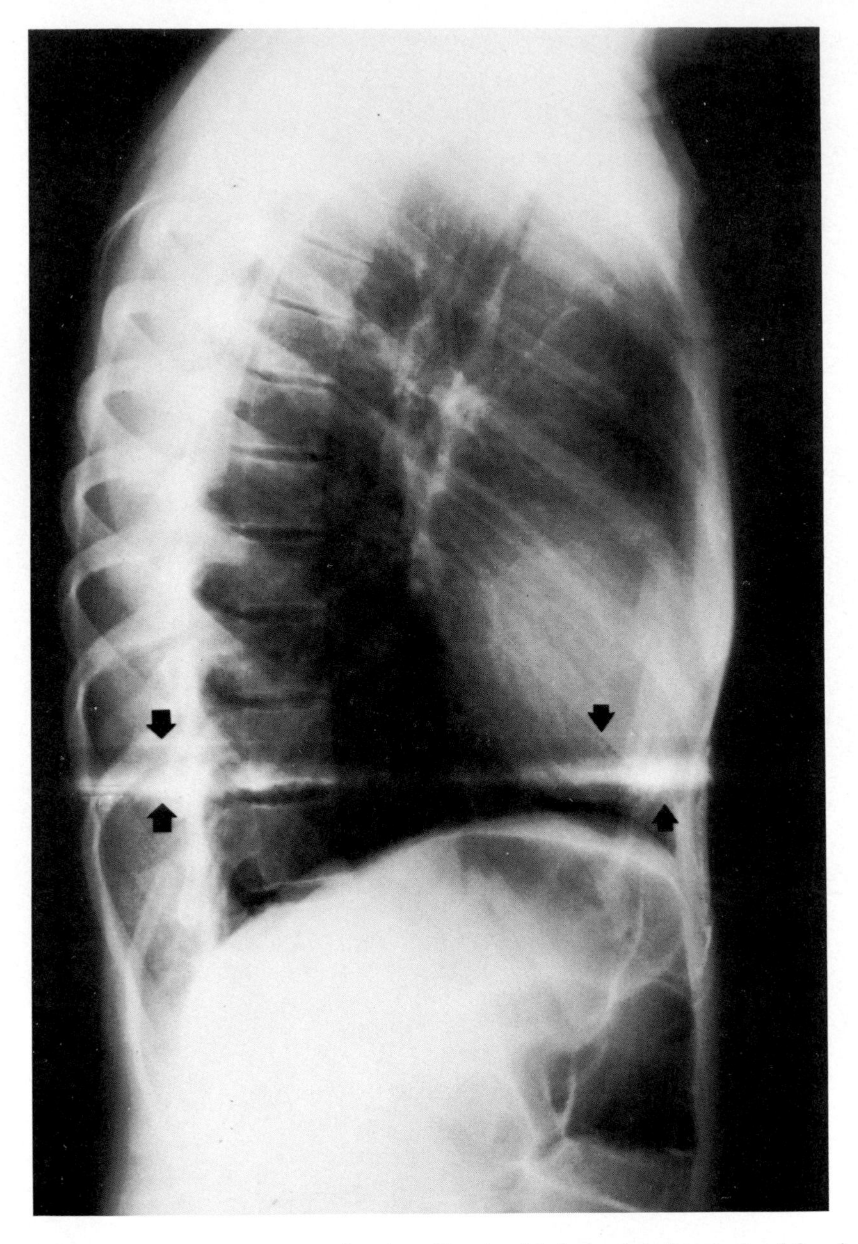

FIG. 3-11. *The pressure-related artifact in this lateral radiograph of the chest* (arrows) *resulted from accidentally closing the drawer of the film bin on the sheet of film.*

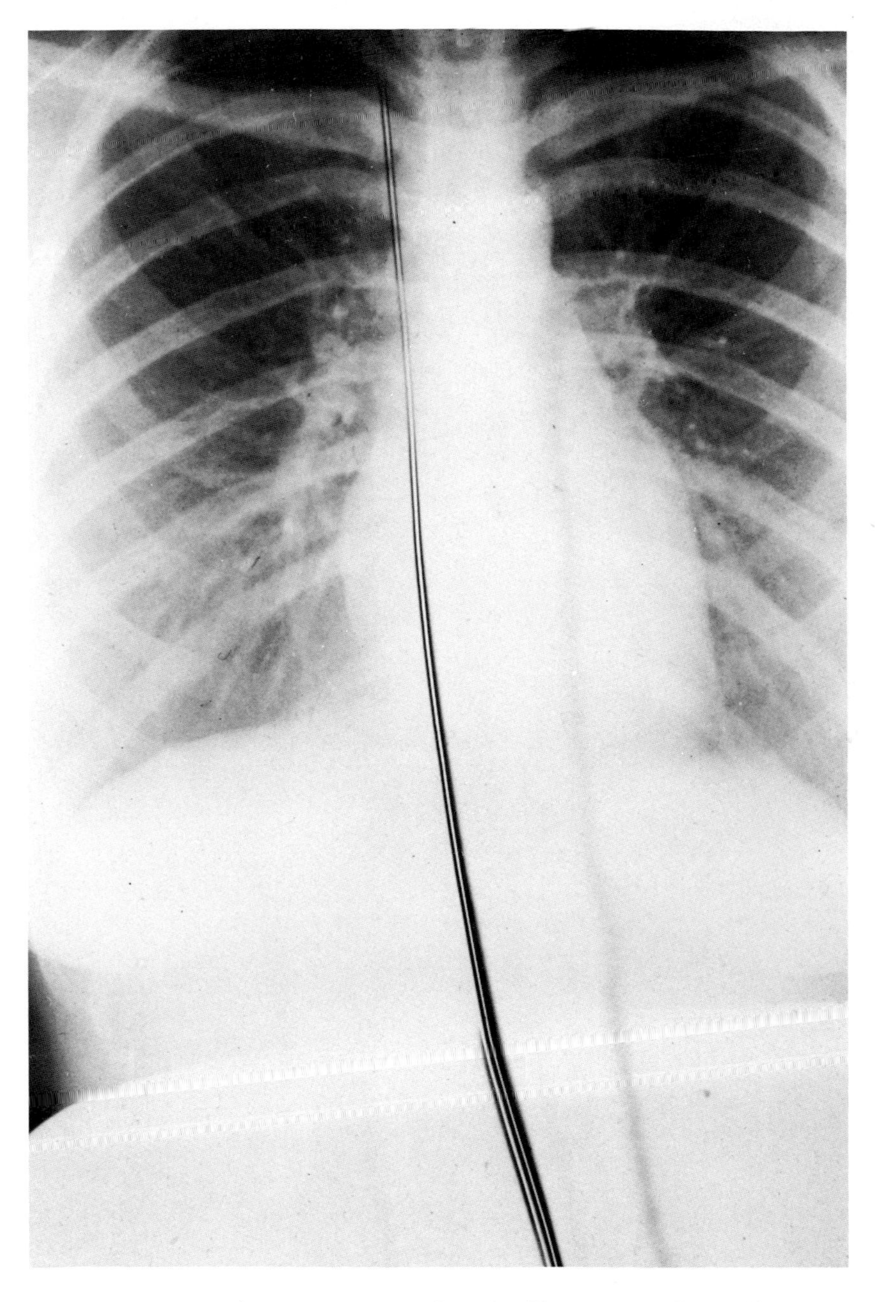

FIG. 3-12. *Excessive pressure exerted on the film can manifest itself in many ways in the radiograph. This artifact occurred when the sheet of film was folded during handling.*

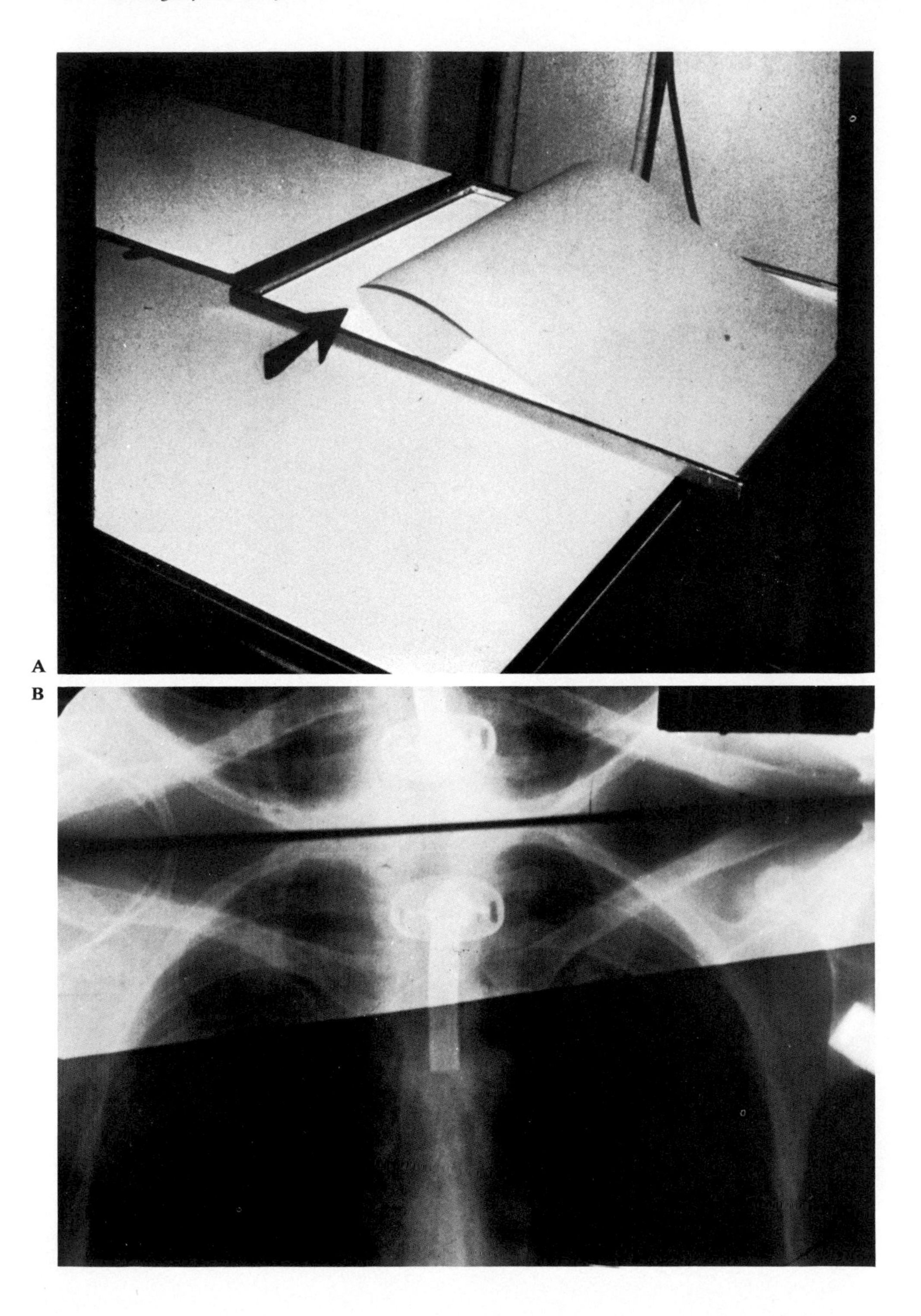
A
B

◁ **FIG. 3-13.** (A) *A sheet of film is sometimes accidentally loaded into the cassette in a folded position. Such a condition was responsible for the appearance of the mirrorlike image in the chest radiograph shown in* B. *The black line at the level of the first segment of the ribs is the point where the film was folded during loading.*

FIG. 3-14. *The multiple image of the thorax in this radiograph resulted from an improper film loading procedure. The film was exposed while folded.*

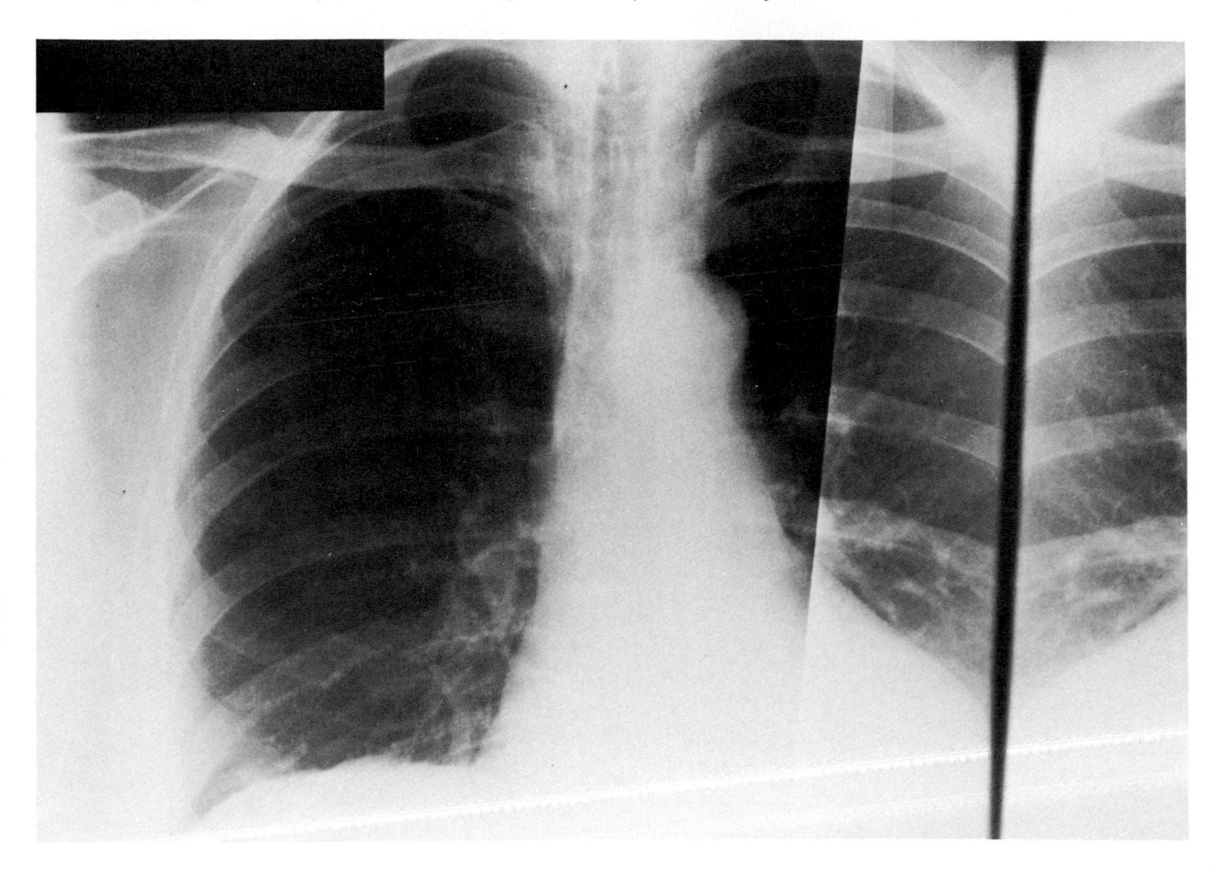

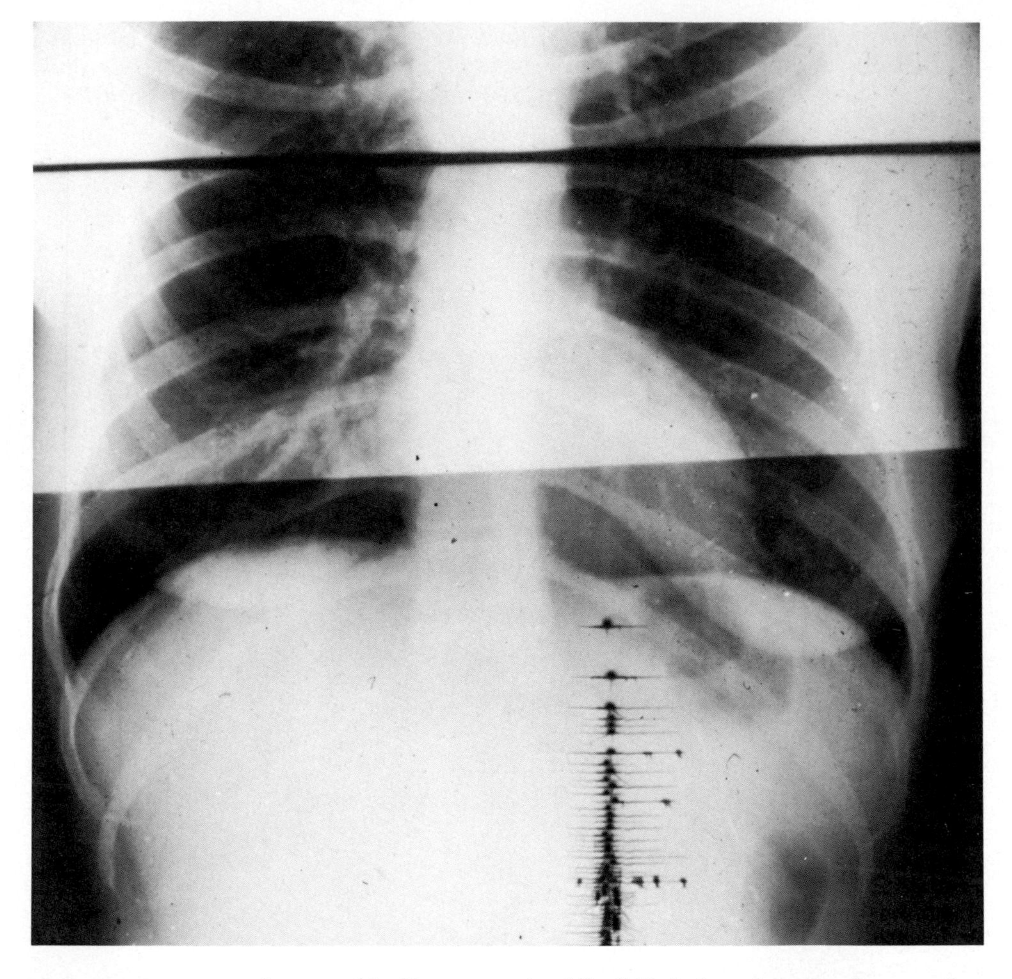

FIG. 3-15. *Not only was this film exposed while folded, but there is also a rather unusual static discharge.*

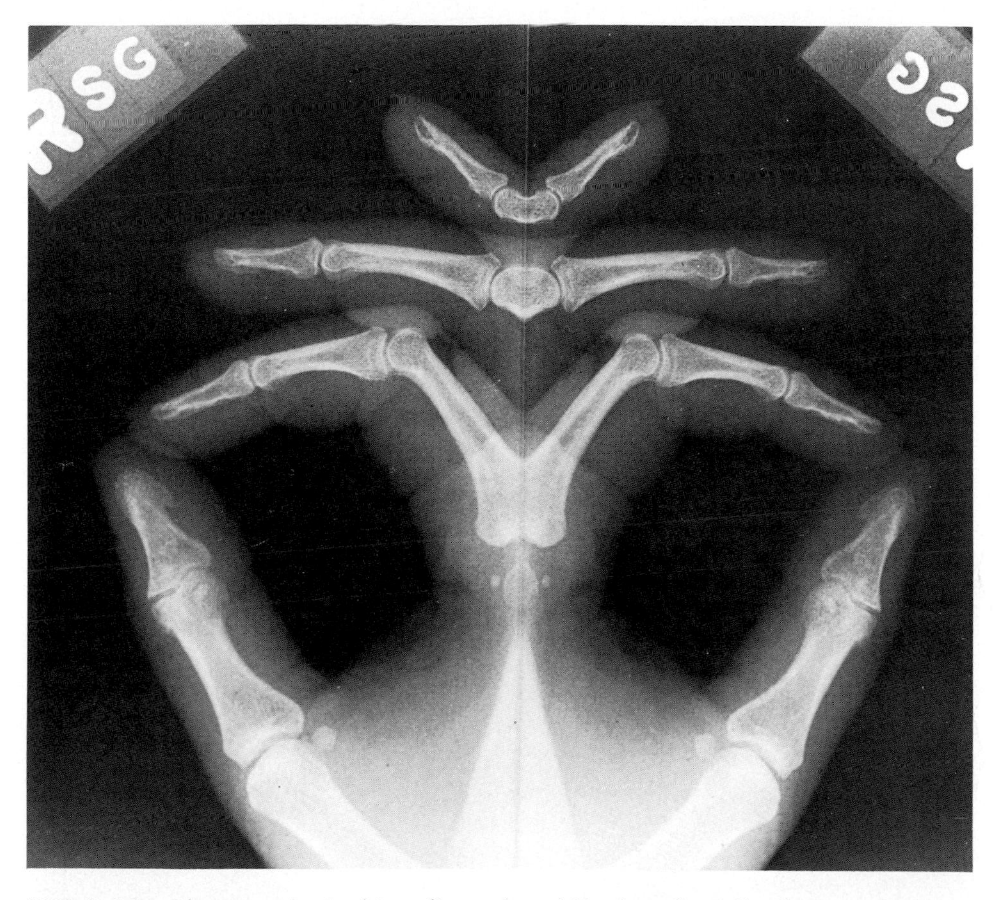

FIG. 3-16. *The extremity in this radiograph could be that of an alien from outer space or that of a rather unusual earthling. However, it is a normal oblique radiograph of the hand made on a film folded in half within the cassette.*

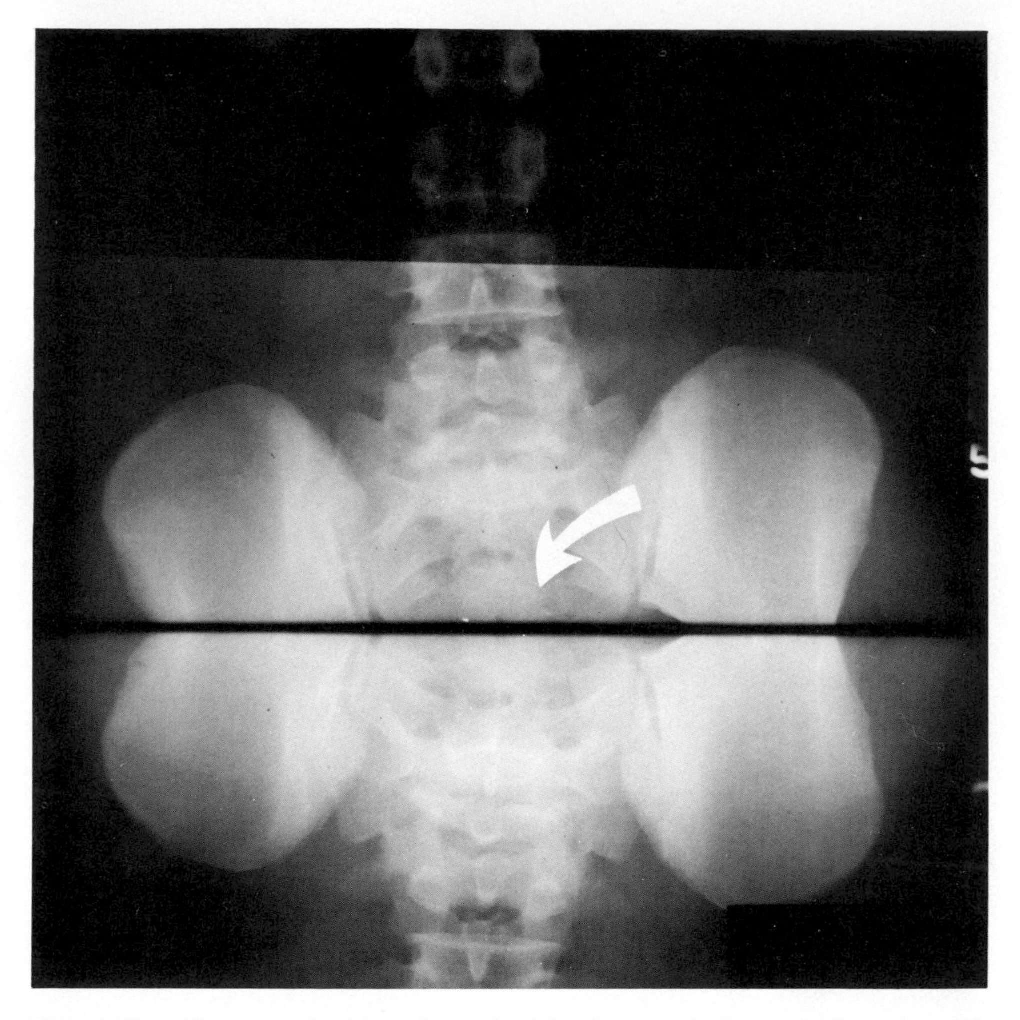

FIG. 3-17. *The* arrow *in this radiograph of the abdomen indicates the line where this film was folded while being loaded into the cassette.*

FIG. 3-18. (A) *The unusual shape of the artifact in this radiograph of the breast resulted from forcing the film against the anterior portion of the chest.* (B) *The improper placement of the packet, as well as the compression of the film by the weight of the breast, was responsible for the appearance of the artifact. (A, courtesy of Robert A. Short, R.T. [R])* ▷

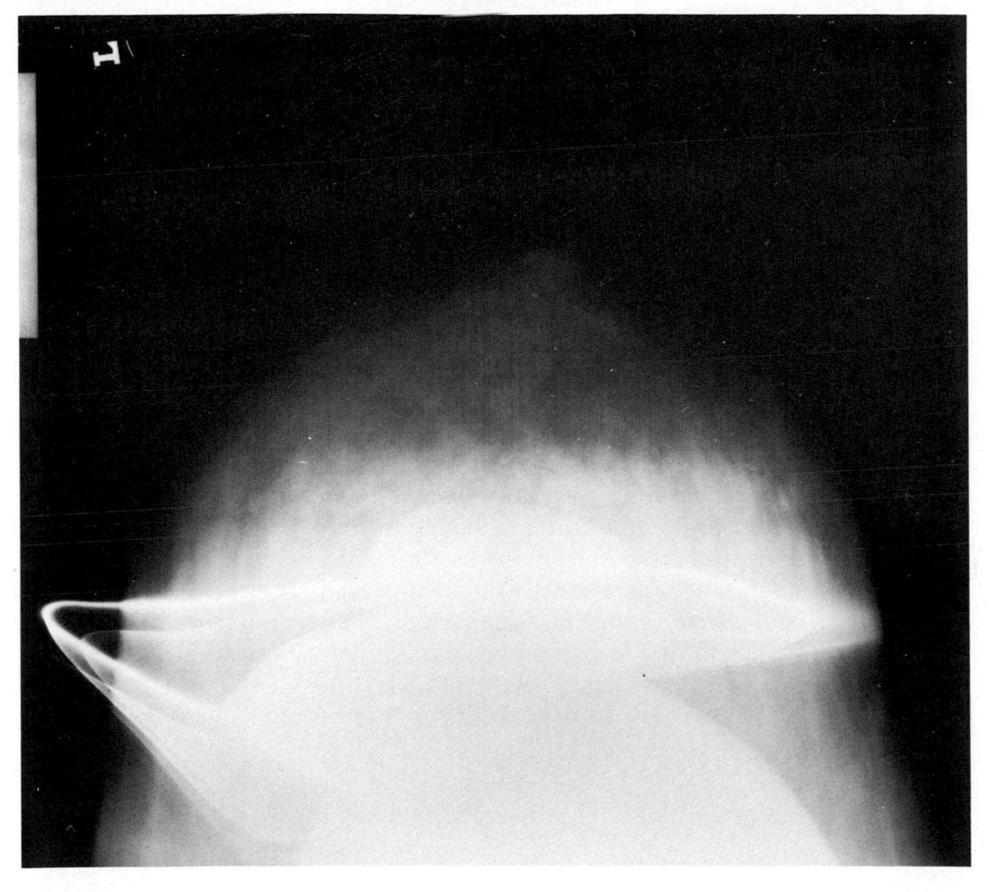

A

B

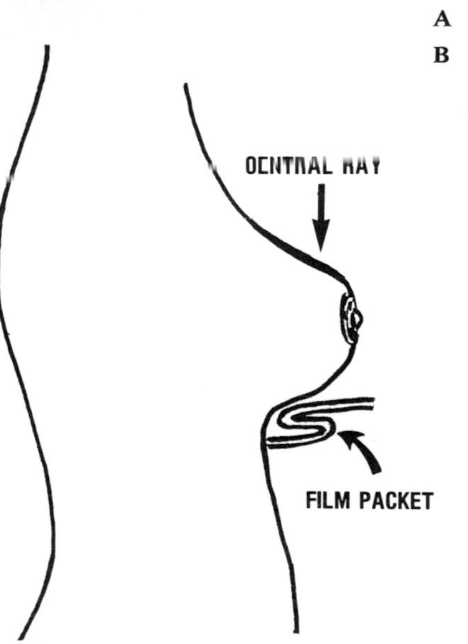

Pressure-Related Artifacts in Cassetteless Film-Transport Systems

One of the most difficult imaging problems I have encountered is attempting to identify the condition responsible for producing an artifact in a cassetteless type of film-transport system. This is due to the fact that such problems often range from pressure occurring in the automated unit to artifacts caused by dirt and dust particles on the surface of the intensifying screens. Although periodic cleaning or replacement of the screens helps to control the appearance of numerous artifacts, pressure marks caused by the operation of the film-transport mechanism or the loading of the supply magazine are most difficult to isolate (Fig. 3-19).

Although it is correct to assume that any type of film can be tried in a bulk film loader, it is incorrect to assume that they all will and should work correctly. This is because not all film is manufactured to the same standards but may vary in dimensions, type of surface coating, and cutting of the corners and sides. Couple these factors with the variation in temperature and relative humidity in the room, and this will have a profound effect on the reliability of film transport.

In order to overcome such problems, the technologist is often instructed either to fan the stack of films or to reverse the direction of each sheet before loading the supply magazine. Such preventive measures may eliminate film jams, but they are often responsible for introducing a number of handling or pressure-type artifacts that can be most difficult to analyze (see Figs. 3-20 and 3-21).

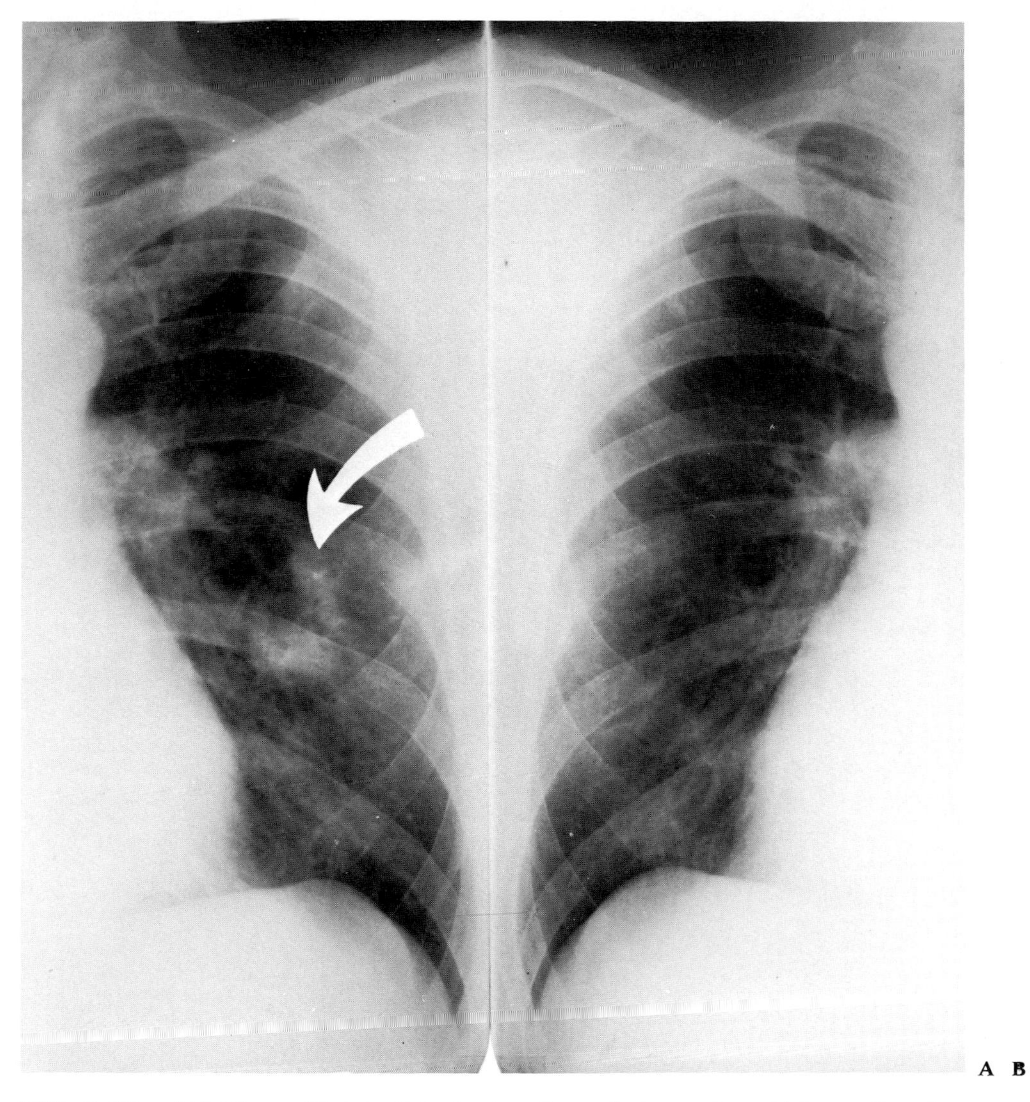

FIG. 3-19. (A) *The artifact in this chest radiograph* (arrow) *could be attributed to numerous imaging conditions. In this instance it was caused by excess pressure on the surface of the film.* (B) *The artifact disappeared in an immediate follow-up radiograph.*

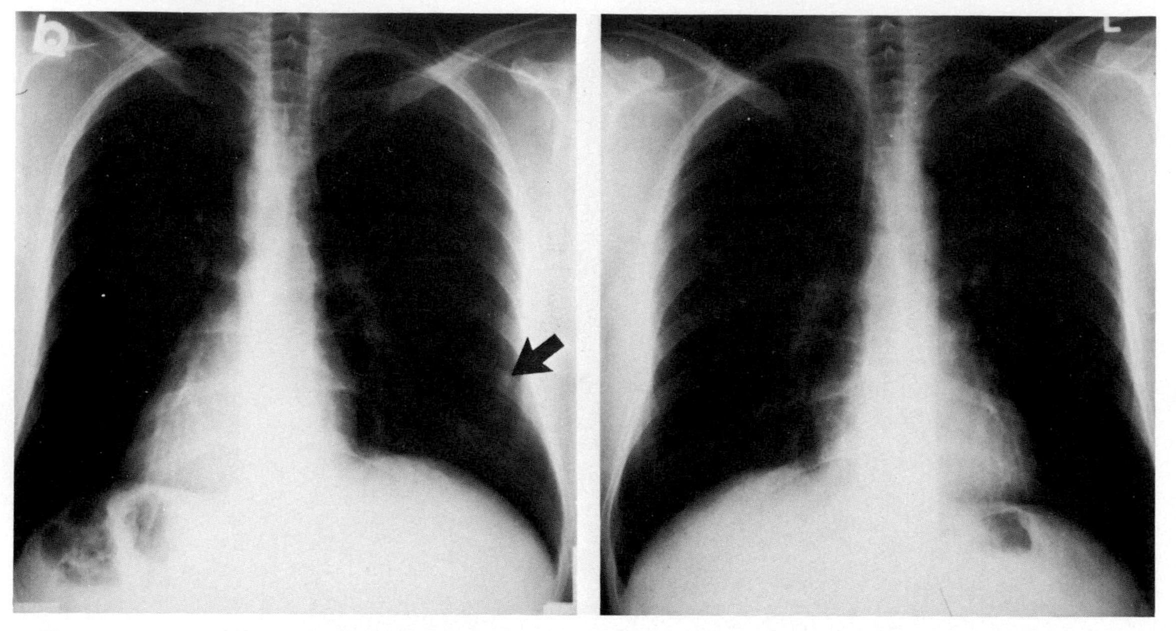

A B

FIG. 3-20. (A) *The radiologist reported that the opacity* (arrow) *in this radiograph of the right hemithorax (shown reversed for comparison) was probably artifactual and suggested a repeat study.* (B) *The artifact disappeared in the follow-up examination. The artifact, which is better seen in the close-up* (C, arrow), *was probably caused by pressure applied to the film during handling.*

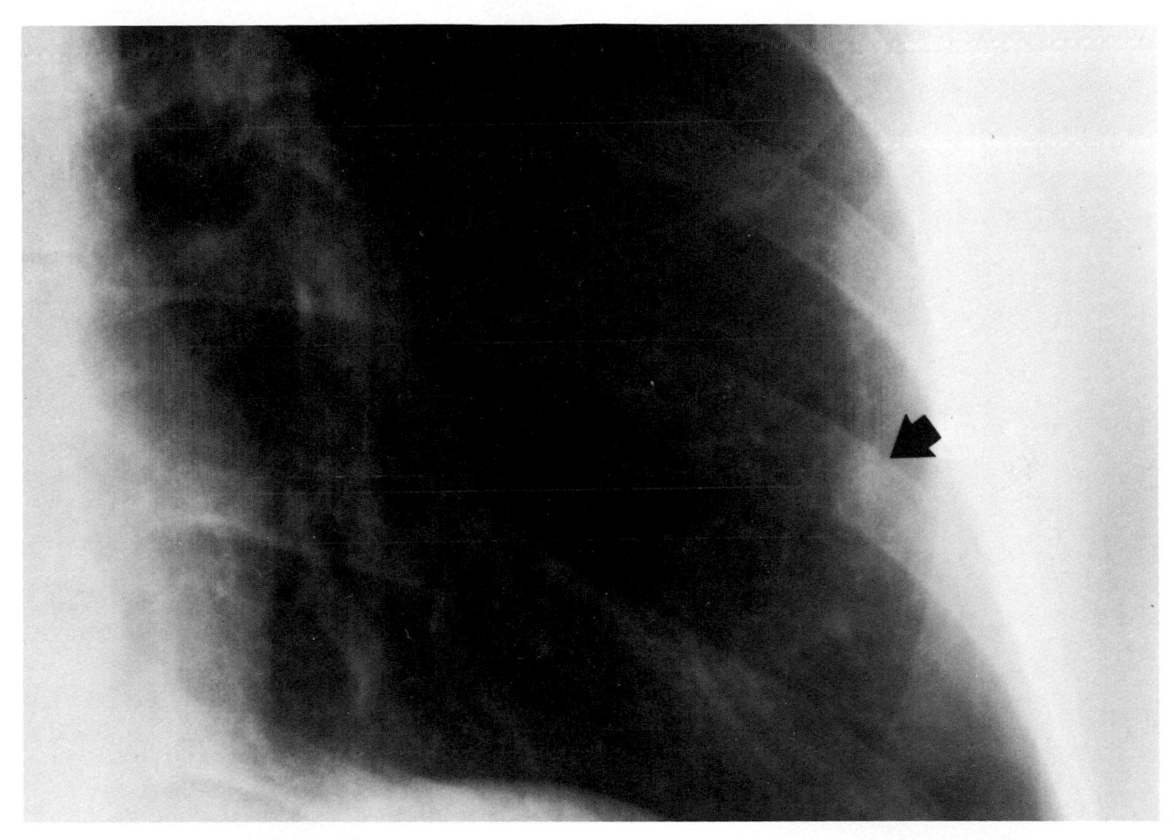

FIG. 3-20. *continued*

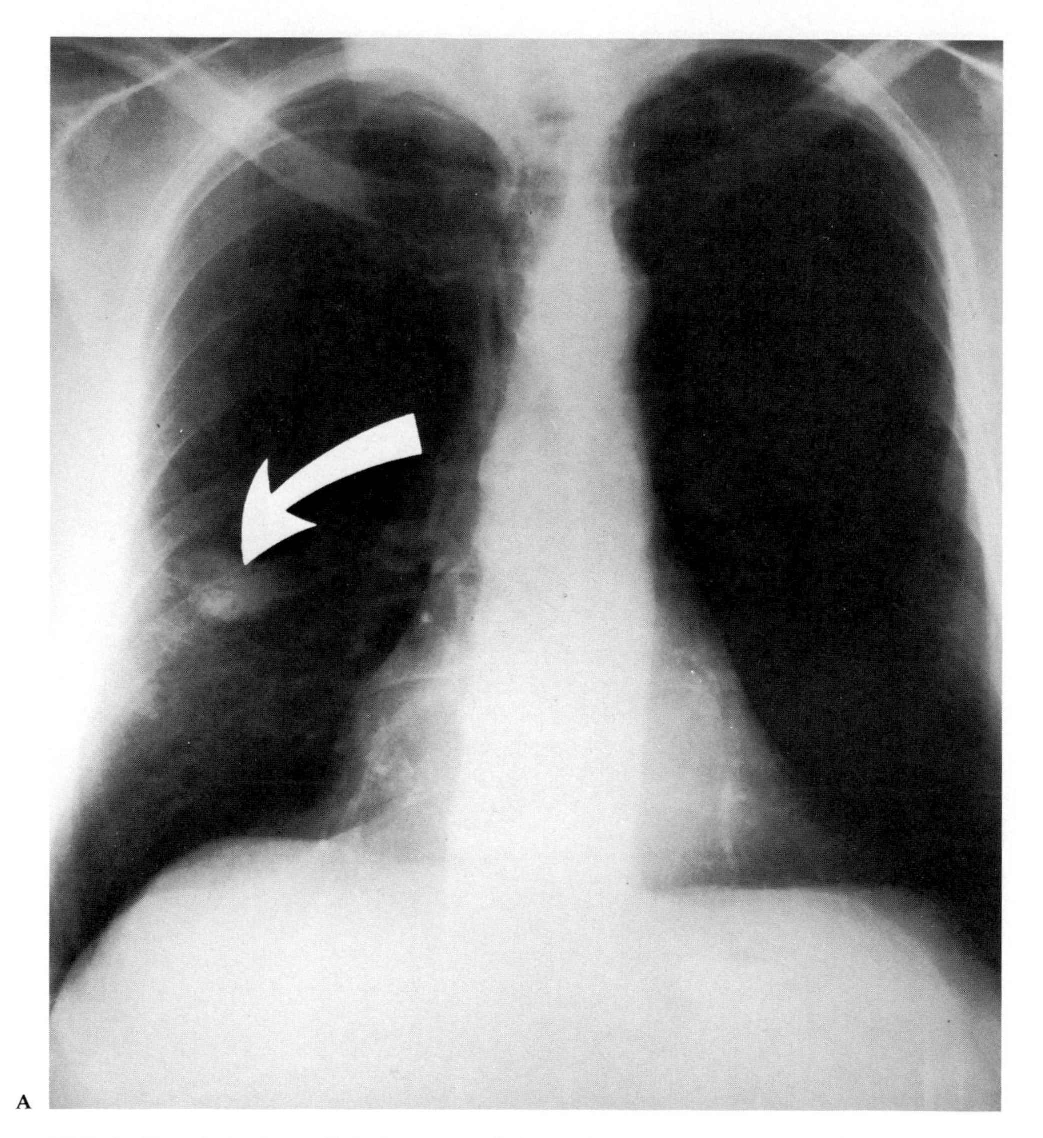

A

FIG. 3-21. (A) *The radiologist reported that a lesion in the right middle lobe* (arrow) *that was noted in previous studies shows little change in size and appearance in this current radiograph.* (B) *In the anterior view of the chest, however, there is an ill-defined nodular density* (arrows) *now projected over the right upper lateral hemithorax.* (C) *The same density* (arrow) *appears in the retrosternal area in the lateral view. The radiologist suggested that the examination be repeated because he felt that the image was artifactual. His suspicion was confirmed. It is interesting to note that when an overlay was made of the anterior and lateral view of the chest, the image of the artifact could be superimposed in each film. Although the exact cause of the artifact is difficult to determine, it probably resulted from pressure exterted on the film while it was being loaded into the chest unit.*

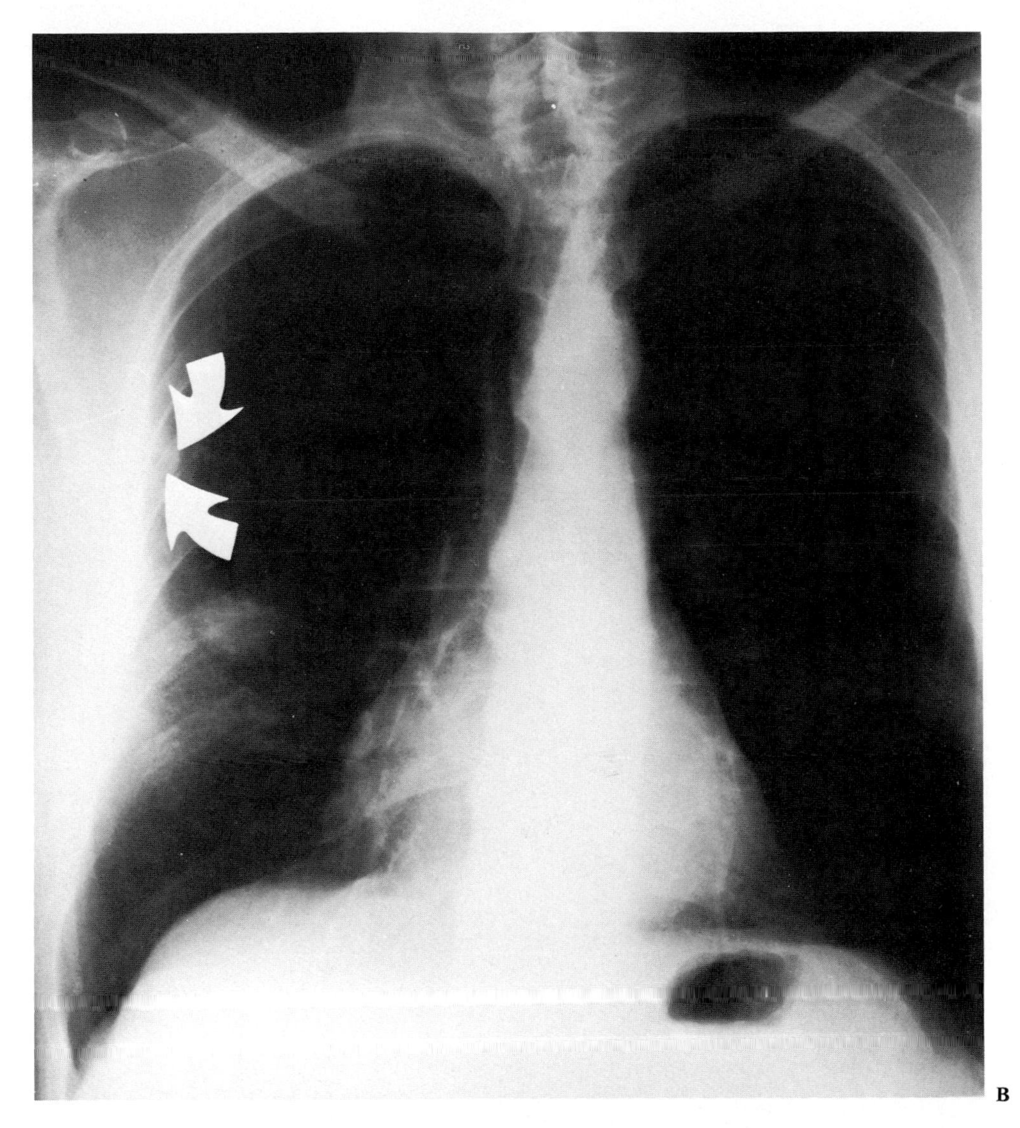

FIG. 3-21. *continued*

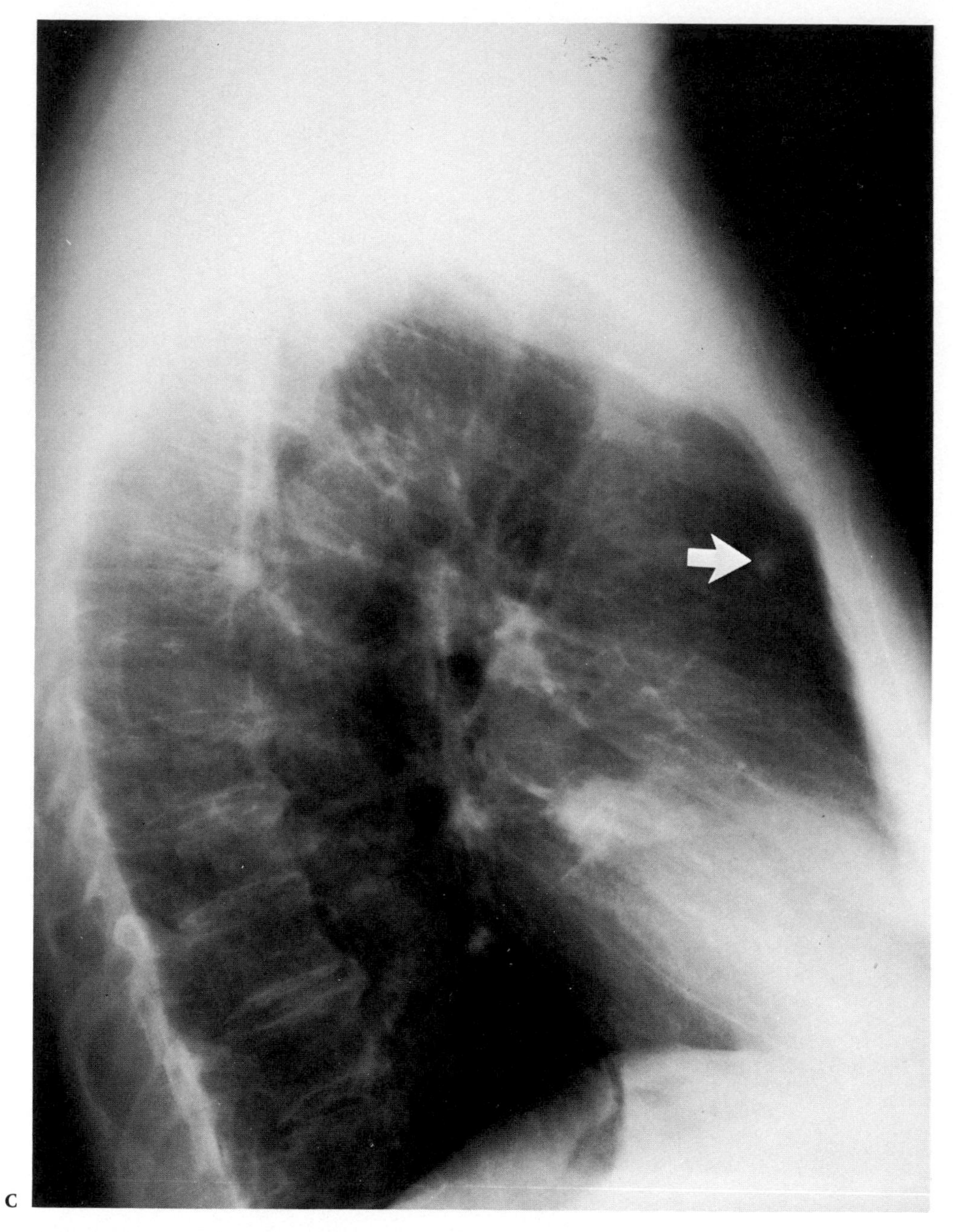

FIG. 3-21. *continued*

References

1. Products for Medical Diagnostic Imaging, Pub. No. M5-15, p 15. Rochester, Eastman Kodak, 1981
2. Thompson TT: Cahoon's Formulating X-ray Techniques, 9th ed, pp 31–33. Durham, Duke University Press, 1979

Suggested Reading

Henny GC: Artifacts in roentgen films. Radiology 24, No. 3:350–356, 1935
Kane IJ: Artifacts simulating minimal tuberculosis. AJR 66, No. 5:803–807, 1951

4 DOUBLE EXPOSURES

In haste it is possible to use a preexposed cassette for a second radiographic examination. This always seems to occur when you are obtaining a radiograph of a very ill patient in the trauma room or at bedside. In order to avoid such problems, you should develop a system whereby unexposed cassettes can be kept separate from those already used for a radiographic examination. Although the identification of such imaging conditions usually does not present a problem, the following examples, which are rather unusual, serve to illustrate the importance of keeping exposed cassettes separate from unexposed ones.

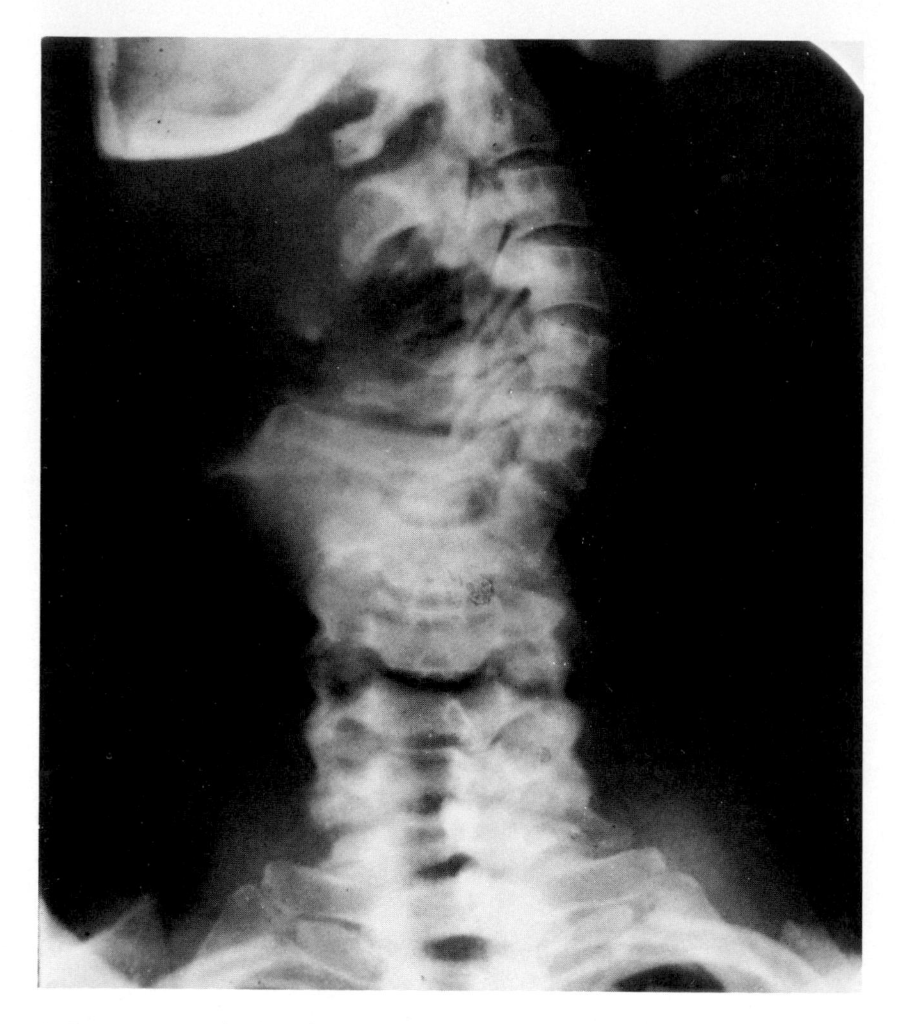

FIG. 4-1. *On viewing this radiograph the emergency-room personnel must have been at a loss to suggest a method of treatment. This unusual image occurred when the same cassette and film were used for both views of the spine.*

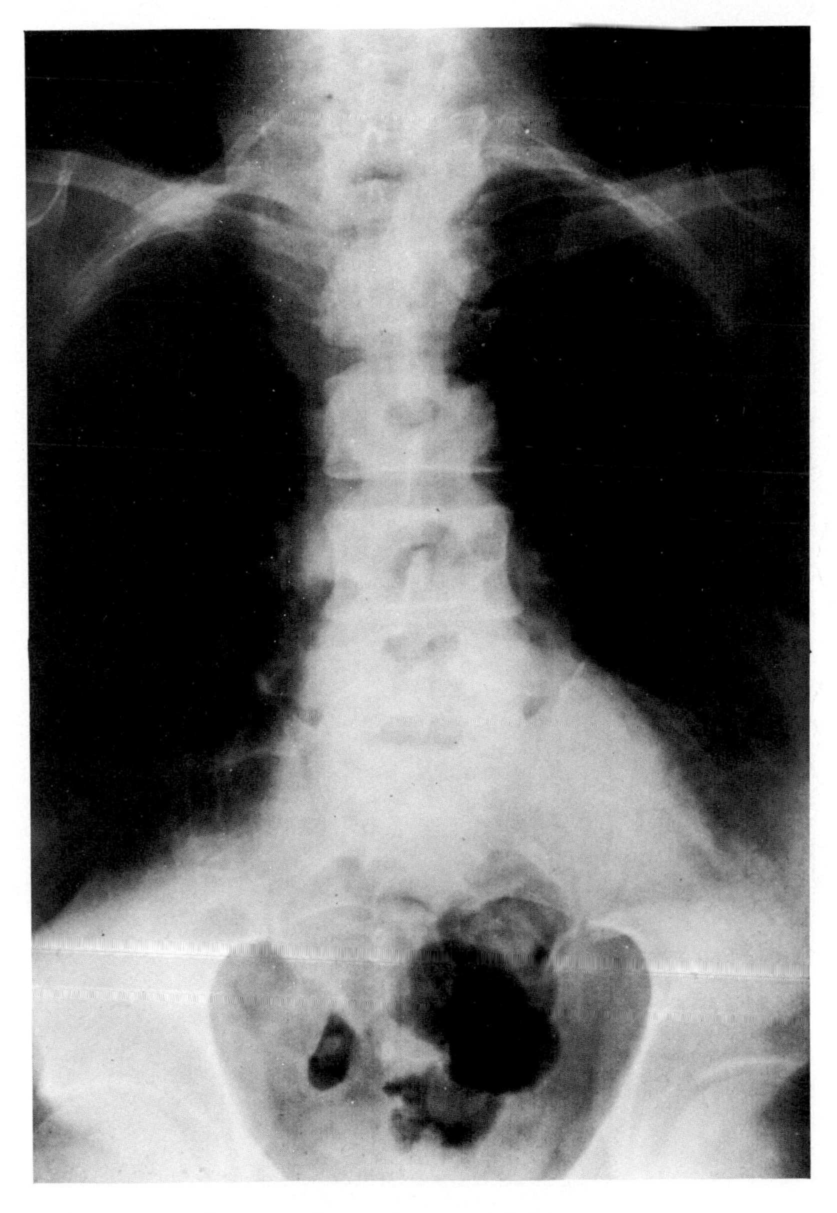

FIG. 4-2. *A chest superimposed over an abdomen.*

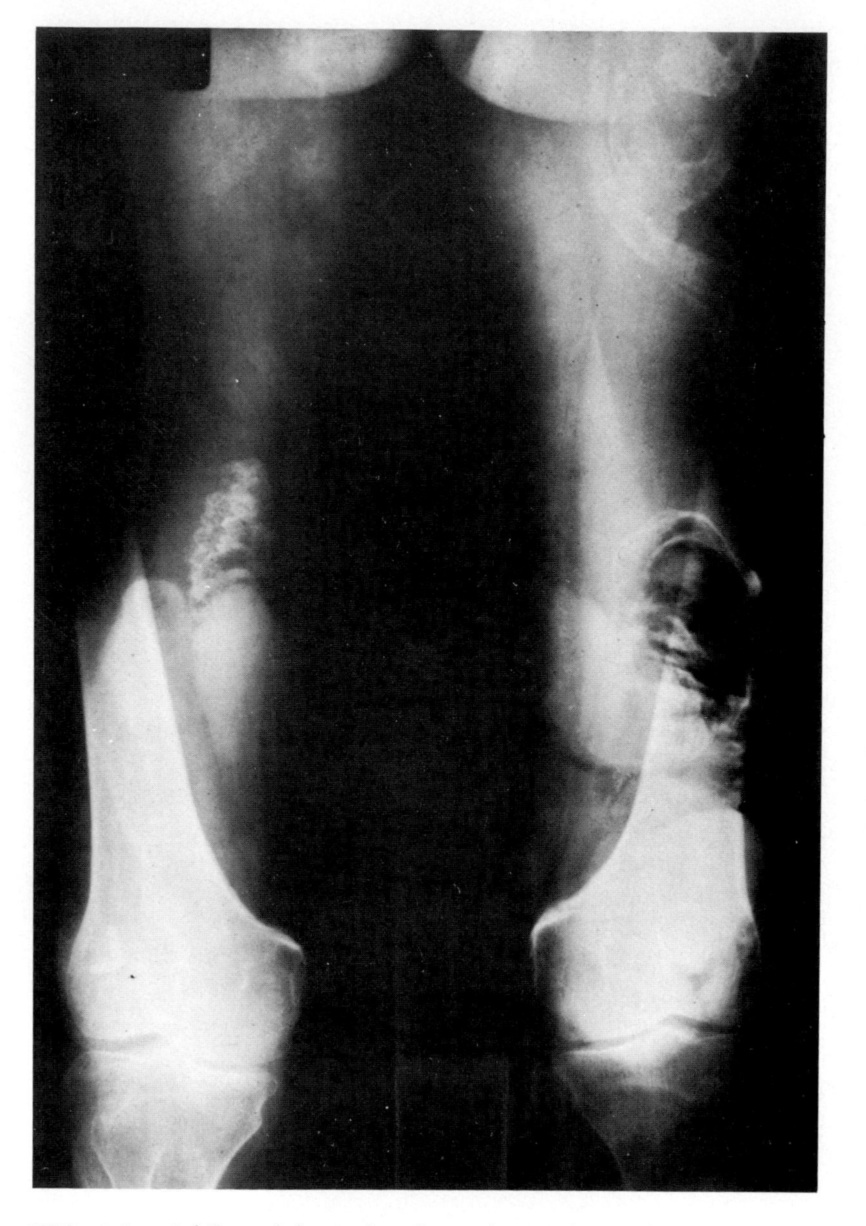

FIG. 4-3. *A bilateral femoral radiograph superimposed over the image of a colon.*

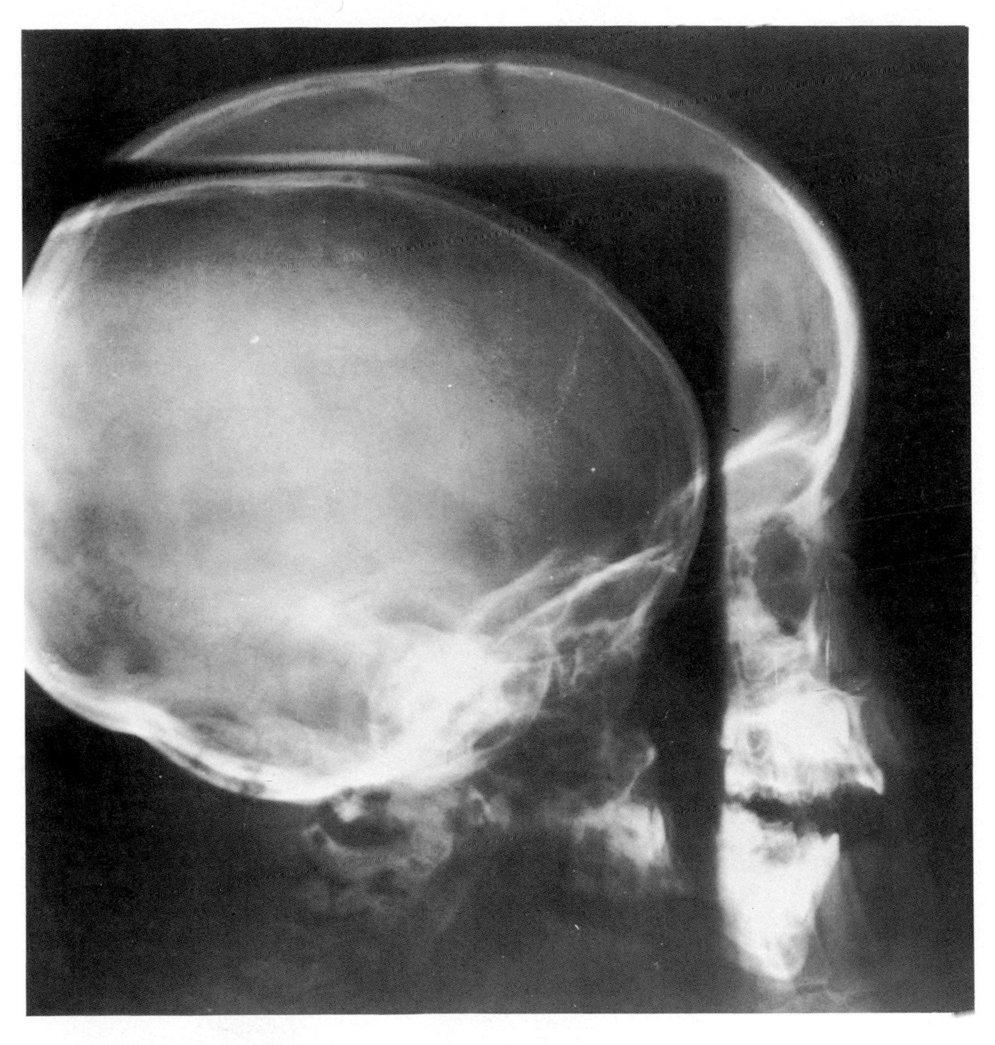

FIG. 4-4. *This double exposure is unusual in that it was made of two different patients positioned in exactly the same manner (Courtesy of Ralph Coates, R.T. [R])*

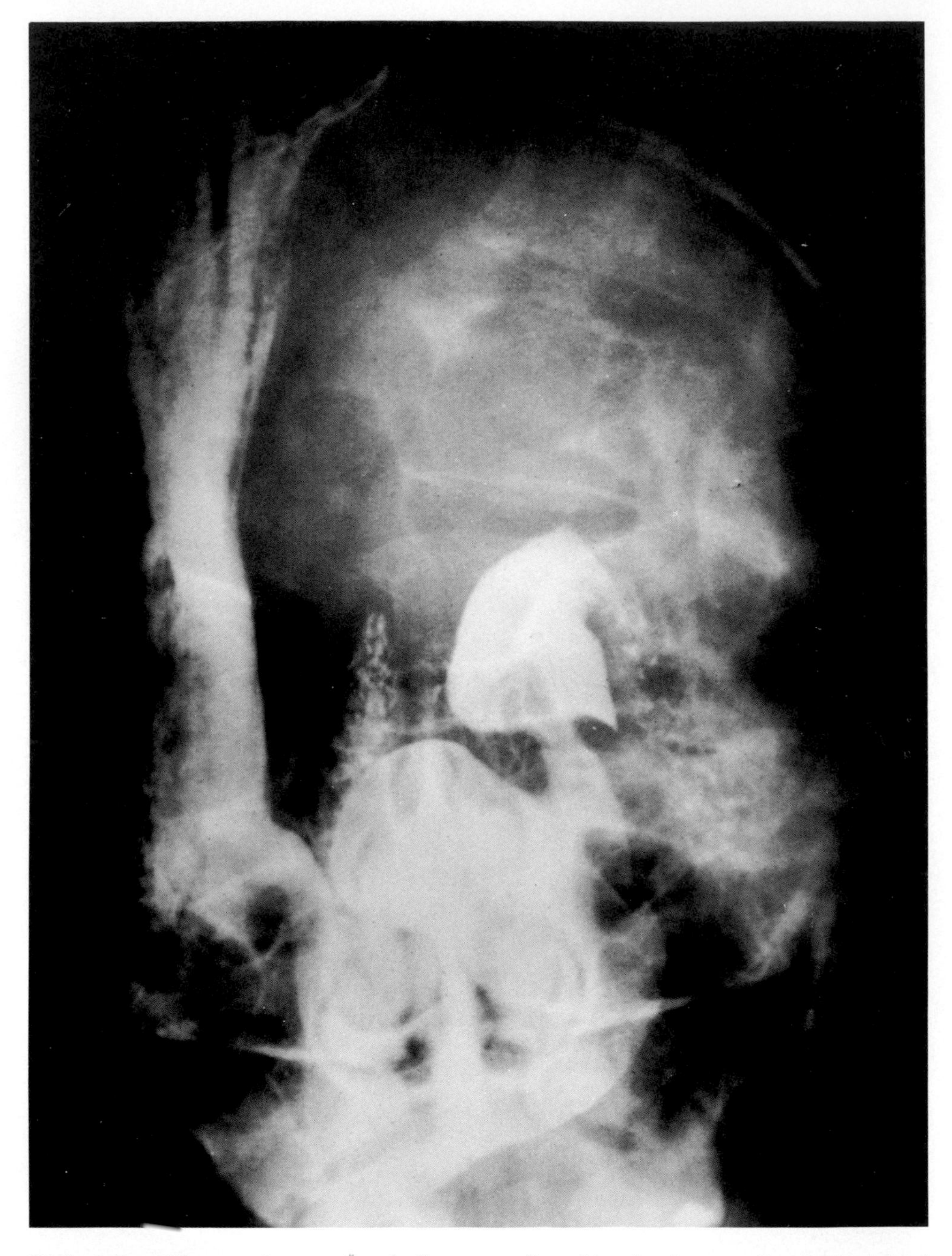

FIG. 4-5. *When an image of a skull appears first thing in the morning in the initial radiograph of a gastrointestinal series, you can feel quite confident in blaming the on-call or night technologist for the problem.*

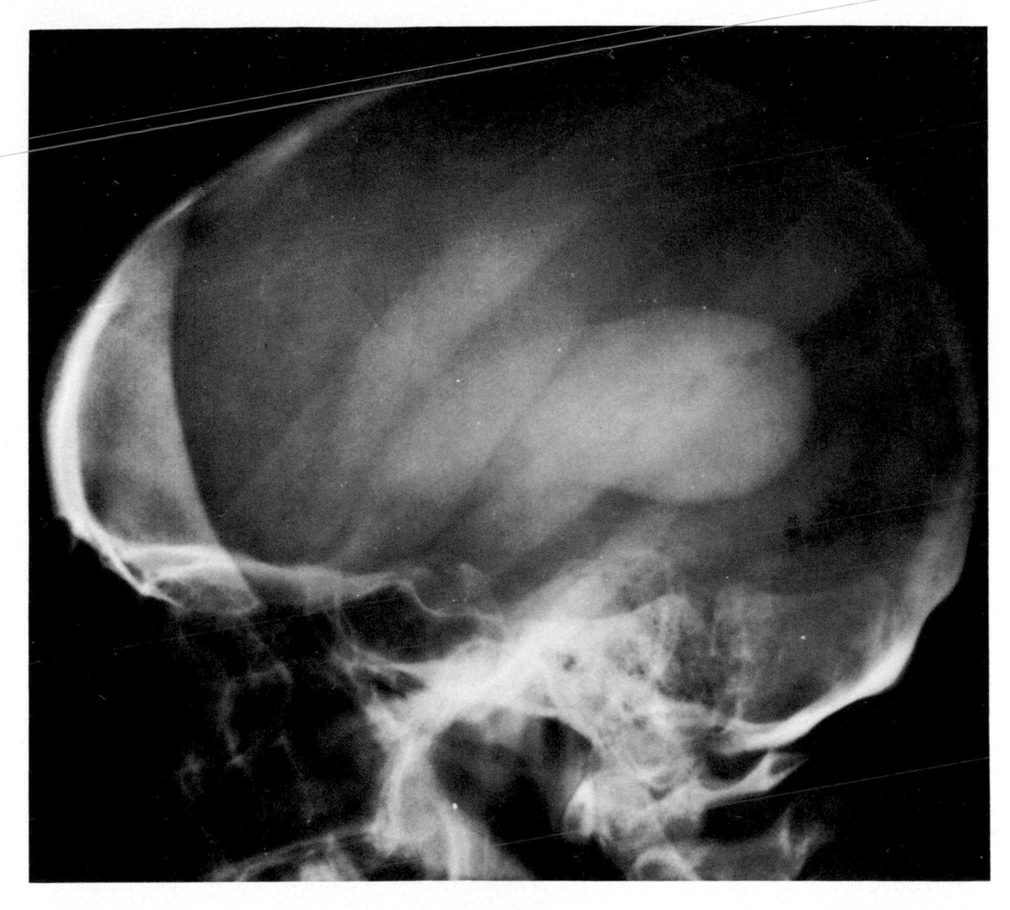

FIG. 4-6. *The appearance of a gallbladder imaged in the skull indicates that an exposed cassette was left in the control area, resulting in a double exposure.*

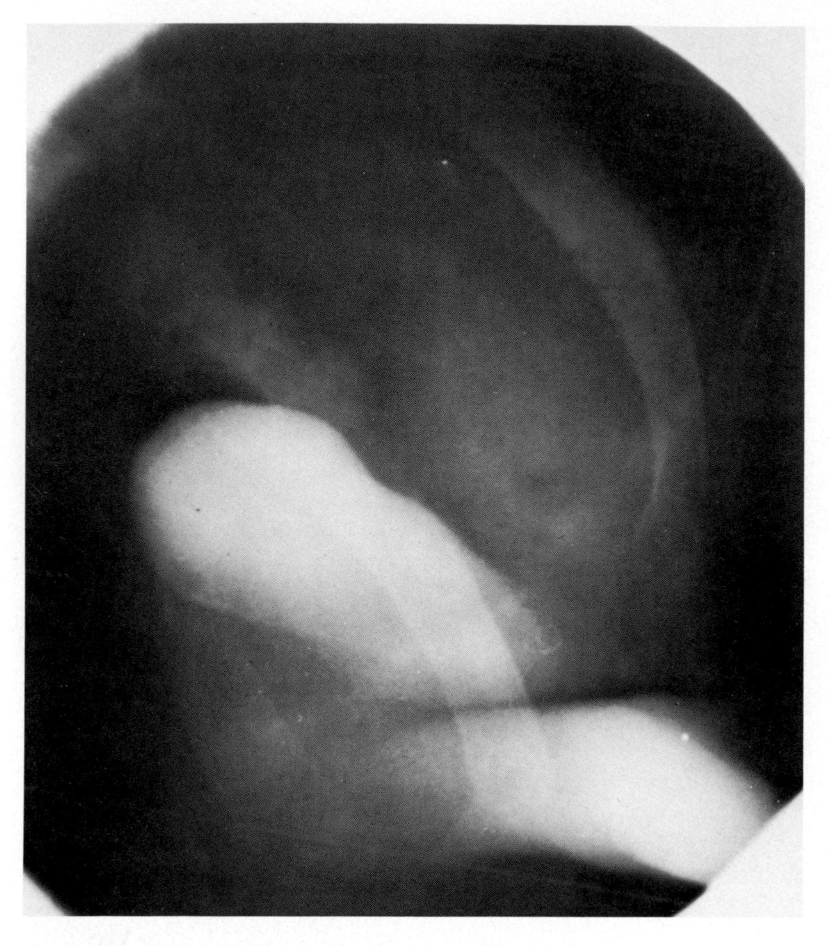

FIG. 4-7. *Double exposure or double gallbladder? Although cases of double gallbladder have been reported in the literature (see Suggested Reading), the peculiar image of the ribs in this radiograph indicates that this is a double exposure.*

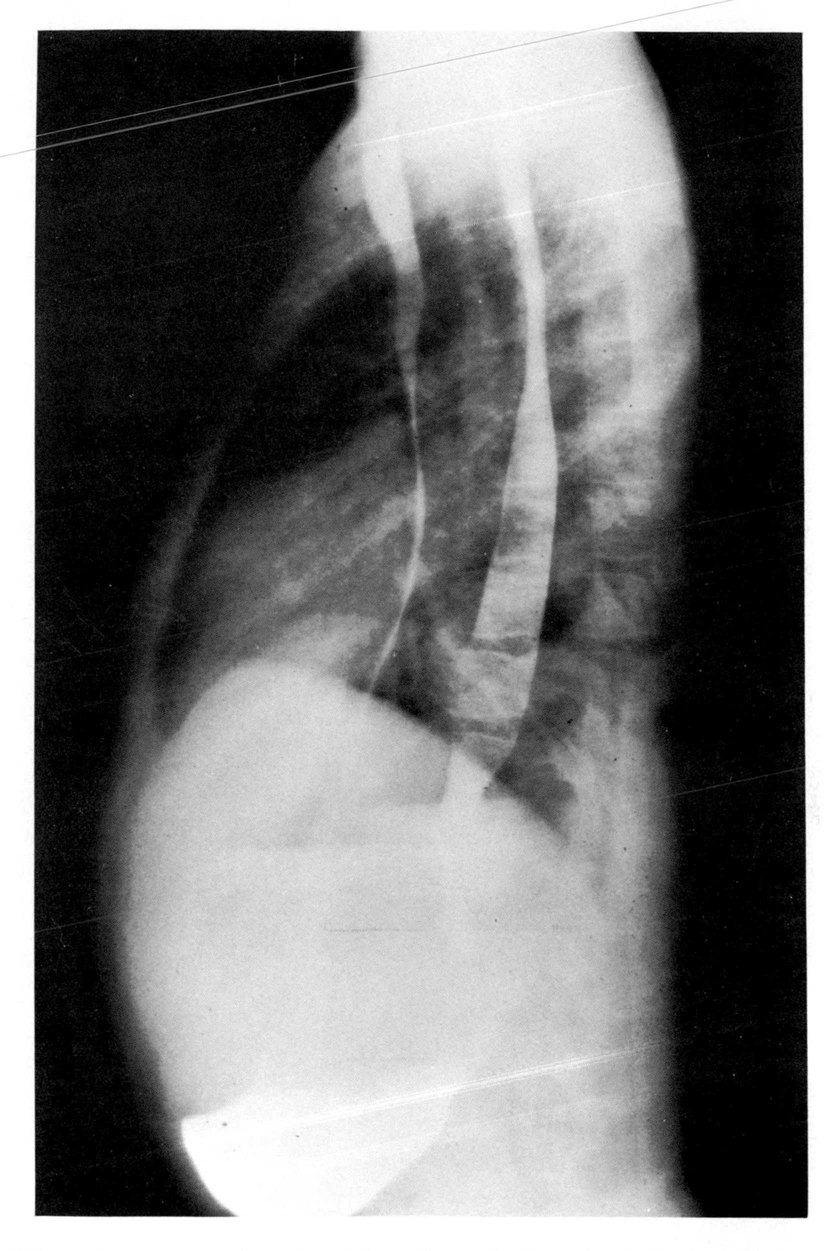

FIG. 4-8. *While in the process of obtaining this radiograph, the technologist thought that the first exposure had not occurred and made another with the same film and cassette. The variation in the appearance of the esophagus is attributed to the repositioning of the patient prior to making the second exposure.*

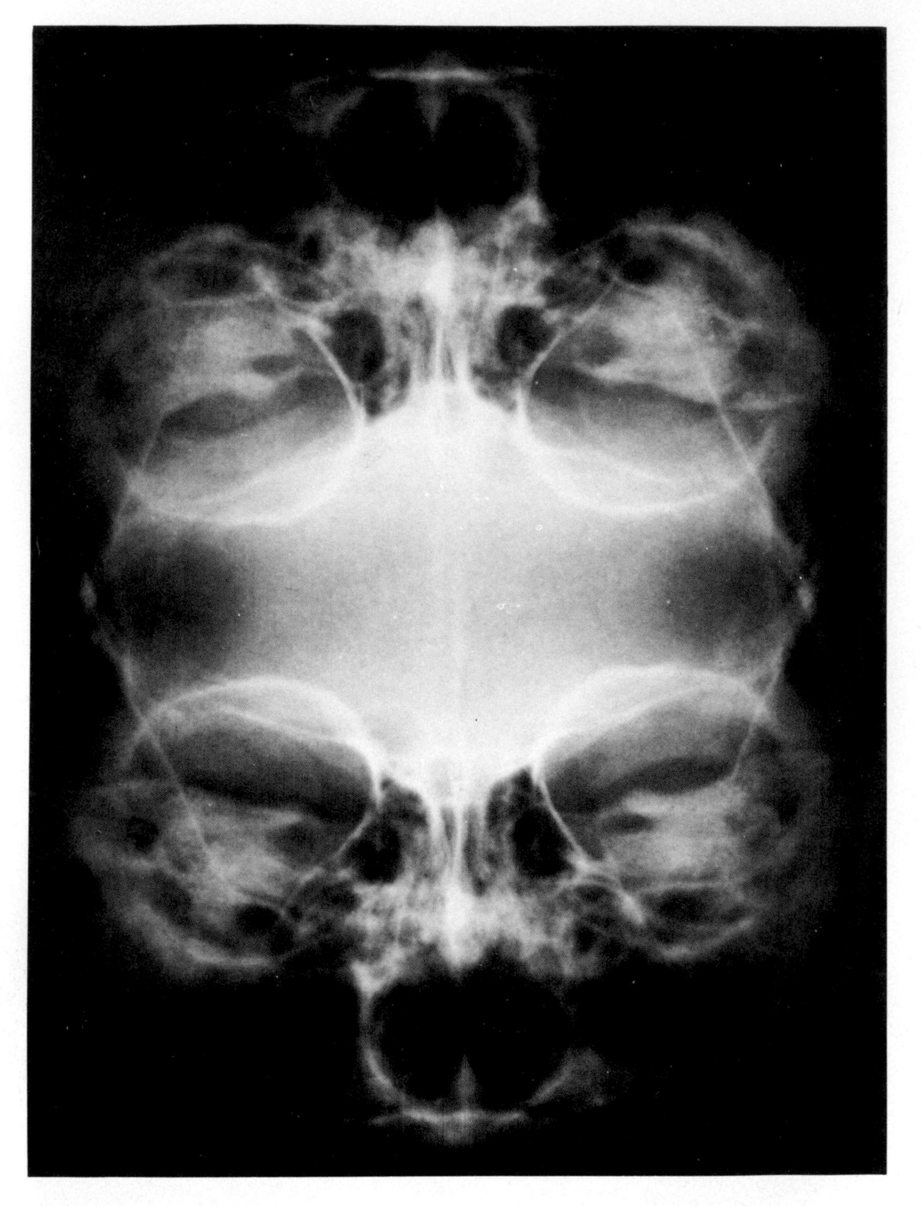

FIG. 4-9. *This unusal image occurred because the same film and cassette were used for both radiographs of a stereo projection of the skull. The inverted appearance of the image indicates that the cassette was repositioned for the second exposure.*

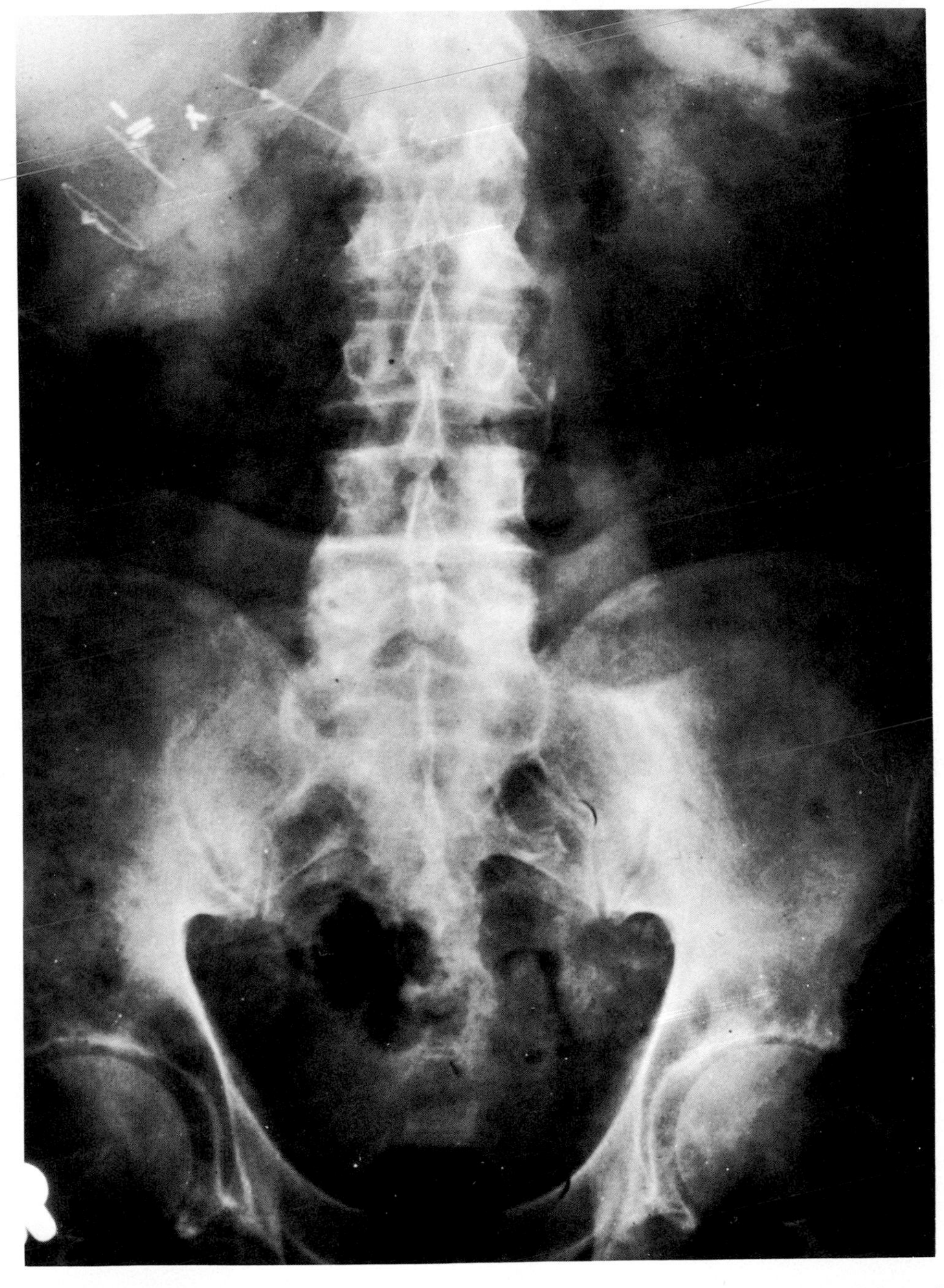

FIG. 4-10. *The image located in the midline portion of the radiograph suggests a bilateral pelvic ilium. The pseudoimage was caused by the compression and superior displacement of fatty tissue in the region of the buttocks (see Suggested Reading).*

Suggested Reading

DOUBLE GALLBLADDER

Barton PB: Double gallbladder, with calculi in one. Med Radiogr Photogr 31, No. 3:127, 1955

Foster DR: Triple gallbladder. Br J Radiol 54(645):817–818, 1981

Kurzwes FT, Cole PA: Triplication of the gallbladder: Review of literature and report of a case. Am J Surg 45(6):410–412, 1979

Ritchie AW, Crucioli V: Double gallbladder with cholecystocolic fistula. Br J Surg 67(2):145–146, 1980

Seabold PS: Double gallbladder, ductular type. Med Radiogr Photogr 31, No. 3:126, 1955

PSEUDO-BILATERAL PELVIC ILEUM

Keates TE: An Atlas of Normal Roentgen Variants That May Simulate Disease, p 105. Chicago, Year Book Medical Publishers, 1975

5 ARTIFACTS CAUSED BY STATIC ELECTRICITY

Origins of Static Electrical Discharges

Before we attempt to control troublesome static artifacts, it may be advantageous to review briefly some of the conditions responsible for static electrical discharges. First of all, it is the natural tendency of all forms of matter to remain electrically neutral. This is accomplished by the movement or migration of electrons from a point of higher electrical potential to one of a lower, negative potential. Such electron movement is likely to produce a static artifact during film handling when dissimilar substances are separated rapidly or rubbed together. Static discharges may or may not produce a visible spark, so static artifacts often appear on the radiograph even though a visible spark was not observed during film handling.[1]

The tree and crown static artifacts shown in Figure 5-1 are thought to be caused by electrical discharges that do not produce visible light. The smudge and spot markings shown in Figure 5-3 are caused by exposure of the film to visible light. In the latter case a spark is produced because of the ionization of the gas molecules in the air next to the film surface.[3]

Although most static artifacts have the appearance of those seen in Figures 5-1 and 5-3, we may encounter some static markings that have a rather bizarre appearance. Such is the case with the artifact shown in Figure 5-2.* This is a very unusual form of static discharge called "swamp static." It results when an electrical charge is released in a manner resembling gas burning in the air or in a swamp. Even though the radiographic appearance of this artifact is unusual, if you employ the same types of preventive measures as for controlling the more common artifacts, the unusual static artifacts will be eliminated as well.

Smudge and spot static are thought to be caused by the exposure of the film to visible light (Fig. 5-4). A discharge is then produced by the ionization of gas molecules in the air next to the surface of the film.[1,6] The periodic use of an antistatic screen-cleaning substance can help to control such problems, but recent advances in the design of the intensifying screens used in angiographic sheet-film changers have reduced the occurrence of such static discharges. These screens are coated on a resin-impregnated paper instead of on a polyester support and then joined to a metal plate that bleeds off the static before it produces an artifact.[2]

*Personal communication: Jon S. Cole and William McKinney, EI du Pont de Nemours & Co, Clifton, NJ, July 17, 1980

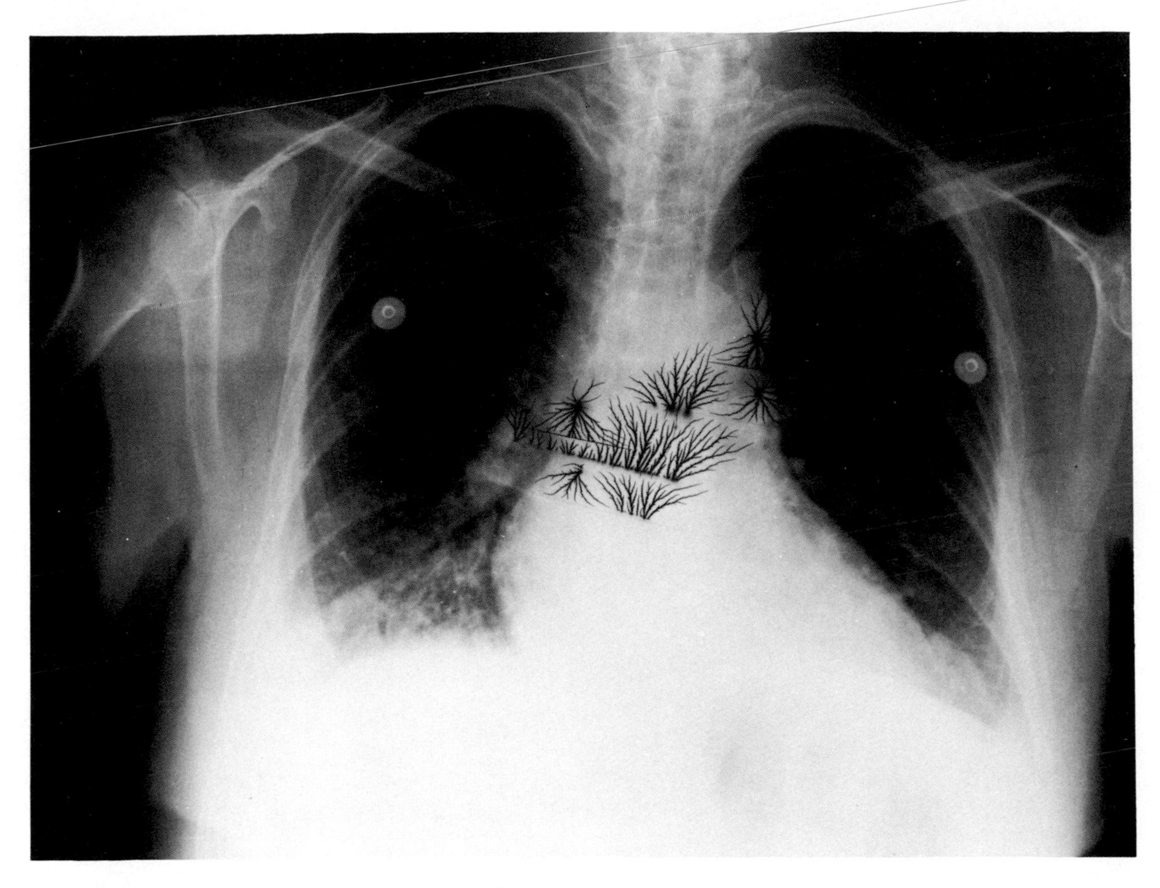

FIG. 5-1. *Crown and tree static artifacts.*

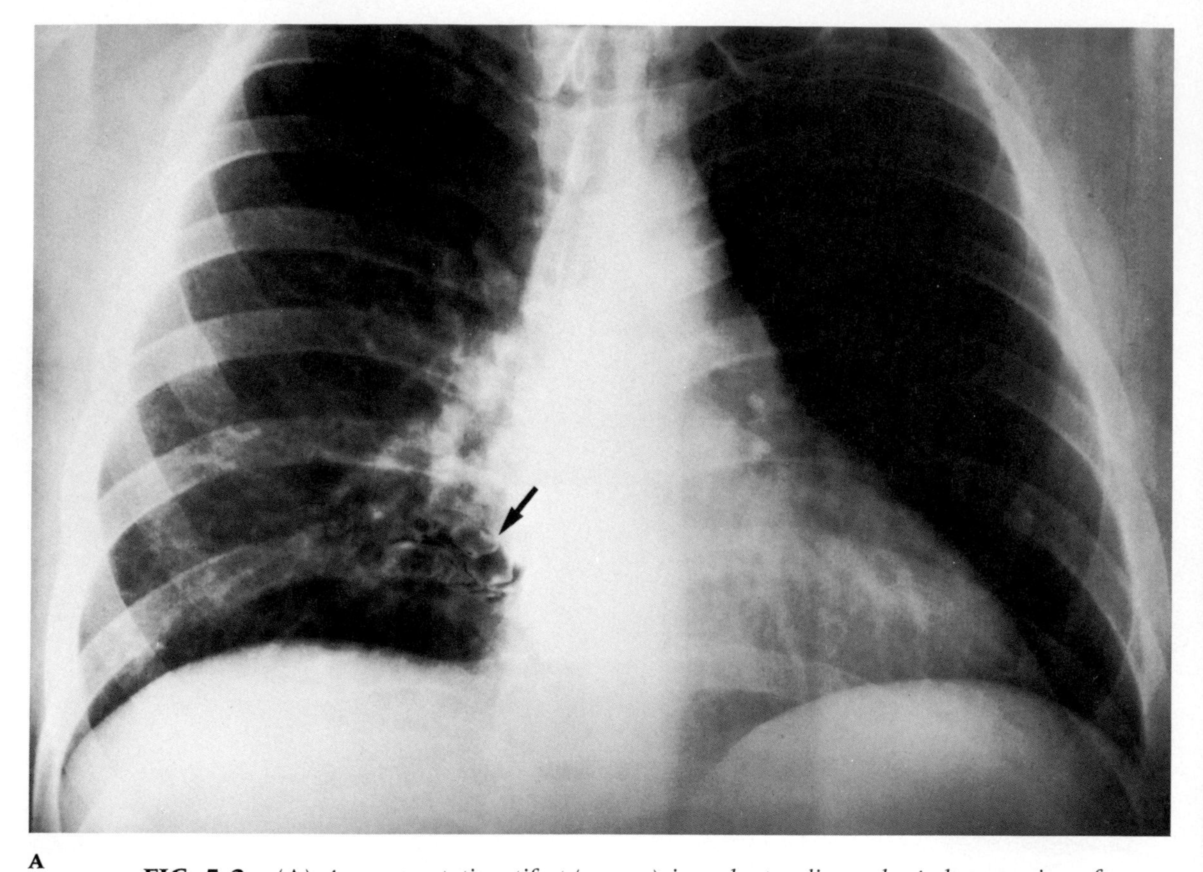

A

FIG. 5-2. (A) *A swamp-static artifact* (arrow) *in a chest radiograph. A close-up view of the artifact is shown in* B (arrow).

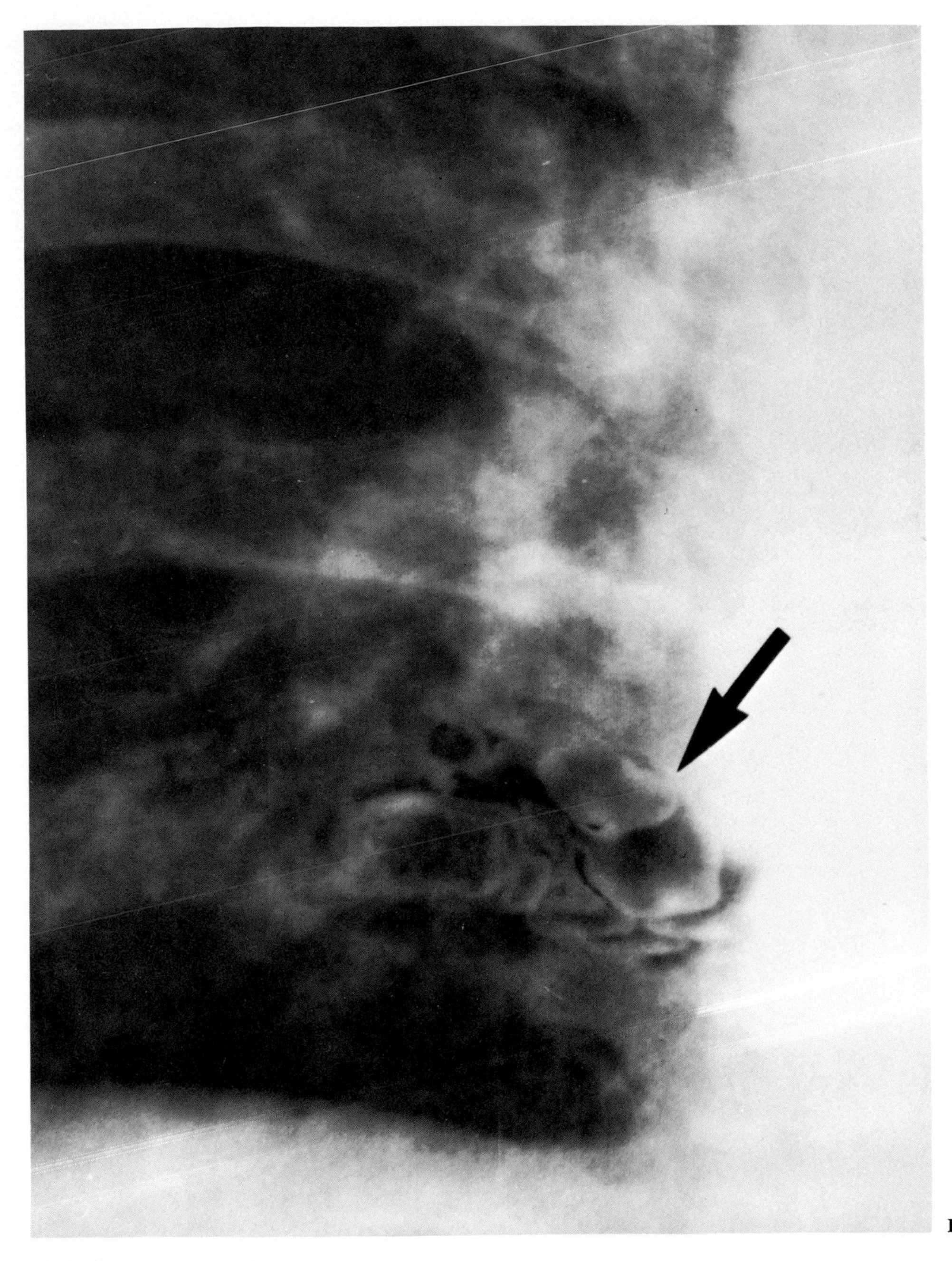

B

FIG. 5-2. *continued*

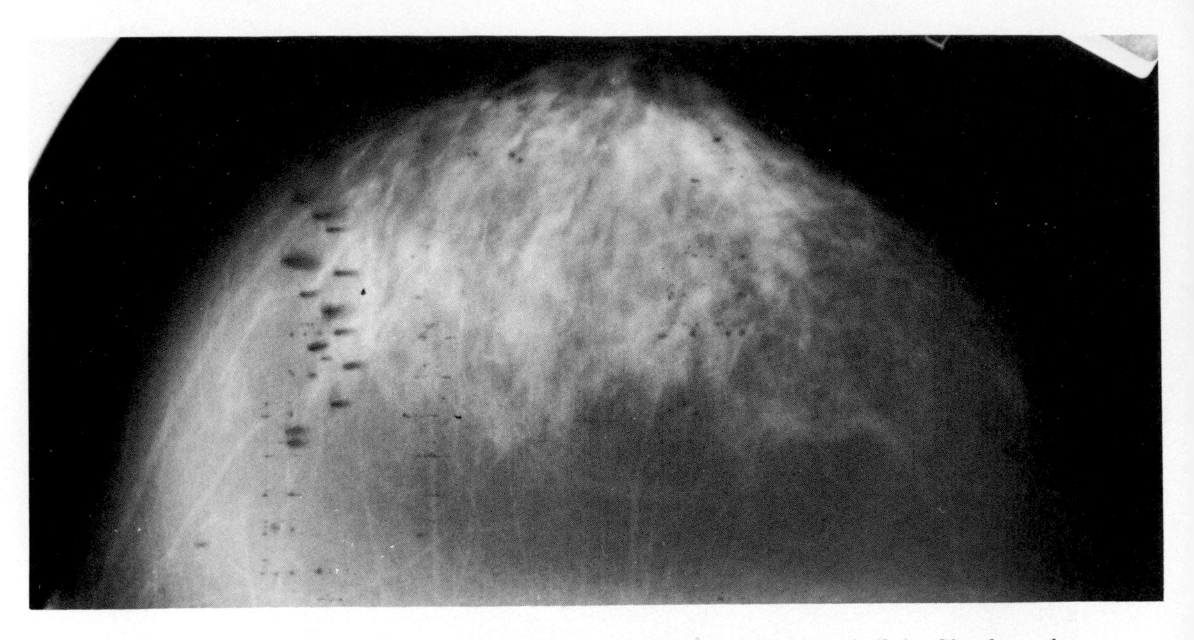

FIG. 5-3. *Smudge and spot static artifacts caused by rapid removal of the film from the intensifying screen in a vacuum cassette.*

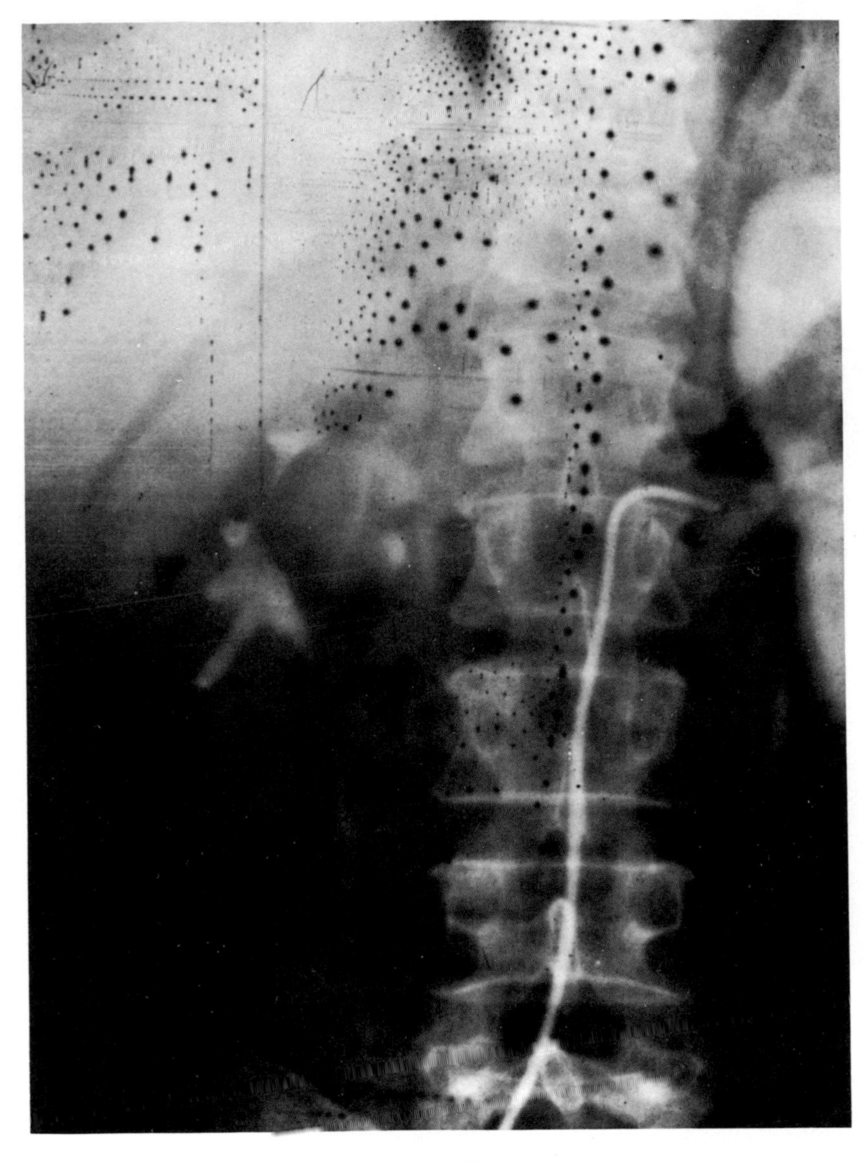

FIG. 5-4. *Smudge and spot static artifacts.*

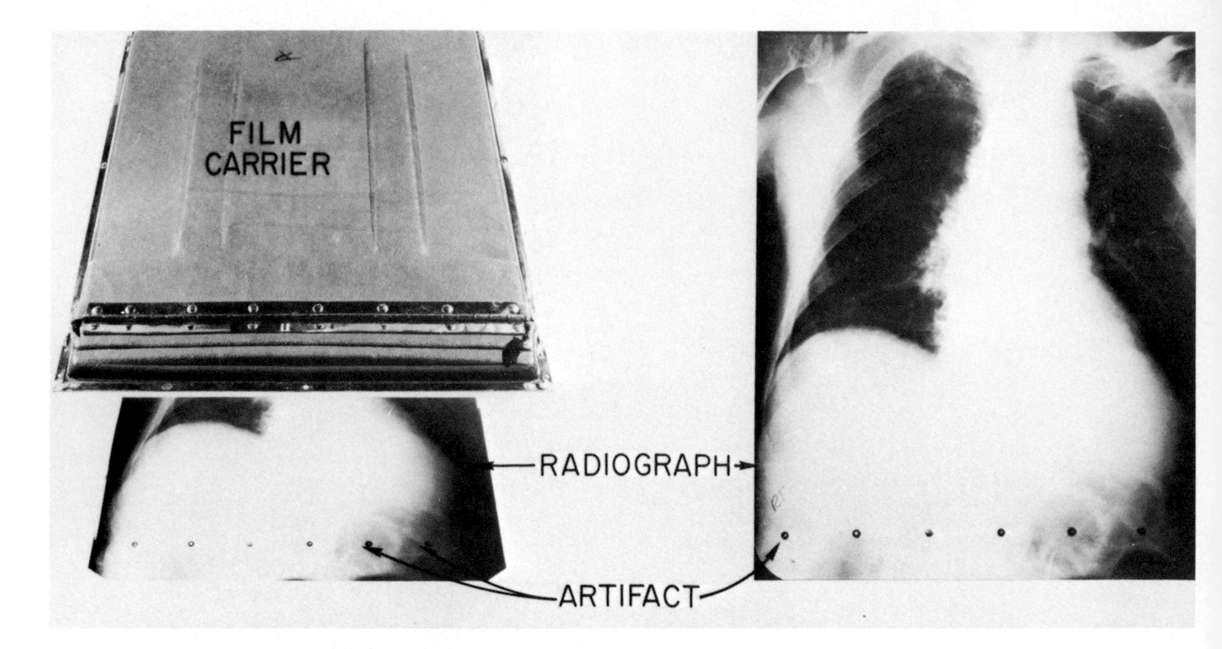

FIG. 5-5. *The static artifacts in this radiograph occurred because the film came into contact with the rivets inside the film carrier. Notice the similarity in the spacing of the artifacts to that of the rivets on the back of the carrier. These carrying cases are excellent, but in time the coating over each rivet disappears, leaving a very prominent surface area. This condition can easily be remedied by covering the surface of each rivet with a strip of a plastic-coated electrical tape. (From Sweeney RJ: On the Technical Side. Radiol Technol 48, No. 1:63–64, 1976)*

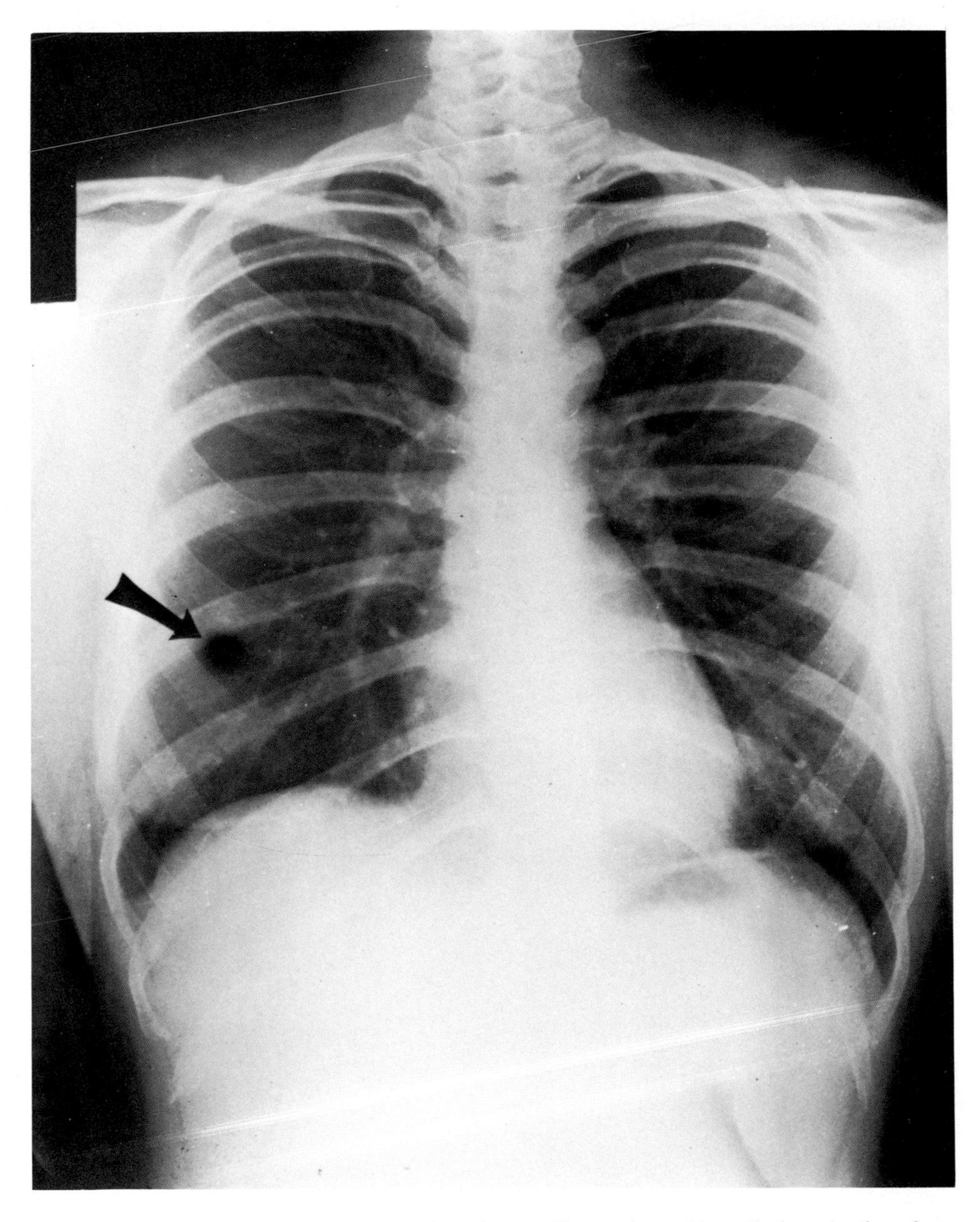

FIG. 5-6. *The artifact* (arrow) *in this chest radiograph could easily be mistaken for a static discharge. In this case, however, the artifact resulted from a radioactive contaminant that became lodged within the film-packing material. Such a condition is uncommon because film manufacturers require samples of packing materials from their suppliers, and each is carefully analyzed for possible impurities.*

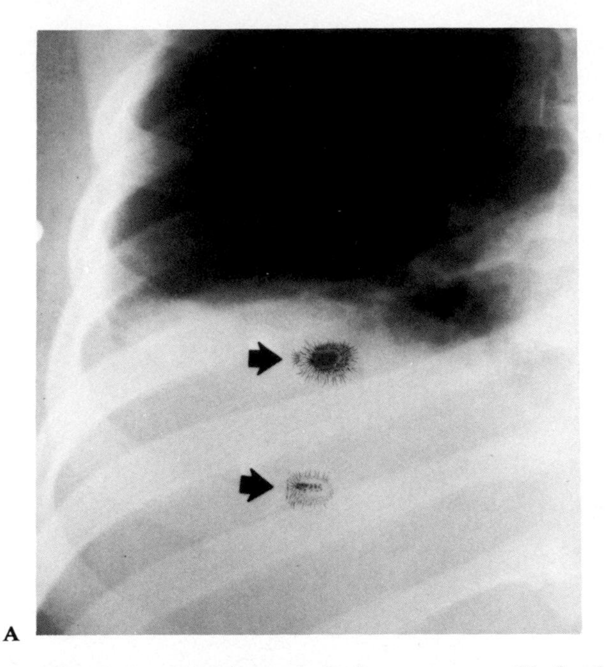

FIG. 5-7. (A) *The static discharges responsible for these artifacts* (arrows) *occurred when the film was being guided across the entrance tray of an automatic processor.* (B) *The configuration of the images indicates that the point of discharge was where the technologist's fingertips came into contact with the film. (A, courtesy of Ralph Coates, R.T. [R])*

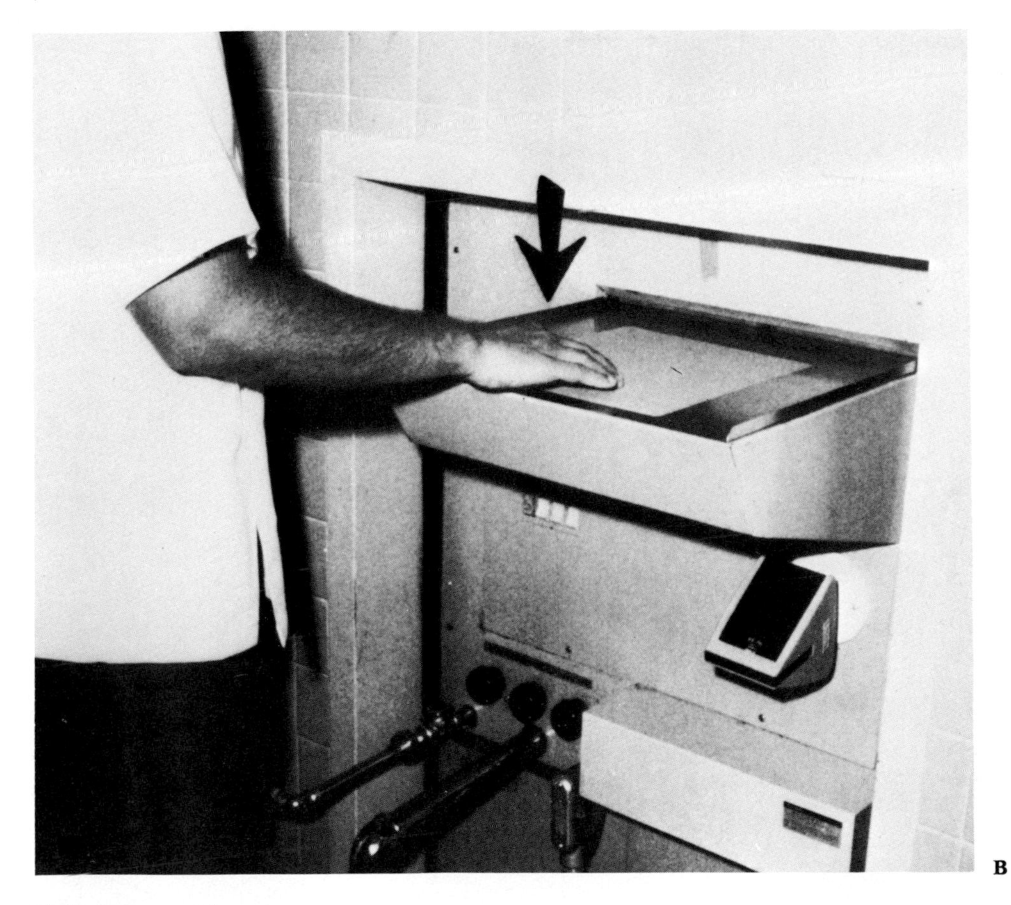

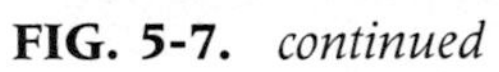

FIG. 5-7. *continued*

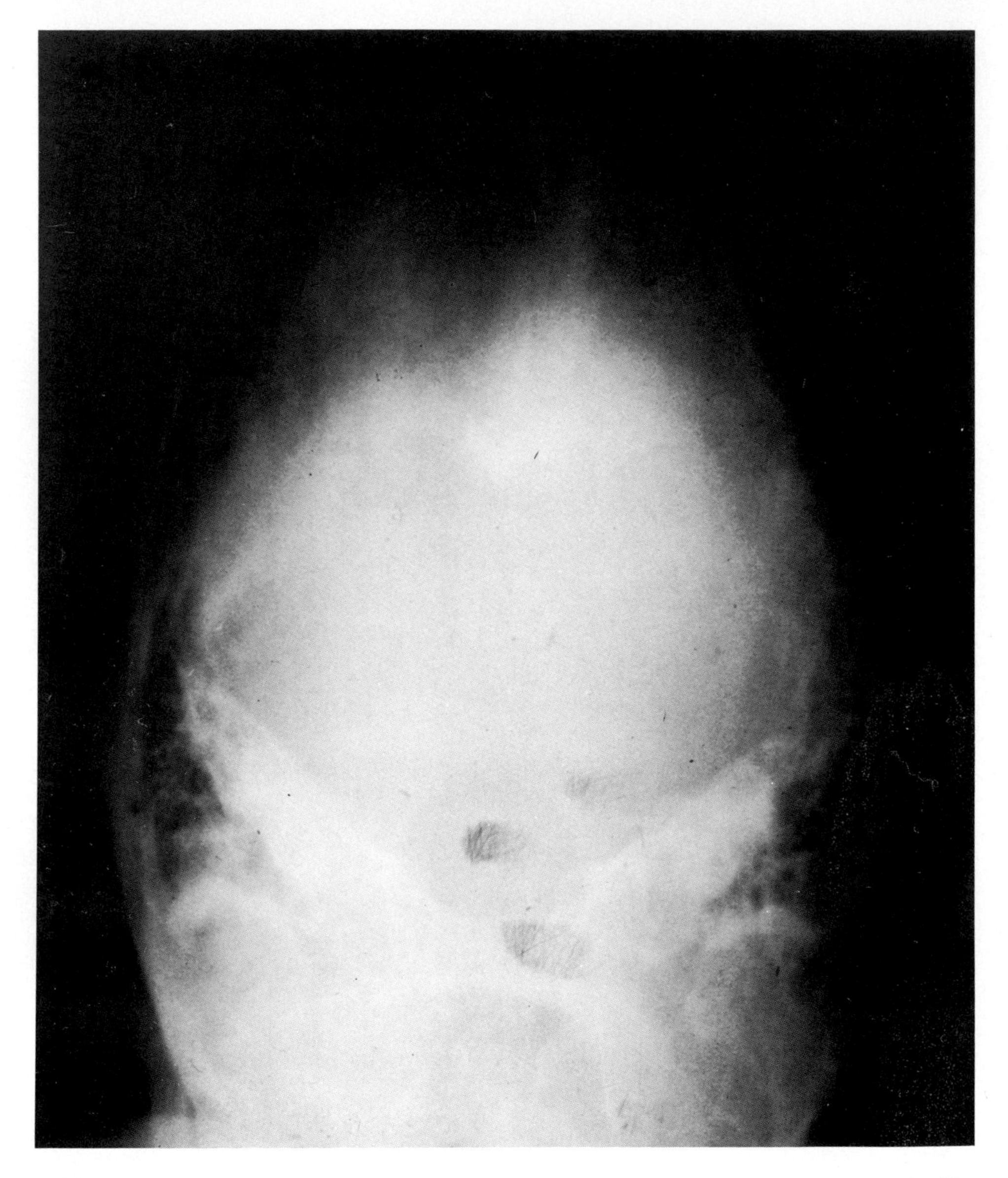

FIG. 5-8. *The abrasion marks in this radiograph resulted from sliding or moving the film across the loading bench or feed-tray assembly of the automatic processor.*

Preventive Measures

In order to dissipate an accumulation of static charges, all radiographic equipment that is in contact with radiographic film should be properly grounded. Although grounding is the simplest solution to static problems, relatively nonconductive materials such as intensifying screens and film present a more difficult problem. Moreover, even though intensifying screens and films are designed to minimize static artifacts, if the surfaces of the screens are not kept free of dust and dirt particles, these particles provide a pathway for the conduction of the electrical discharge.

In addition, commercial intensifying-screen cleaners contain antistatic agents that inhibit such discharges. Consequently, if a periodic screen-cleaning procedure is followed, it will not only help to eliminate the static problem, but in addition, it will remove the dust and dirt particles that are responsible for the appearance of other types of artifacts.

A word of caution concerning the method and type of screen-cleaning agent used for this procedure: The technologist should consult with the manufacturer's technical representative for his recommendations on cleaning intensifying screens. Various types of screen-cleaning agents may leave residues that can seriously affect the light transmission of some intensifying screens.

TEMPERATURE AND RELATIVE HUMIDITY

The technologist should make every effort to control the temperature and relative humidity in the radiology department. All factors considered, one of the most successful ways of minimizing static problems is to maintain a high relative humidity in order to increase the moisture content of all surfaces, including clothing, floors, radiographic film, and accessories. This added moisture in turn increases the conductivity of these surfaces, which decreases their tendency to accumulate static charges. Most authorities recommend that radiographic film be stored at a temperature between 65°F and 75°F (13°C and 24°C) at a relative humidity of 50%.[1,3–5,7,8] However, as with screen-cleaning procedures, one should always consult the manufacturing representative on the use of their specific products.

References

1. Darkroom Technique for Better Radiographs Processed Manually or Automatically, Pub. No. A-61109, p 4. Wilmington, EI du Pont de Nemours & Co, 1961
2. Eastman Kodak: Tech Talk No. 7. Med Radiogr Photogr 56, No. 2, 1980
3. Processing and Handling of Medical X-ray Films, Pub. No. XM3-3E, p 4. New York, Fuji Photographic Film, 1978
4. Products for Medical Diagnostic Imaging, Pub. No. M5-15, p 15. Rochester, Eastman Kodak, 1981
5. Roderick JF, Sutherland B: The Static Problem in the Darkroom. Wilmington, EI du Pont de Nemours & Co, 1958
6. A Study of the Static Electricity Problem in the X-ray Darkroom. Wilmington, EI du Pont de Nemours & Co. Reprinted from X-ray Technician 23:a, 343–348; b, 365, 1952
7. Thompson TT: A Practical Approach to Modern X-ray Equipment, pp 7–8. Boston, Little, Brown & Co, 1978
8. Thompson TT: Cahoon's Formulating X-ray Techniques, 9th ed, pp 30–36. Durham, Duke University Press, 1979

6 ARTIFACTS RELATED TO IMPROPER CONDITIONS OF THE DARKROOM AND INTENSIFYING SCREENS

Maintaining Darkroom Cleanliness

Back when all films were manually processed, time was somehow allotted for cleaning the darkroom and processing accessories. However, today the emphasis has shifted to the care and maintenance of the automatic processor, and darkroom cleanliness is often ignored. In fact, in many departments the film-handling areas are considered off limits to housekeeping personnel. Although it is important not to allow anyone to clean these rooms without proper supervision, it is extremely important that film-handling areas be kept clean in order to minimize the possibility of numerous artifacts. Consequently, the technologist should develop a schedule in order to see that these areas are inspected and cleaned periodically.

IDENTIFICATION CARDS

Using identification cards in conjunction with a photographic printer is a conventional method of identifying radiographs. However, allowing identification cards to accumulate on the loading bench in the darkroom is responsible for their accidental entry into the cassette during the film-loading procedure (Fig. 6-1). In some instances the card may have multiple carbon inserts for use by file-room personnel and others. These inserts, too, may cause a variety of artifacts if not handled properly (Fig. 6-2).

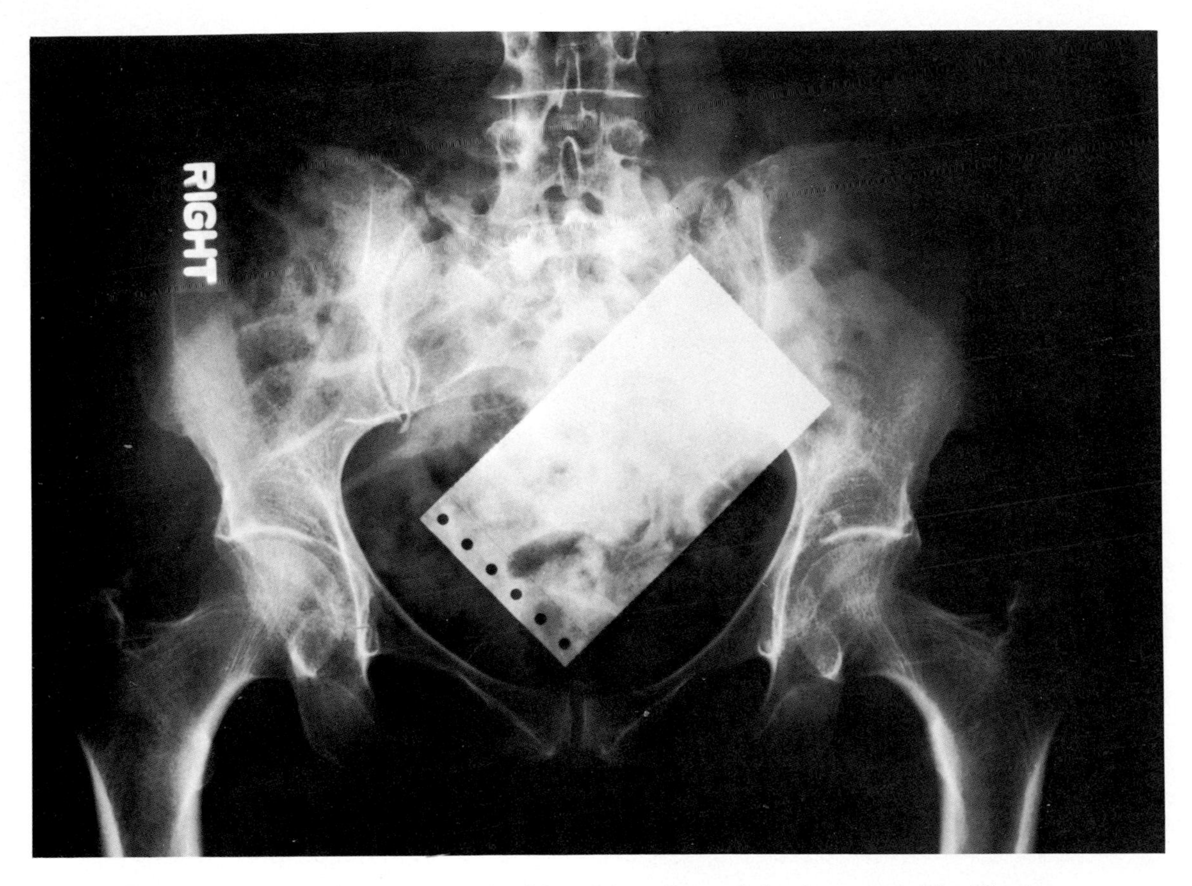

FIG. 6-1. *The artifactual image in this pelvic radiograph is of a patient identification card that somehow found its way into the cassette during film loading.*

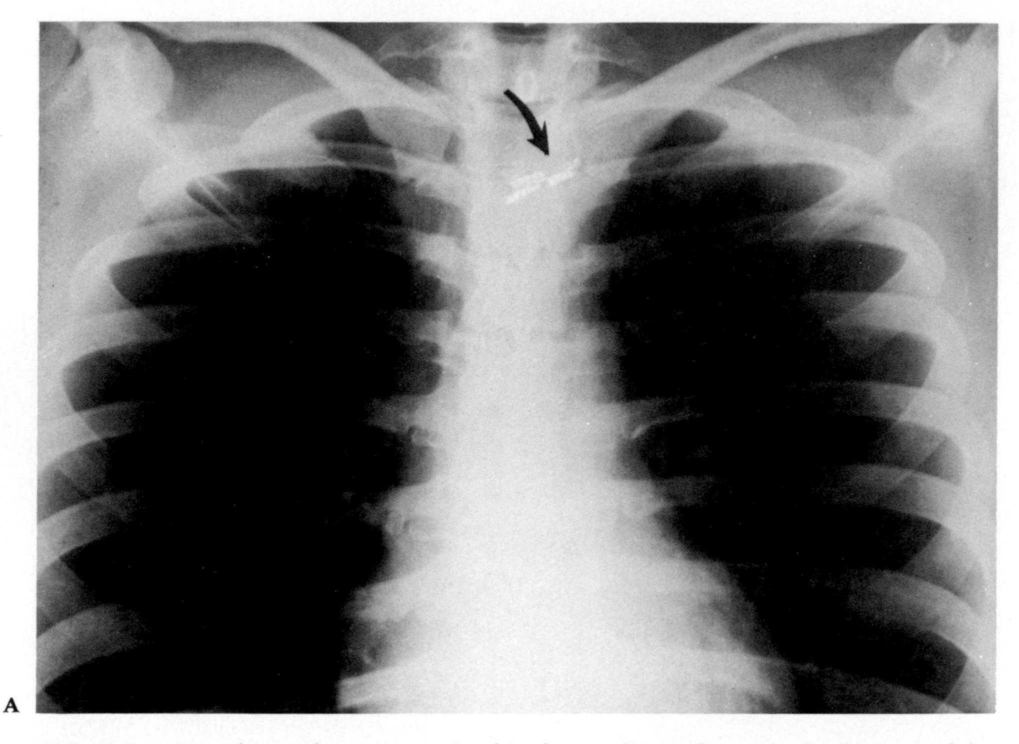

FIG. 6-2. (A) *The artifact* (arrow) *in this chest radiograph occurred when one of the carbon inserts from an identification card* (B) *accidentally found its way into the cassette during film loading. When the cassette was closed, the raised lettering on the carbon was imprinted on the surface of the intensifying screen. Following discovery and removal of the carbon paper, the cassette was put back into use. It was used until the imprint was discovered during a visual inspection of the intensifying screen.*

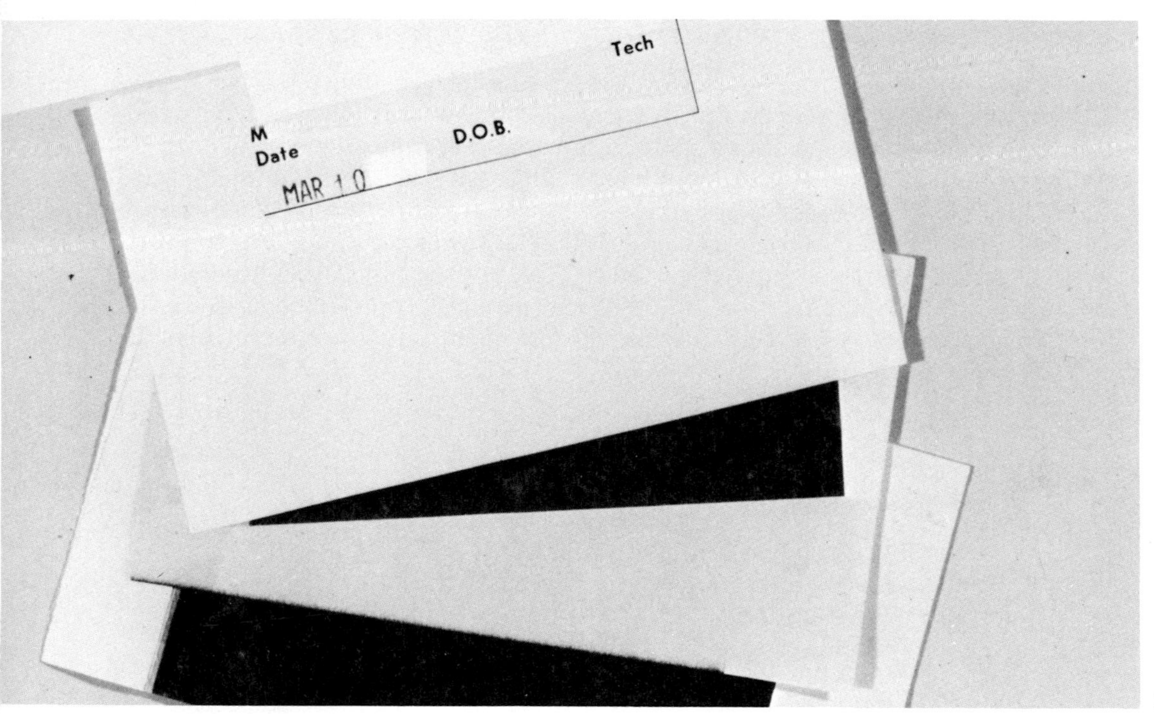

FIG. 6-2. *continued*

Causes and Control of Intensifying-Screen Artifacts

When you use an intensifying screen, the screen's light-emitting phosphor coating is responsible for most of the density seen in the radiograph. However, if the screen is damaged or dust and dirt are on its surface, the proper amount of light does not reach the film. Such a condition produces white spots or specks on the finished radiograph (Fig. 6-3). In order to illustrate this point, the surface of an old intensifying screen was deliberately damaged (Fig. 6-4). The damaged portions appear as white areas on a radiograph of the screen (Fig. 6-4, *B*), just as they would in an actual radiographic image. These problems are generally costly, not only as a result of repeat studies but also because of the expense incurred in the replacement of the screens.

Screen-related problems can be reduced or eliminated by adhering to the following guidelines:

Periodically inspect and clean screen surfaces.

If you are responsible for loading and unloading cassettes, avoid wearing rings and bracelets that can scratch the surface of the screen.

Exercise greater care when loading and unloading cassettes.

Never leave cassettes open when they are not in use.

Be more critical of all types of artifacts when you check radiographs.

Establish a regular schedule for cleaning intensifying screens.

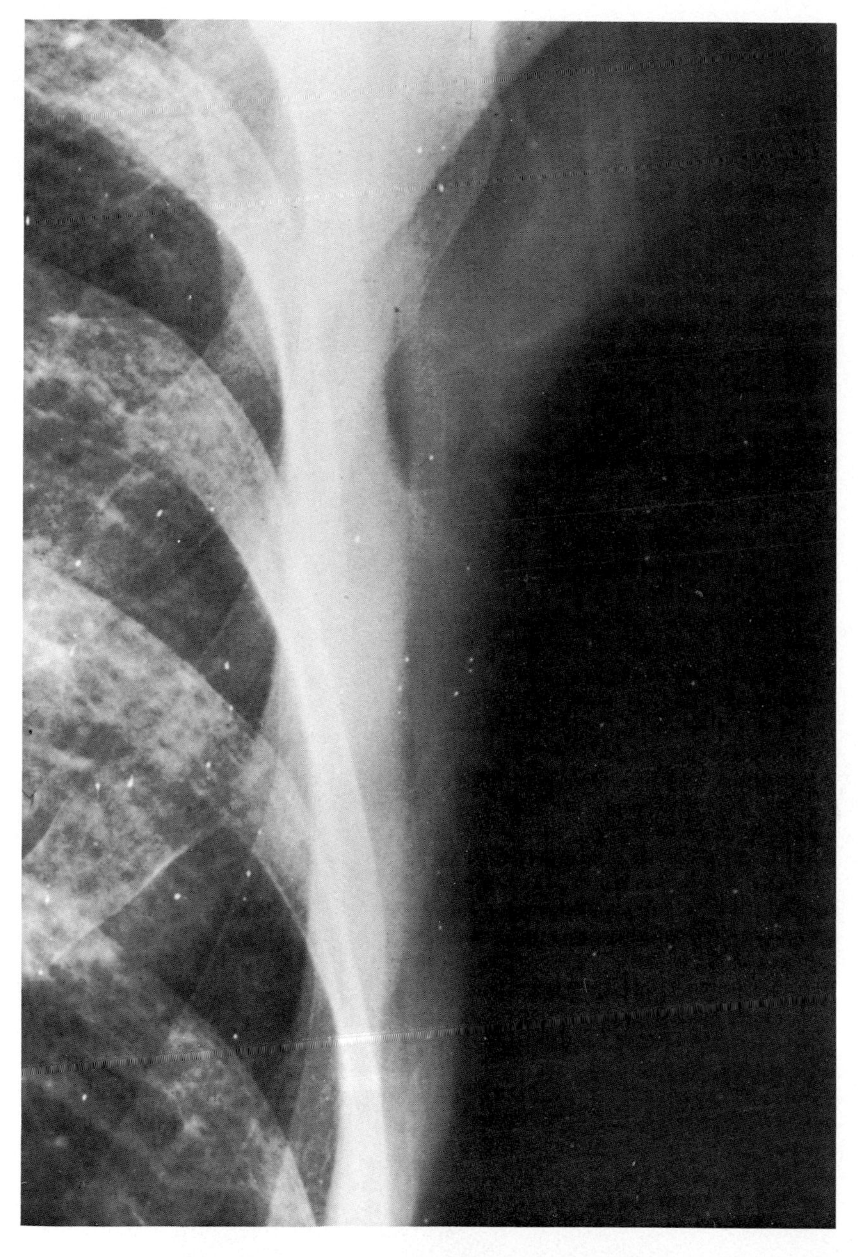

FIG. 6-3. *The multiple specks of minus density seen in this radiograph were caused by the presence of dust and dirt particles on the surface of the intensifying screen.*

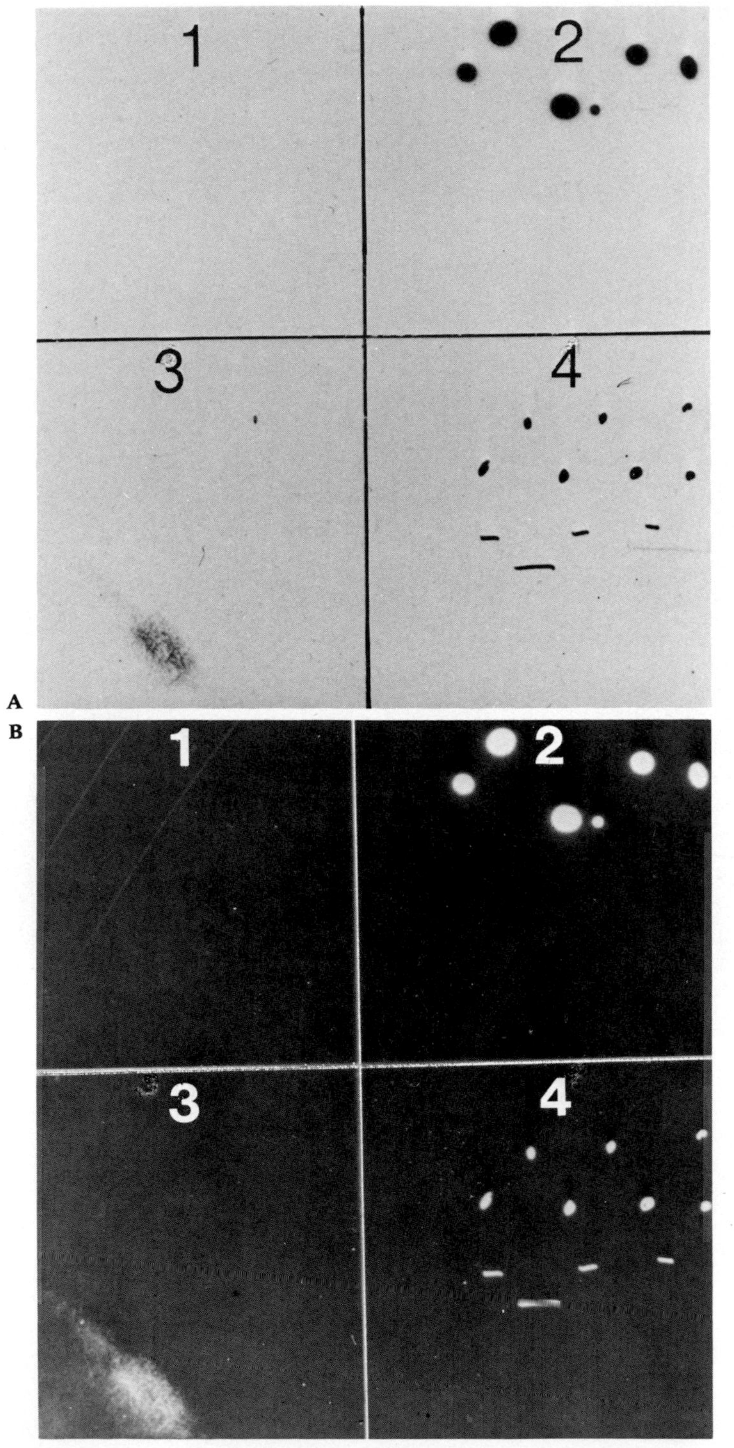
A
B
1
2
3
4
1
2
3
4

◁ **FIG. 6-4.** (A) *A photograph of a deliberately damaged intensifying screen shows* (1) *a deeply scratched area,* (2) *stains from developer chemicals,* (3) *cigarette ashes, and* (4) *marks from a felt tipped pen.* (B) *The damage seen in* A *appears as marks of minus density in a radiograph of the screen.*

FIG. 6-5. *The minus-density image* (arrow) *in this radiograph is of a piece of adhesive tape that was located on the surface of the intensifying screen.*

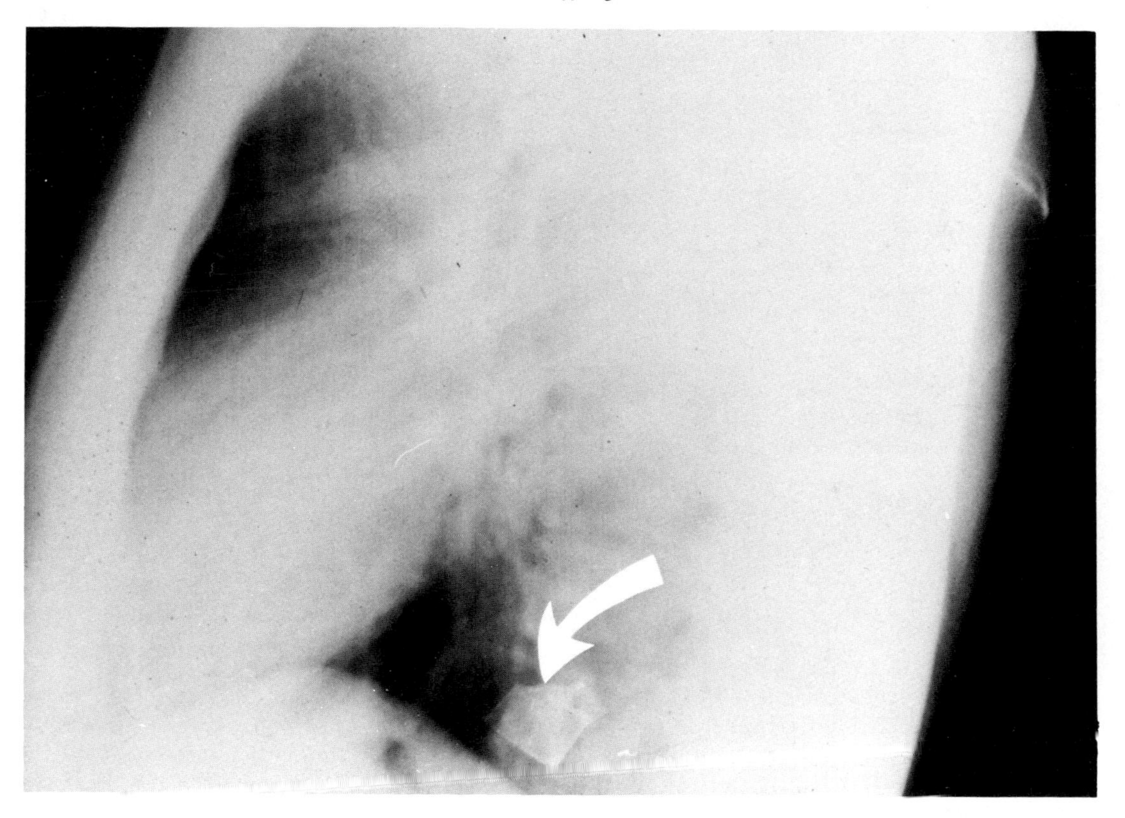

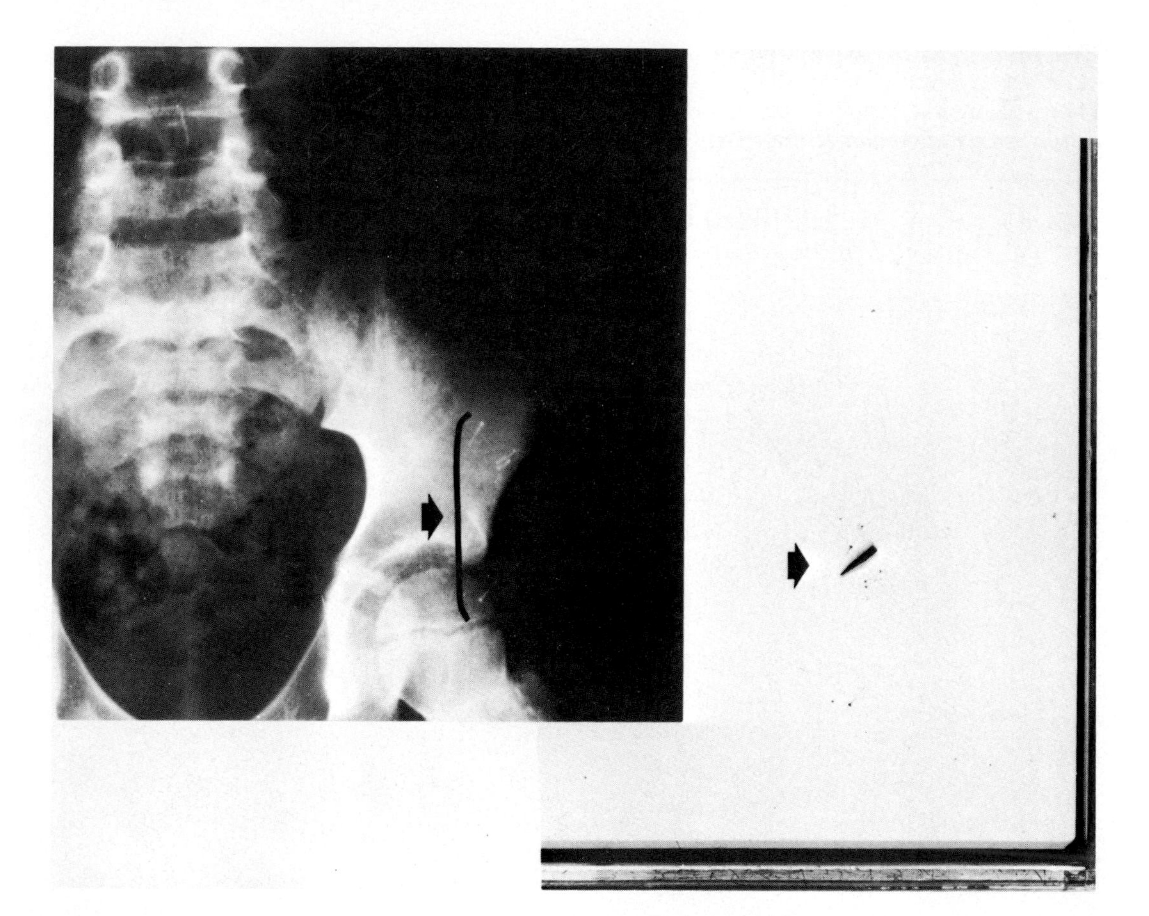

FIG. 6-6. *The radiograph* (left) *demonstrates several pinlike configurations* (arrow) *in the region of the hip. While inspecting the cassette, the technologist discovered that a pencil tip had become embedded in one of the intensifying screens* (right, arrow). *The numerous depressions in the screen indicate that the tip of the pencil must have been in the cassette for some time. Situations such as this can be avoided by following the suggestions outlined on pp. 118 and 120. (See Sweeney RJ: On the Technical Side. Radiol Technol 47:31–32, 1975)*

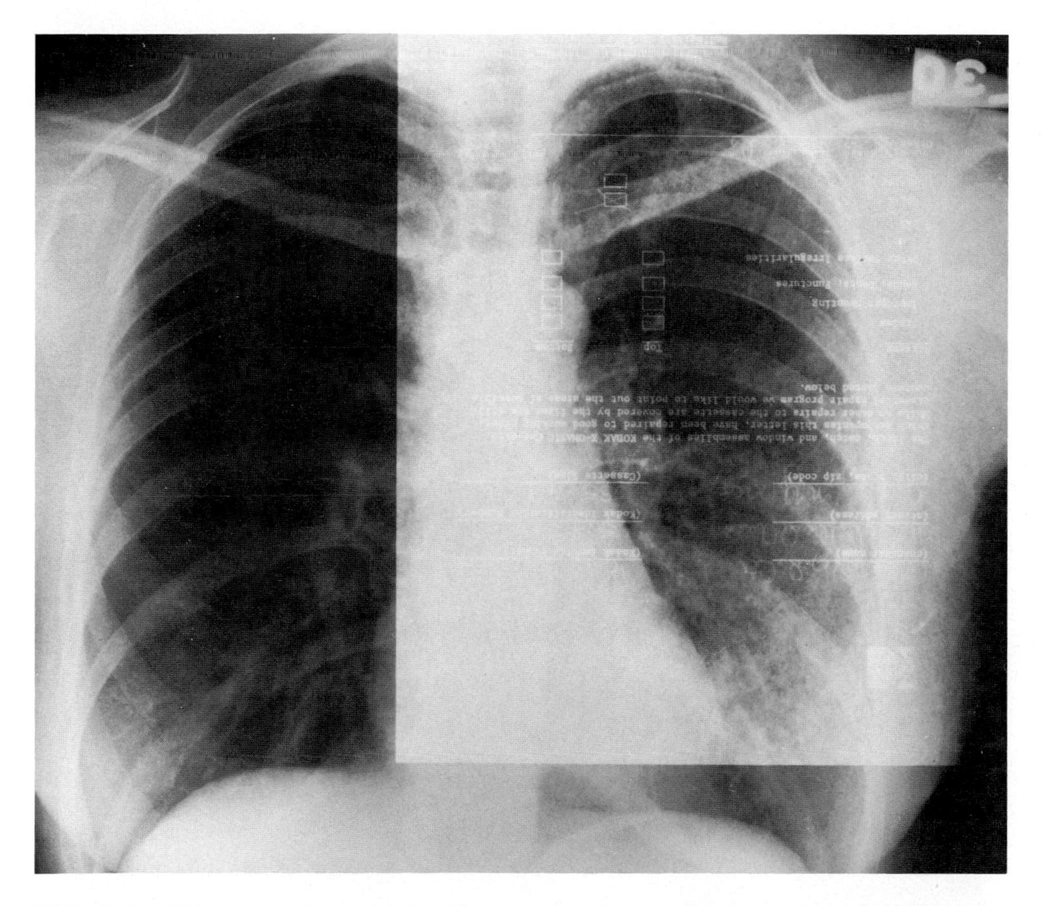

FIG. 6-7. *When cassettes and intensifying screens become damaged, they should be returned to the manufacturer for repair or replacement. This chest radiograph demonstrates the image of a repair-analysis form that was not removed from inside the cassette following its return from the manufacturer.*

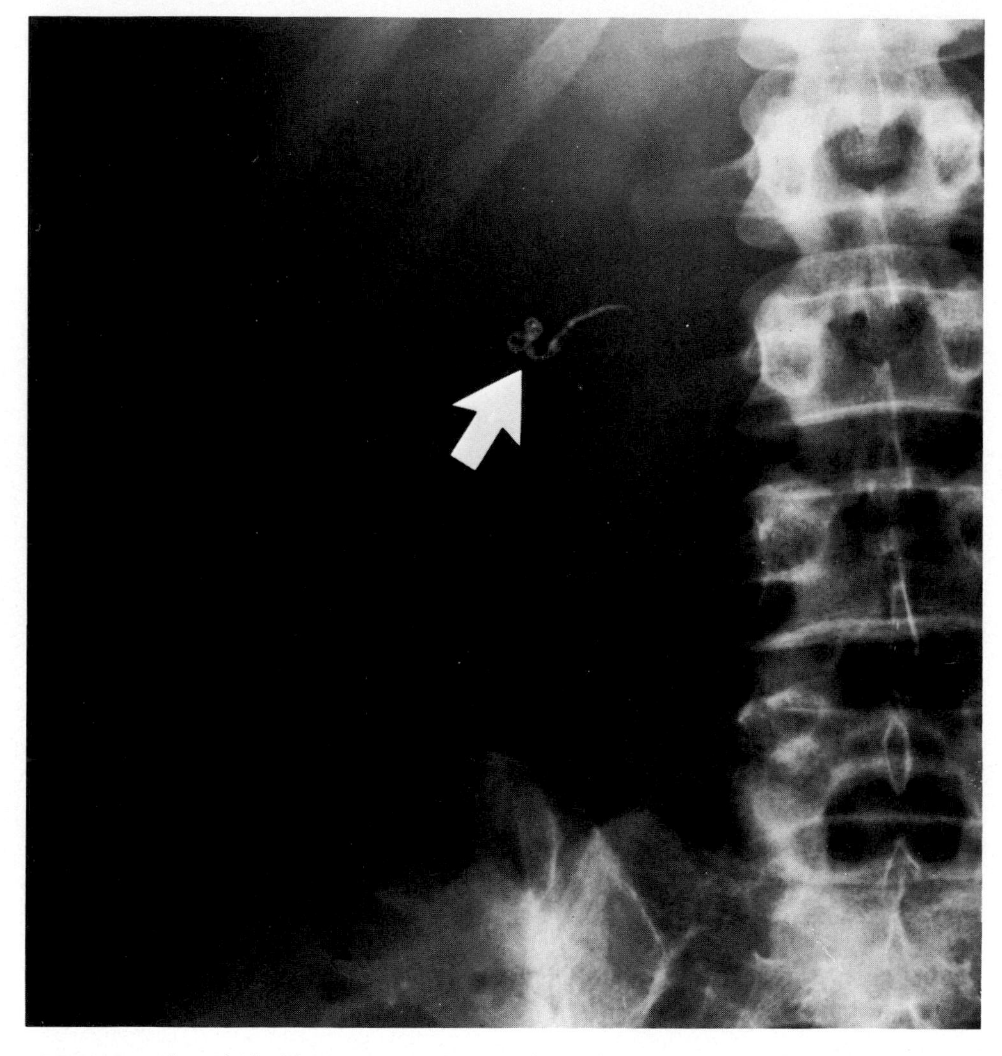

FIG. 6-8. *The calcific-like image* (arrow) *in this radiograph is not a foreign body but an artifact on the surface of the intensifying screen. It is a small strand of cotton fiber that prevented the light from the screen from reaching the film.*

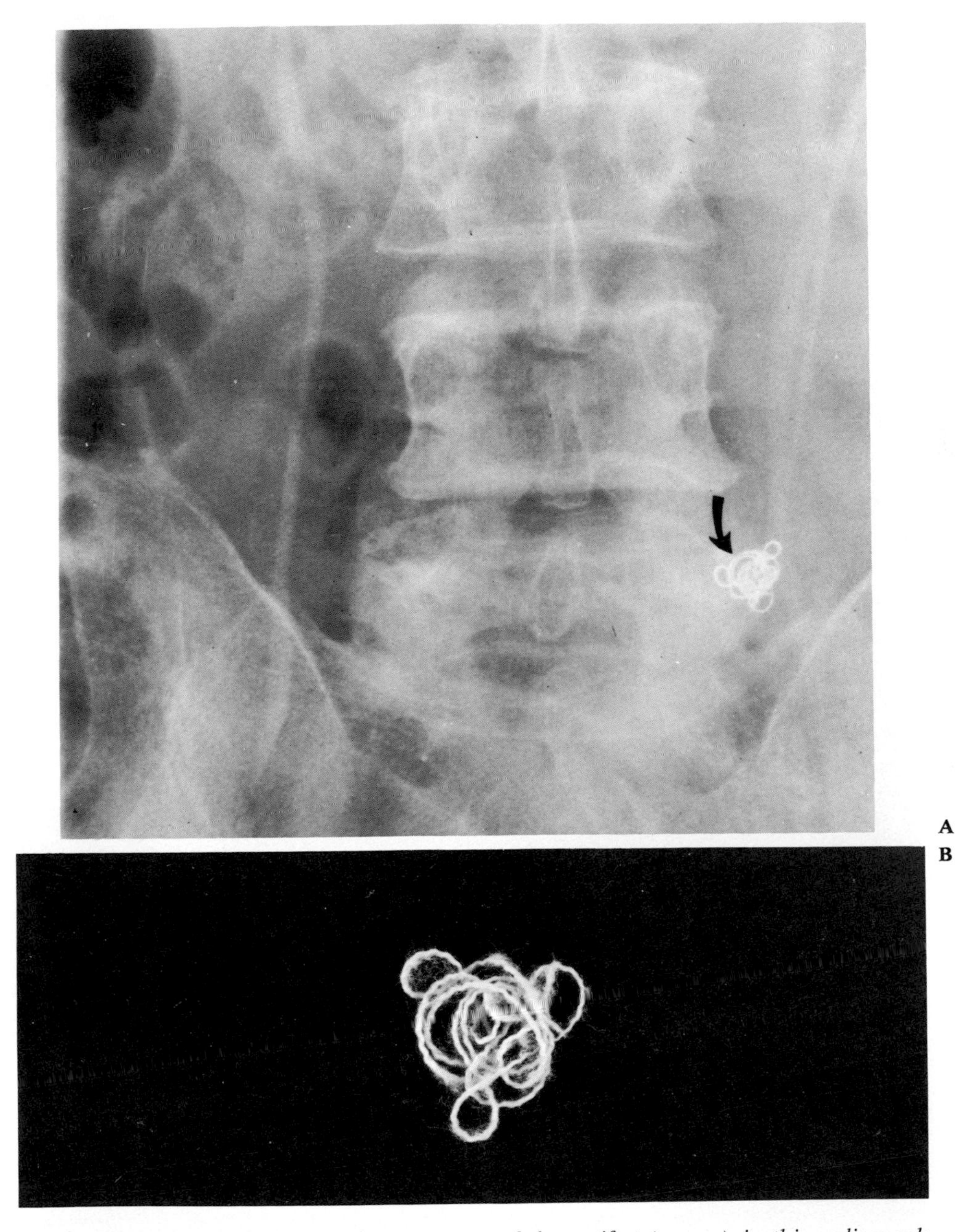

FIG. 6-9. (A) *The ornamental appearance of the artifact* (arrow) *in this radiograph suggests that it is an image of a piece of jewelry. However, in this instance the condition responsible for the artifact was determined to be a small wad of cotton fiber* (B) *located on the surface of the intensifying screen.*

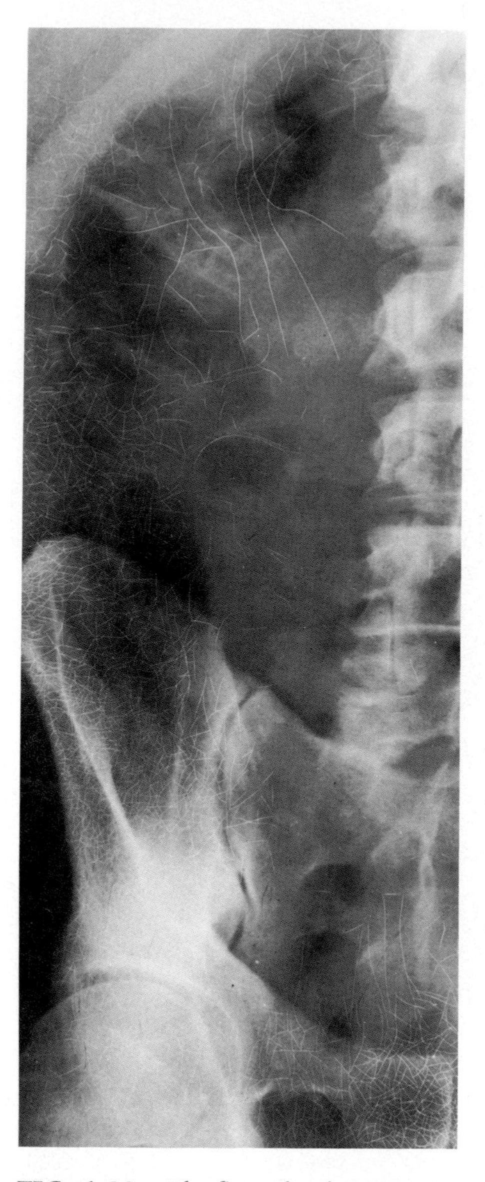

FIG. 6-10. *The fine cobweb appearance of this radiograph could easily be mistaken for reticulation (see Fig. 12-15). However, this artifact is known as* screen craze *and is attributed to a breakdown or cracking of the surface of very old intensifying screens. This example is rather unusual in that a screen that exhibits this degree of deterioration is usually replaced long before such extensive cracking is detected.*

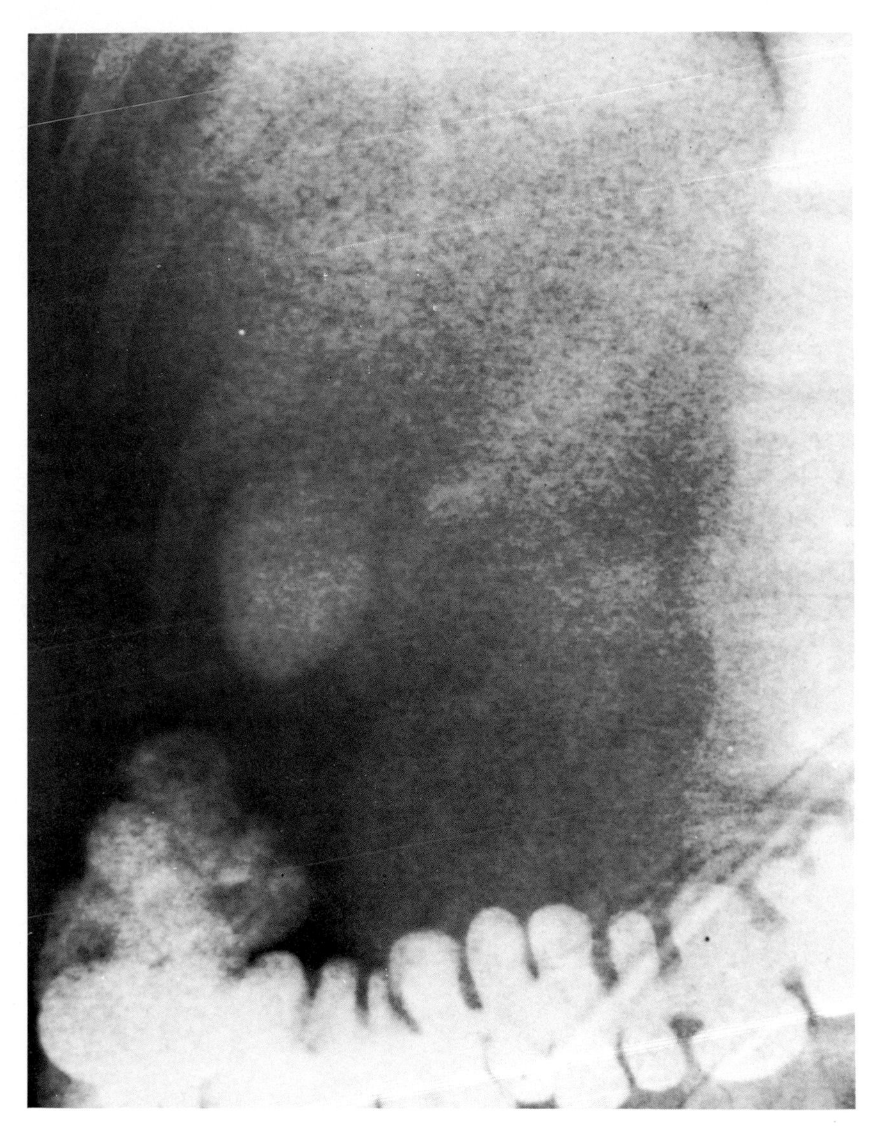

FIG. 6-11. *Most radiographic film is now supplied without interleaving paper, although some manufacturers still provide film packaged in this manner on a special-order basis. These sheets of paper must be removed when the film is exposed with intensifying screens, otherwise it will result in the mottled pattern seen in this radiograph.*

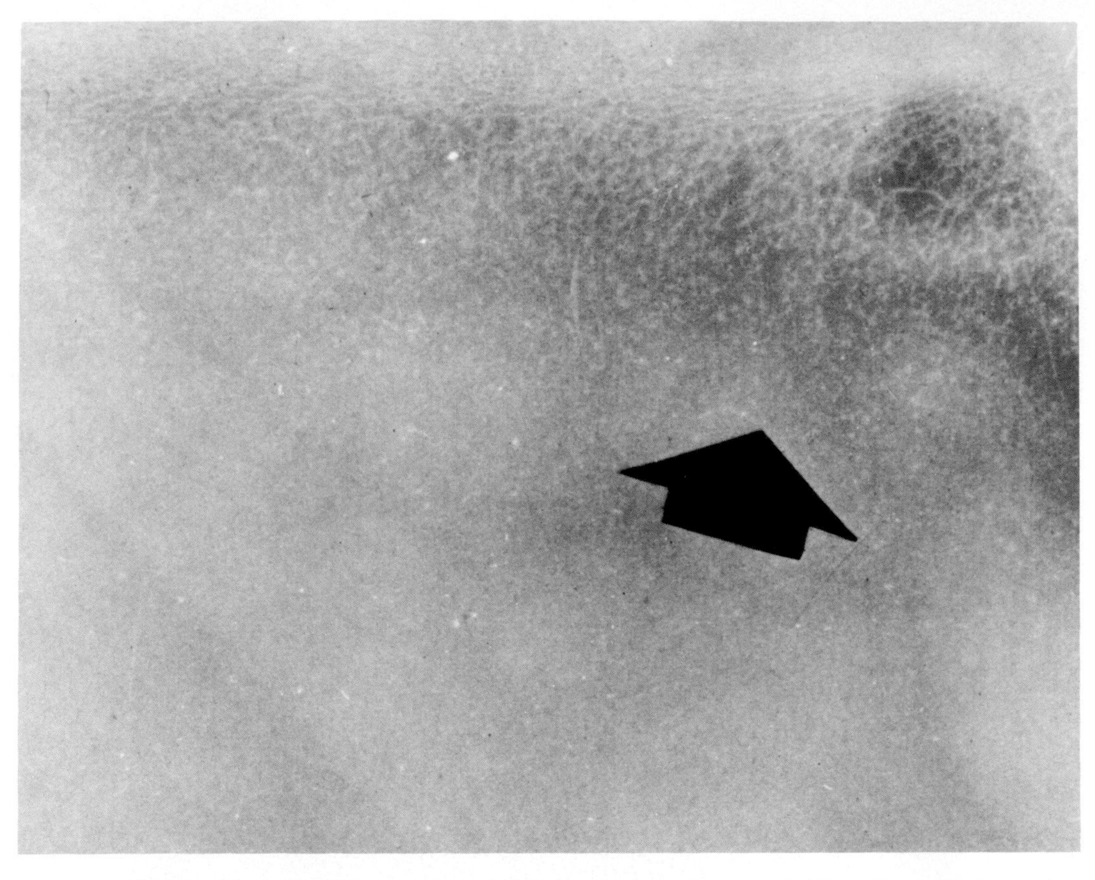

FIG. 6-12. *The filigree pattern* (arrow) *along the border of this radiograph is another example of screen craze, which is caused by the breakdown or cracking of the surface of the intensifying screen.*

FIG. 6-13. (A) *A photograph of the surface area of an intensifying screen used in a cassetteless radiographic table demonstrates the border outlines* (arrows) *of various sizes of film.* (B) *These lines* (arrows) *become progressively more pronounced until the screen must be replaced. The two large circular images are photographic artifacts caused by light reflected in the attempt to demonstrate the lines.*

Inspecting and Cleaning Intensifying Screens

A periodic inspection of the surface of the intensifying screens to detect dirt and dust particles responsible for the appearance of various artifacts is very important. In some instances, a normal visual inspection fails to demonstrate foreign substances and defects in the surface of the screen. This is where the use of an ultraviolet Screen-Check Lamp can be most useful (Fig. 6-14). If any foreign substances are found, the technologist should clean the screen thoroughly and allow it to air-dry before loading the next film (Fig. 6-15). If the condition of the intensifying screen is still in question following inspection and cleaning, a uniform exposure should then be made on a sheet of radiographic film that is sufficient to produce a medium density of 1.0 to 1.5. The processed film can then be examined for screen-related artifacts.

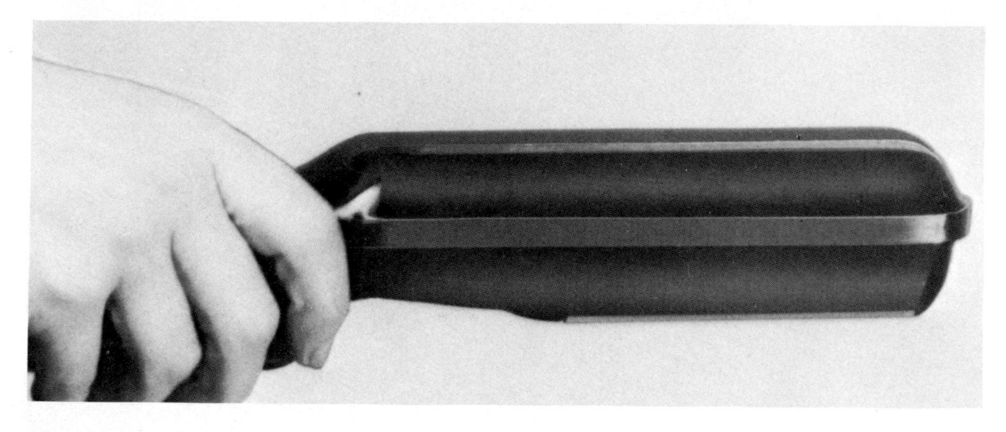

FIG. 6-14. *An ultraviolet lamp can detect foreign objects and defects in an intensifying screen that may be invisible to the naked eye. (Courtesy of Nuclear Associates, Carle Place, NY)*

FIG. 6-15. *The multiple stains in* A *and* B *resulted from an improper intensifying-screen cleaning procedure. The technologist did not allow the surface of the screens to air-dry sufficiently before loading a film between them. Consequently, the moisture on their surface was responsible for presensitizing the film emulsion prior to exposure and processing. (*B, *courtesy of Ralph Coates, R.T. [R])* ▷

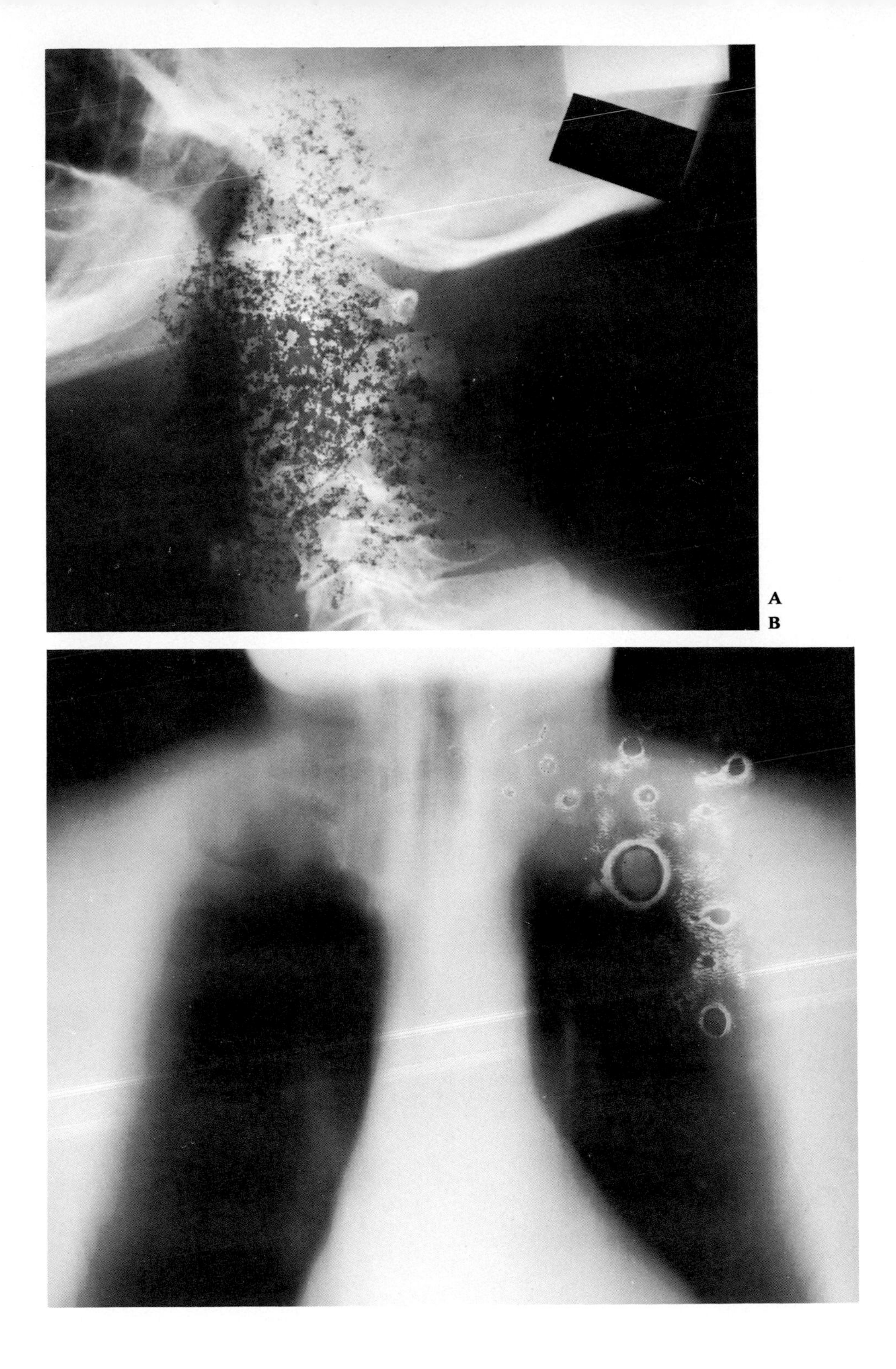
A
B

Avoiding Damage to Intensifying Screens

Intensifying screens and cassettes are expensive, and under no circumstances should anyone attempt to mount the screens in the cassette without understanding the importance of following the manufacturer's instructions. Whenever possible, consult your technical representative for assistance, or have the screens mounted in the cassette by the manufacturer. To do otherwise is to risk damaging costly equipment (Figs. 6-16 and 6-17).

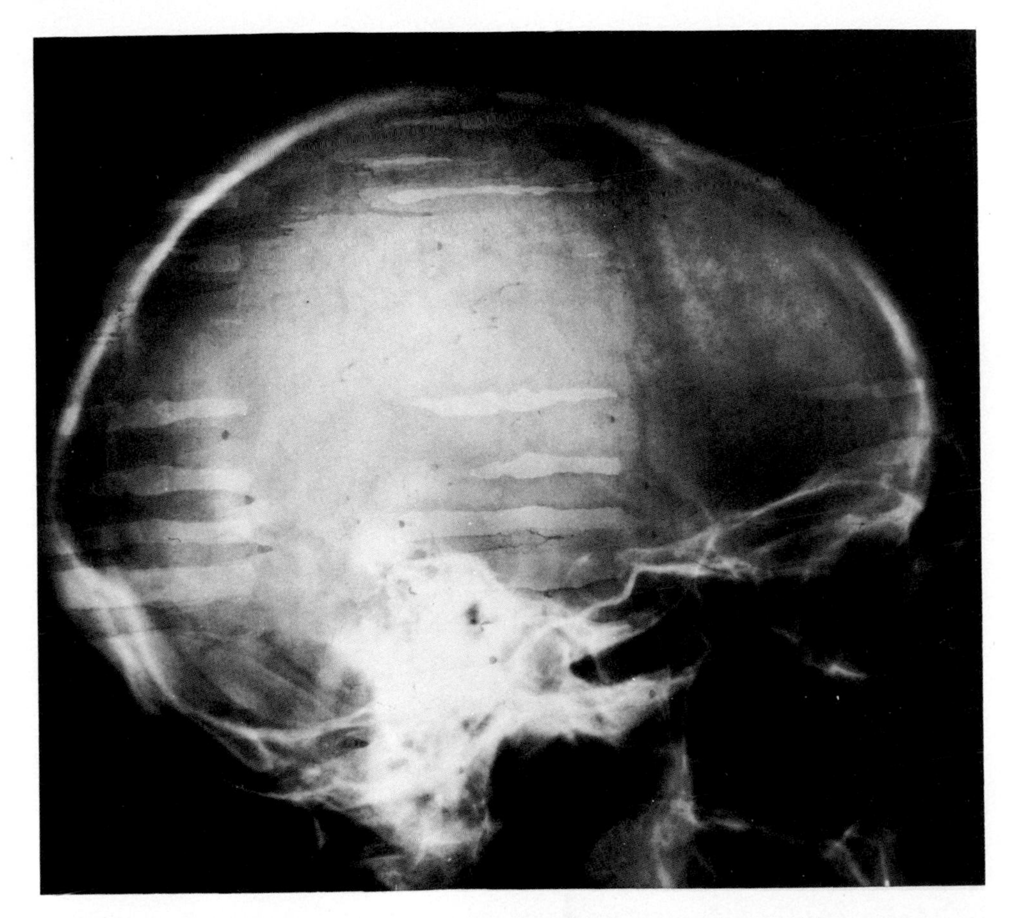

FIG. 6-16. *The artifacts in this radiograph of the skull were due to an improper screen-mounting procedure. The technologist had accidentally placed the surface of the intensifying screen in contact with some residual adhesive located on the countertop used to mount the screens. As a result, the adhesive damaged the phosphor layer, resulting in the bandlike artifacts.*

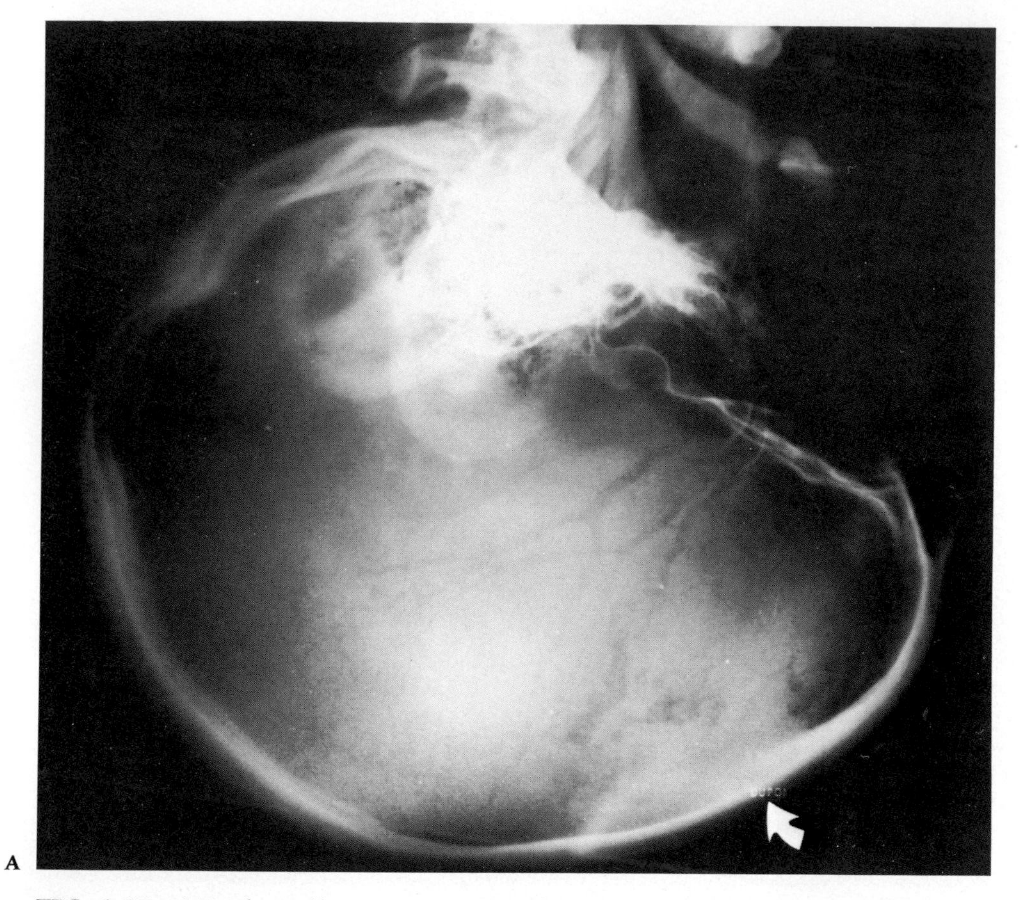

A

FIG. 6-17. (A) *The artifact* (arrow) *seen in this radiograph of the skull (shown inverted in* A *and* B *and enlarged for clarity in* B*) occurred when a portion of the coating on the surface of the intensifying screen was damaged as a foreign particle was removed from inside the cassette.* (C) *Notice that a portion of the stenciled lettering has peeled away from the surface of the screen and is positioned in the manner in which it was imaged in* A *and* B.

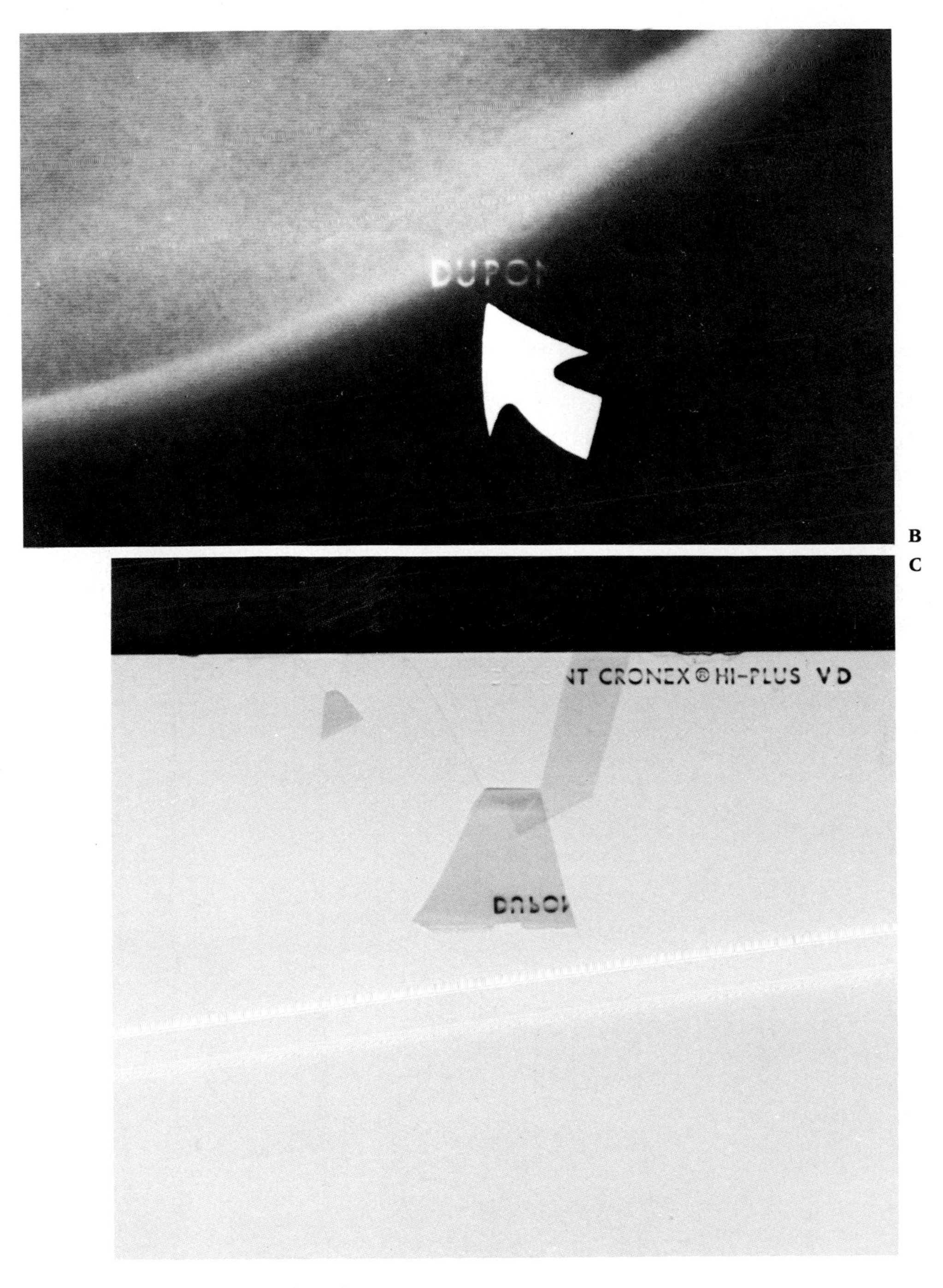

FIG. 6-17. *continued*

Labeling Cassettes and Intensifying Screens

Cassettes and intensifying screens can be identified in a number of ways in order to enable you to locate the cassette whenever you suspect an artifact. One of the more popular methods is to use commercially available decals or labels (Fig. 6-18). Another is to engrave the identification along the edge of the cassette (Fig. 6-19, *A* and *B*). Obviously, a corresponding number should be indicated on the surface of the intensifying screen. This is usually done with an opaque marking pen, but unfortunately, repeated cleaning of the screen diminishes the opacity of the lettering to a point at which it becomes impossible to identify the legend. In order to overcome such problems, you can cover the lettering with a piece of polyester tape or use pressure-sensitive numbers brushed with a thin coat of clear fingernail polish (Fig. 6-19, *C*). A word of caution: Always consult the screen manufacturer before attempting any of these techniques.

A method of labeling that I have found to work well in the identification of the cassette is to glue small lead numbers near the edge of the foam backing under the intensifying screen (Fig. 6-19, *D*). This would, of course, require that the cassettes be ordered without screens in order to affix the lead numbers. Any type of identification mark, however, might not appear on the radiograph if the region in question is overexposed or excluded owing to collimation. (An exceptional case in point is illustrated in Chap. 9, Fig. 9-8.)

If cassettes are used in various locations throughout the department, it may be difficult to locate a particular cassette when you suspect an artifact problem. By posting a "wanted" notice in the processing area, you can locate the wandering cassette in a relatively short time (Fig. 6-20).

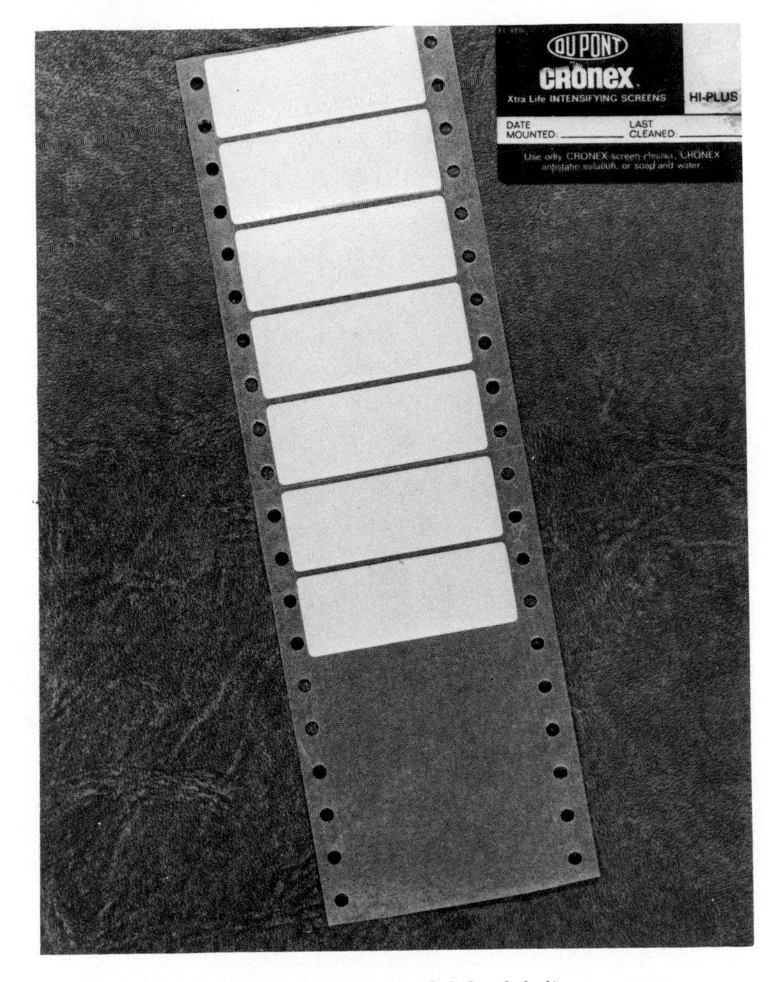

FIG. 6-18. *Commercially available labels are often useful for labeling cassettes.*

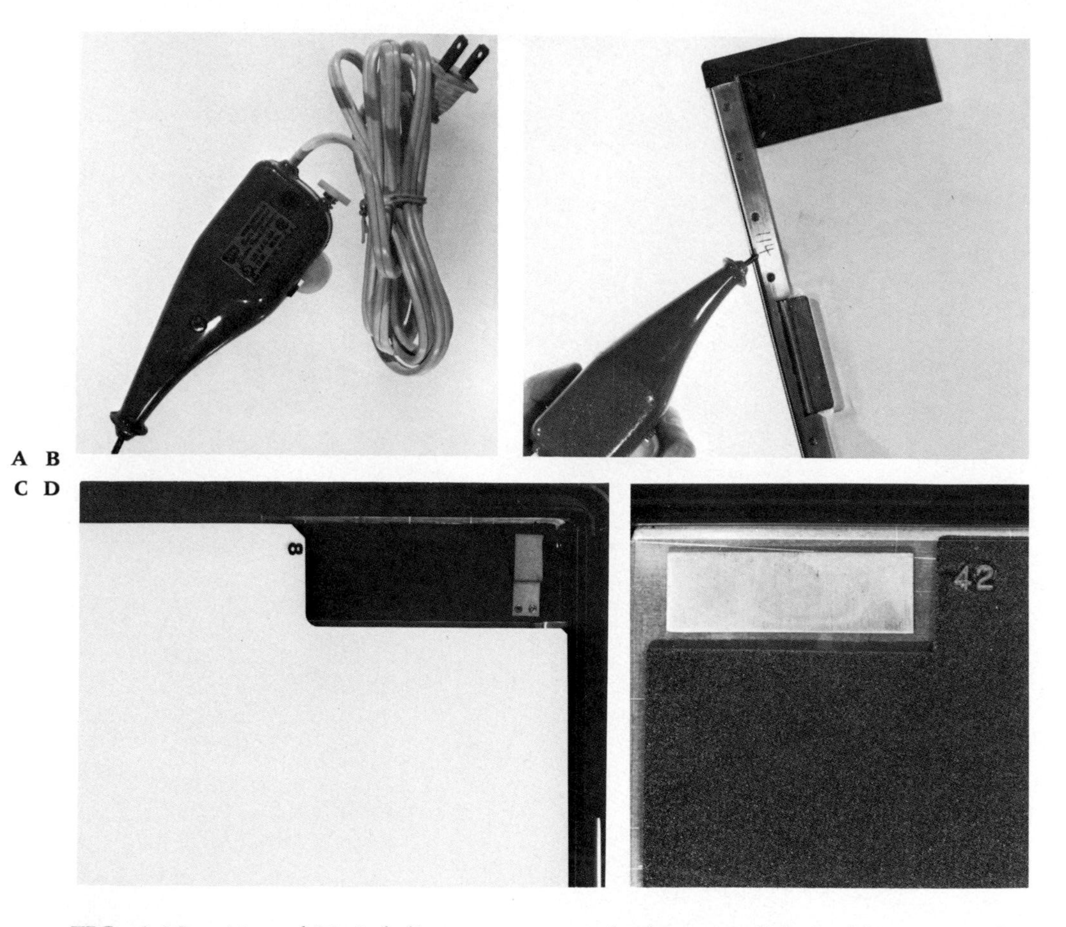

FIG. 6-19. (A *and* B) *Labeling cassettes can also be accomplished with an engraving tool. Using pressure-sensitive labels coated with clear fingernail polish* (C) *and small lead numbers glued to the back of the screen* (D) *are other useful methods of labeling intensifying screens.*

WANTED

CASSETTE NUMBER: ______ SIZE ______

PROBLEM: ______

FIG. 6-20. *One easy means of tracking down a troublesome cassette.*

Film–Screen Contact Problems

If the surface of the film and intensifying screen are not in direct contact, light diverges from the phosphor, producing an unsharp image (Fig. 6-21). Some of the conditions responsible for screen–film contact problems are as follows:

The cassette frame is cracked.

The cassette is warped.

There is a foreign substance under the surface of the intensifying screen.

There are broken hinges or a broken latch on the cassette.

The mounting substances behind the surface of the screen may be somewhat irregular. This could result in a wrinkling or depression in the surface of the intensifying screen.

Air may be trapped between the surface of the film and the intensifying screen.

The weight of the patient may be responsible for compressing the surface of the cassette when it is placed directly under the patient during bedside examinations.

When you suspect poor screen contact, you can use a wire-mesh test pattern to evaluate the questionable cassette (Fig. 6-22). Following exposure and processing, areas of poor contact will appear blurred (Figs. 6-23 and 6-24). Poor contact is easier to detect if you view the test film while standing a few feet away from the surface of the illuminator.

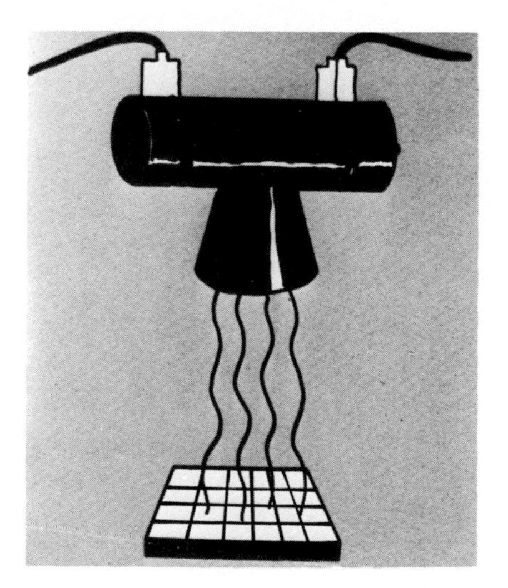

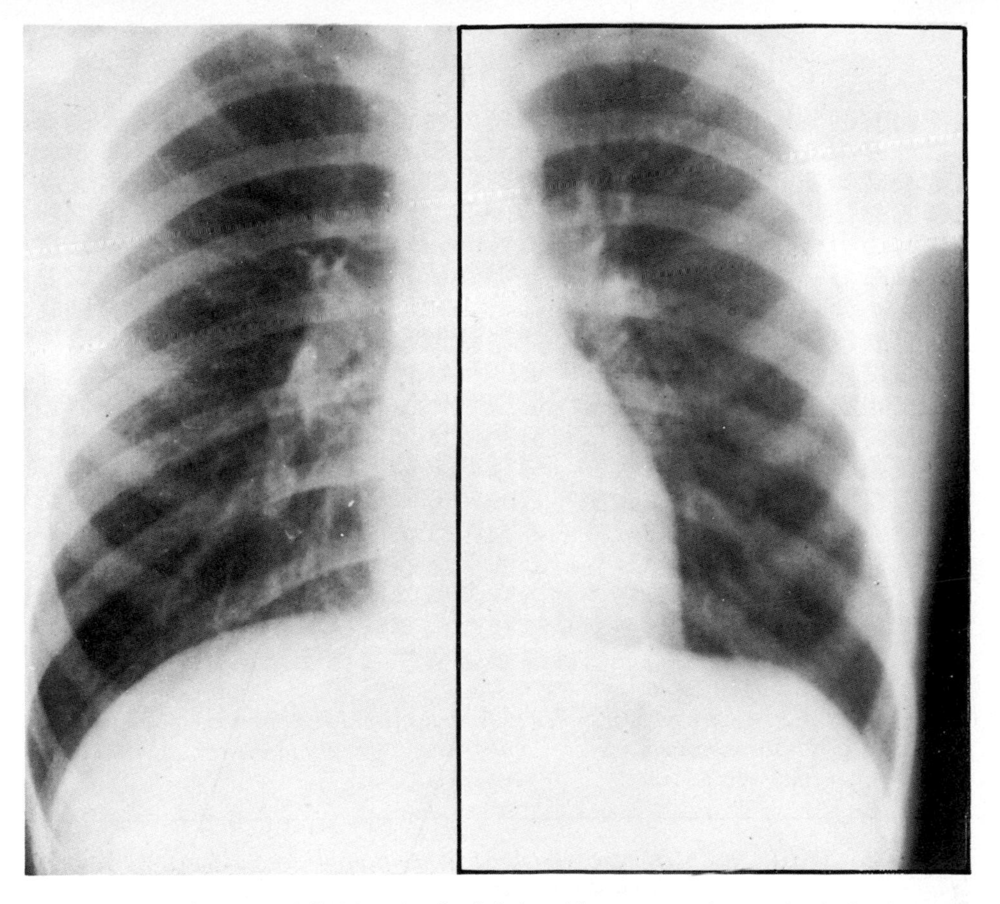

FIG. 6-21. *The poor definition in the left hemithorax was due to the lack of contact between the surface of the film and the intensifying screen.*

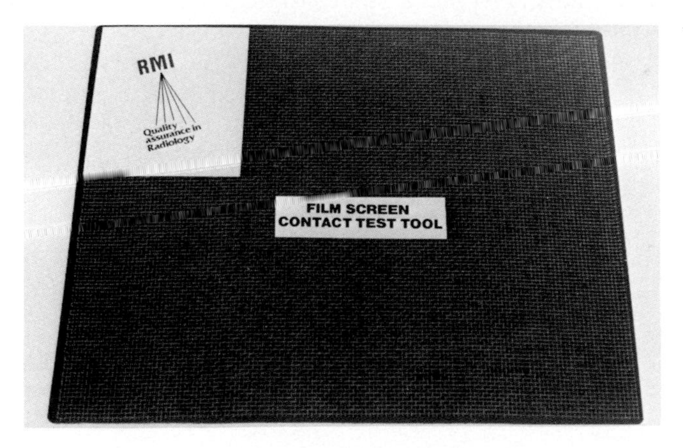

FIG. 6-22. *The RMI film–screen contact test tool, which consists of a plastic-encased wire-mesh pattern of 3 lines per centimeter (8 lines per inch), is used to inspect cassettes for film–screen contact. (Courtesy of Radiation Measurements, Inc., Middleton, WI)*

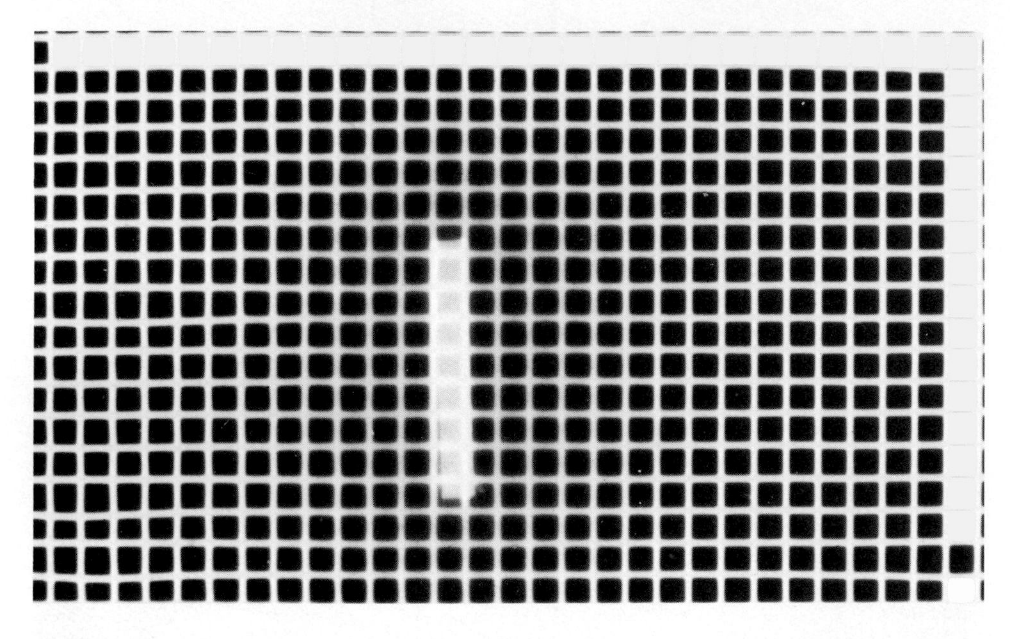

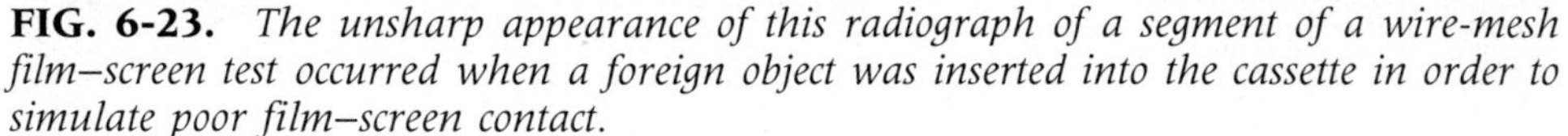

FIG. 6-23. *The unsharp appearance of this radiograph of a segment of a wire-mesh film–screen test occurred when a foreign object was inserted into the cassette in order to simulate poor film–screen contact.*

FIG. 6-24. (A) *The unsharp image of the plastic beads in this radiograph resulted from small pieces of cardboard that were deliberately placed into the cassette in order to simulate poor contact. Very often, the lack of image sharpness may be incorrectly attributed to patient motion, and the problem may go undetected until a contact test is performed.* (B) *A radiograph made of the plastic beads after the pieces of cardboard were removed from the cassette is much sharper. (From Sweeney RJ: Image clarity and detail perception in the radiograph. Radiol Technol 46, No. 6:443–451, 1975)* ▷

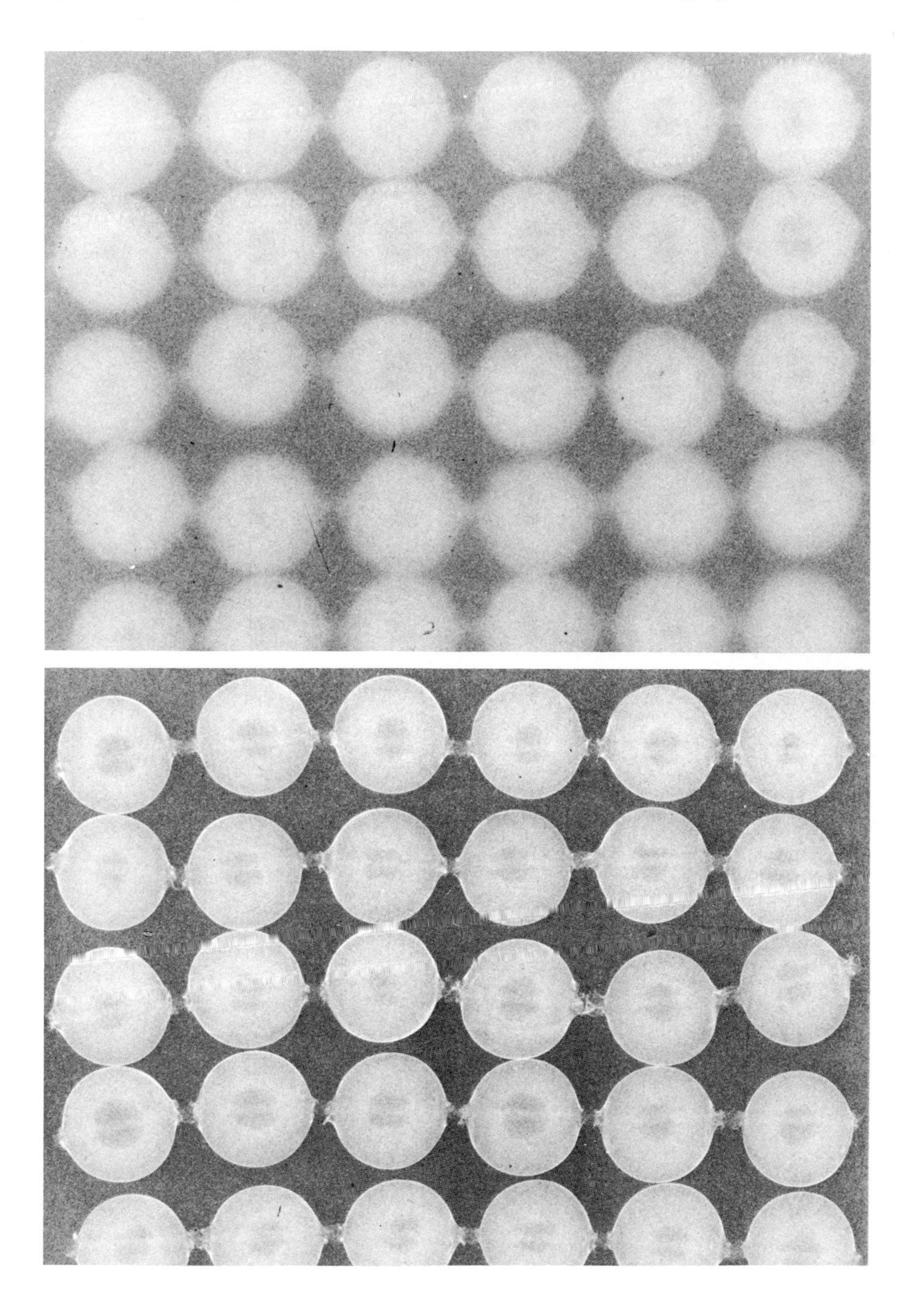

Suggested Reading

American Association of Physicists in Medicine: AAPM Report No. 4, Basic Quality Control in Diagnostic Radiology, pp 31–33, Chicago, November 1977

Ardran GM, Crooks HE, James V: Testing x-ray cassettes for film–intensifying screen contact. Radiography XXXV, No. 414: 143–145, 1969

Bushong SC: Radiologic Science for Technologists, 2nd ed, pp 225–226. St Louis, CV Mosby, 1980

The Care and Use of Intensifying Screens, Pub. No. A-38660, pp 17–19. Wilmington, EI du Pont de Nemours & Co, 1966

Cullinan JE, Cullinan AM: Illustrated Guide to X-ray Technics, 2nd ed, p 61. Philadelphia, JB Lippincott, 1980

Eastman TR: Radiographic Fundamentals and Technique Guide, pp 63–65. St Louis, CV Mosby, 1979

Products for Medical Diagnostic Imaging, Pub. No. M5-15, p 31. Rochester, Eastman Kodak, 1981

Selman J: The Fundamentals of X-ray and Radium Physics, 6th ed, pp 291–292. Springfield, IL, Charles C Thomas, 1979

Thompson TT: Cahoon's Formulating X-ray Techniques, 9th ed, p 64. Durham, Duke University Press, 1979

FILM–SCREEN CONTACT TEST

Gray JE, Winkler NT, Stears J, Frank ED: Quality Control in Diagnostic Imaging. a, pp 51–52; b, pp 55–58. Baltimore, University Park Press, 1983

7 ARTIFACTS CAUSED BY FAILING TO ATTIRE THE PATIENT PROPERLY

The Importance of Effective Communication

Because our failure to communicate can lead to confusion and numerous problems, nowhere is effective communication more vital than in the field of medicine. In radiologic technology, specifically, many repeat studies are needed because directions were not followed or because we failed to understand a request for a specific procedure. Often our communication is misunderstood not so much because of what we say but rather because of how we say it—the tone of our voice as well as the kind of interest demonstrated during the conversation.

In considering the difficulty we as technologists have in understanding some of the technical and medical terms used in our profession, just think how bewildering all of this must seem to the patient. Moreover, even simple instructions are often misunderstood by the patient, and the result is a repeat study.

In most instances the patient is not at fault, and perhaps he is not even aware of the importance of following certain directions. Consequently, it is our responsibility to ensure that the patient understands his role in preparing for certain aspects of the procedure. In order to eliminate confusion when communicating with patients, avoid talking too fast, use plain words, and speak in short, concise sentences.

Artifacts resulting from inappropriate patient attire often have their root in the patient's misunderstanding of instructions given by the technologist. By avoiding confusion in communication, you can also avoid a large number of artifacts, which have been with us since the very first radiographs were pulled from the developing tray.

• *Historical Note*

Having discovered the effect of x-rays on a photographic plate, Wilhelm Röntgen, the father of modern radiology, recorded images of various structures, including his wife's hand (Fig. 7-1).[6] Röntgen made a photographic positive print of this plate as well as eight others, which he sent to his colleagues along with a copy of his first paper on x-rays.[4] The radiograph of his wife's hand was made in December 1895 and required an exposure time of 15 minutes.[3] Not only was this undoubtedly the first radiograph of a human subject, but the presence of the ring on the subject's finger indicates that Röntgen was also responsible for producing the first artifact ever recorded in a radiograph.

Over the years, a number of common objects have appeared in various studies because the patient was not properly attired for the examination. Although recognizing many of these objects is not a problem, I have selected a few that I feel are very unusual. The figures that follow illustrate the wide range of artifacts that can result from failing to convey to the patient exactly how to prepare for the examination.

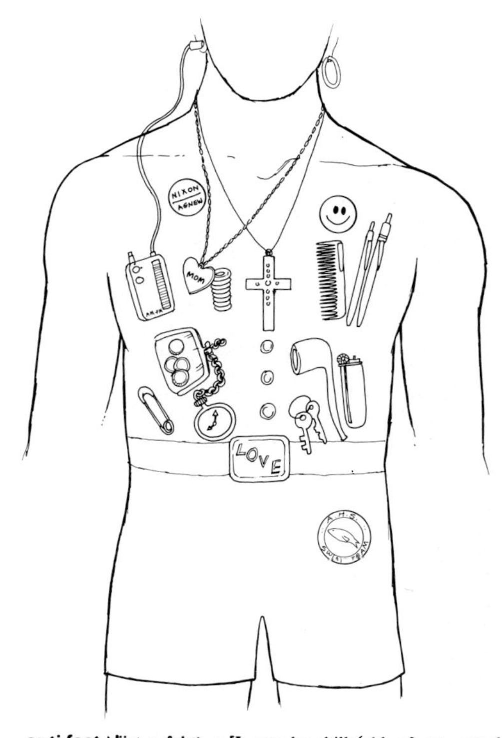

ar·ti·fact \\'ärt-ə-ˌfakt\\ *n* [L *arte* by skill (abl. of *art-*, *ars* skill) + *factum*, neut. of *factus*, pp. of *facere* to do — more at ARM, DO] **1 a :** a usu. simple object (as a tool or ornament) showing human workmanship or modification **b :** a product of civilization (an ~ of the jet age) **c :** a product of artistic endeavor **2 :** a product (as a structure on a prepared microscope slide) of artificial character due to extraneous (as human) agency — **ar·ti·fac·tu·al** \\ˌärt-ə-'fak-chə(-wə)l, -'faksh-wəl\\ *adj*

(By permission. From Webster's New Collegiate Dictionary, © 1981 by Meriam–Webster Inc., Publisher of the Merriam–Webster Dictionaries.)

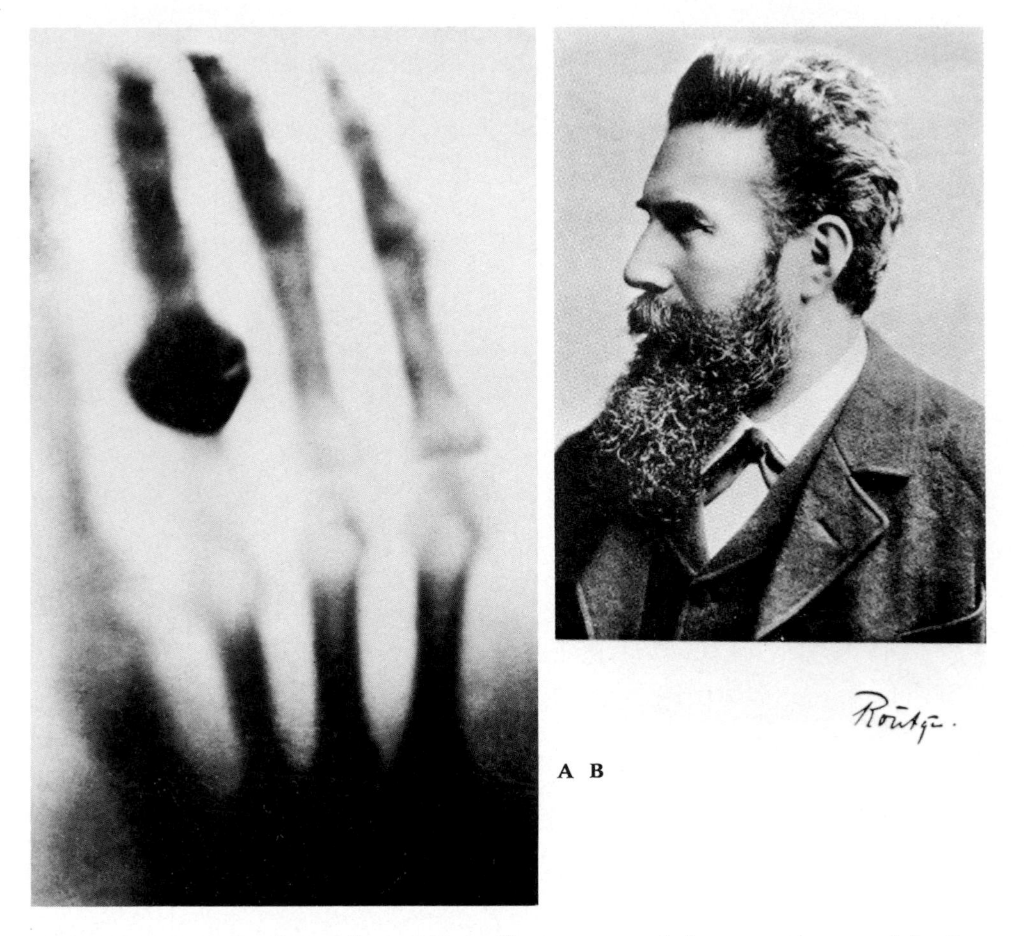

FIG. 7-1. (A) *The first radiographic artifact ever recorded appears in one of the first radiographs ever made.* (B) *The person responsible for recording the artifact in* A *was none other than Professor Röntgen himself. (*A, *courtesy of Eastman Kodak. Copyright © by Eastman Kodak Company, Rochester, NY)*

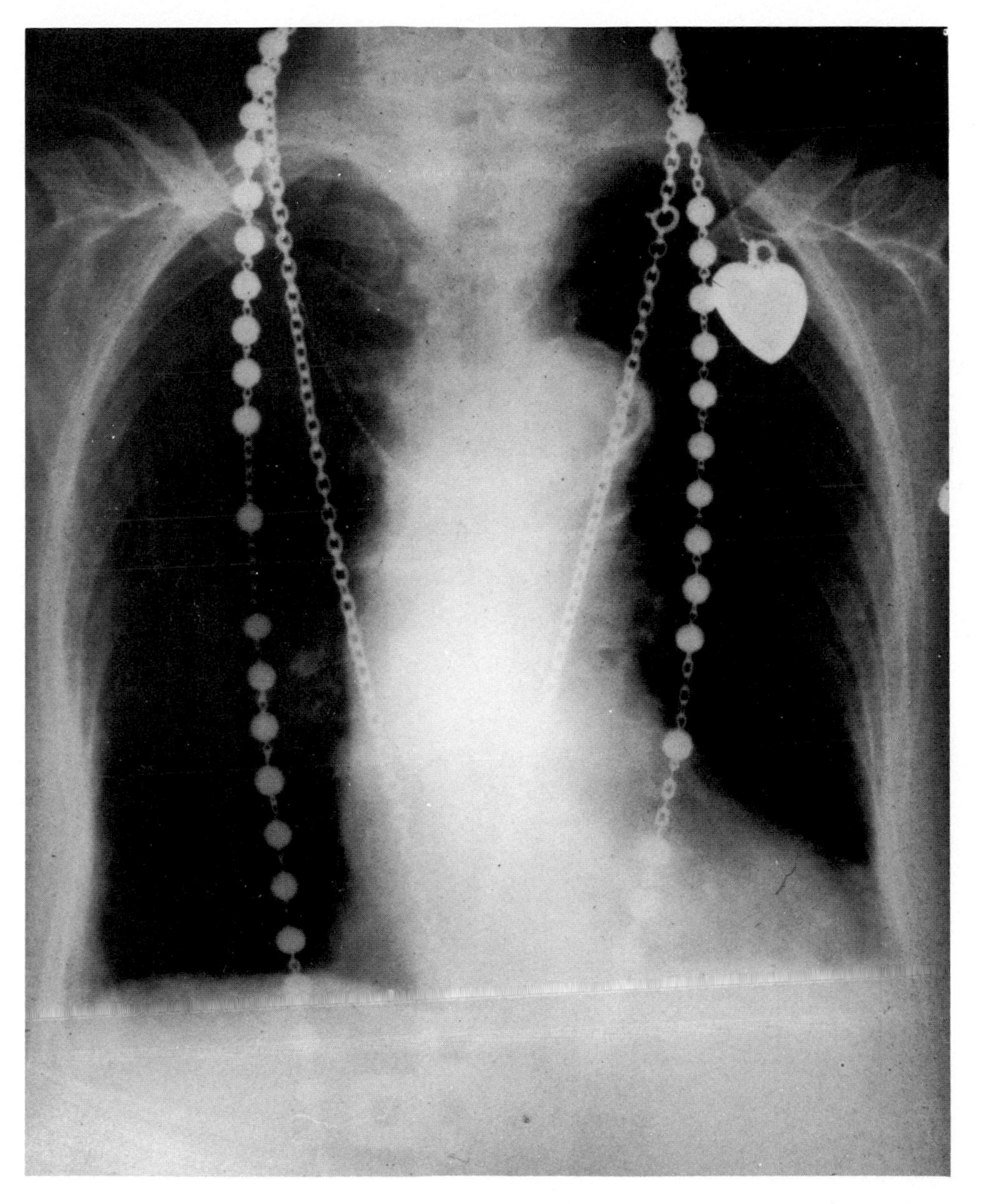

FIG. 7-2. *Instructions such as "Take off everything but your underwear and put on the gown" can often lead to confusion and repeat studies. Notice that in addition to the locket and chain the patient was wearing a rosary, a necklace, and a cross for a radiographic examination of her chest.*

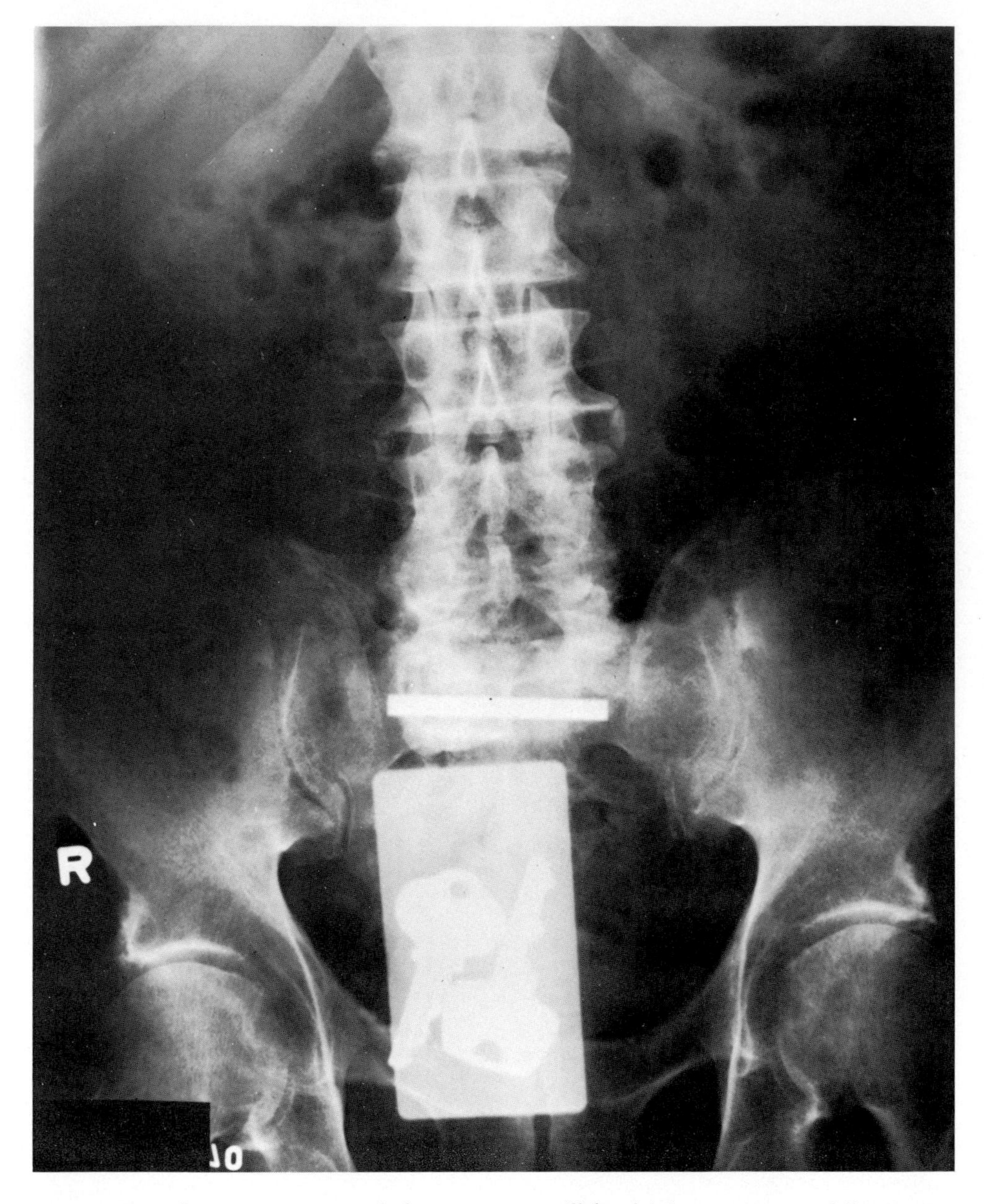

FIG. 7-3. *If you are concerned that someone will be driving your car while you are hospitalized, this may be the answer. These car keys are located within a metal container attached to the inside of the patient's underwear.*

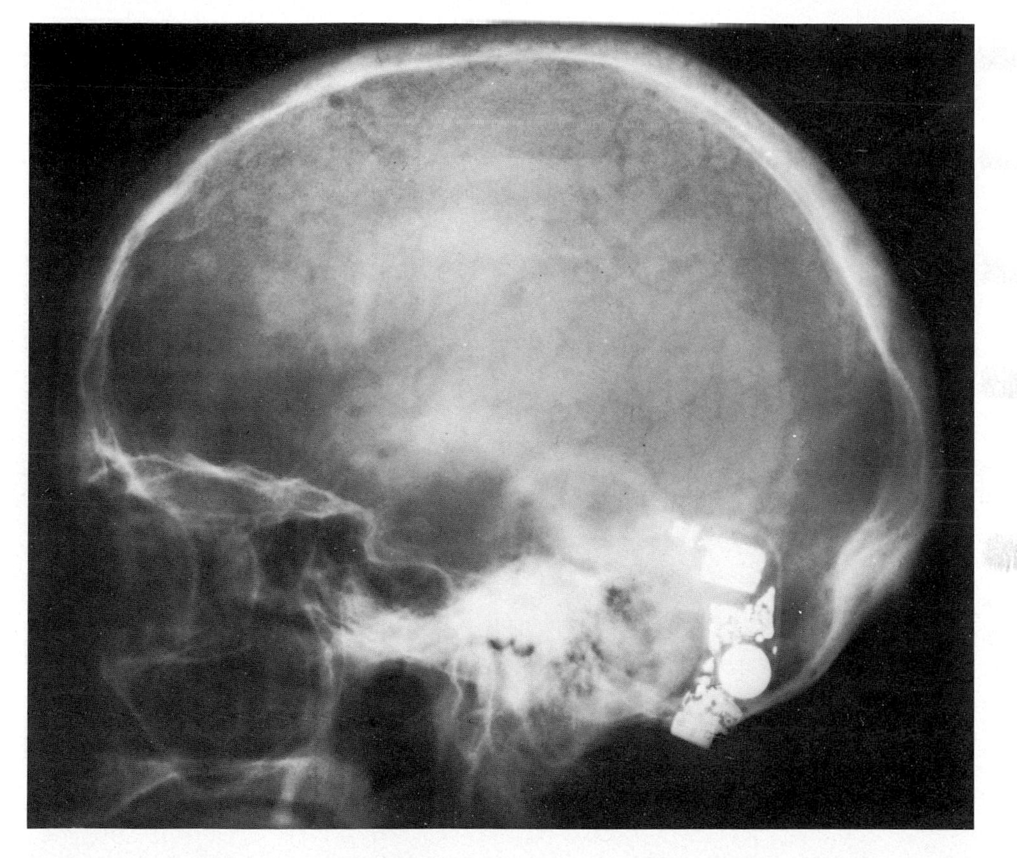

FIG. 7-4. *The electrical components visible in this radiograph of the skull are located within a hearing aid accidentally left on the patient.*

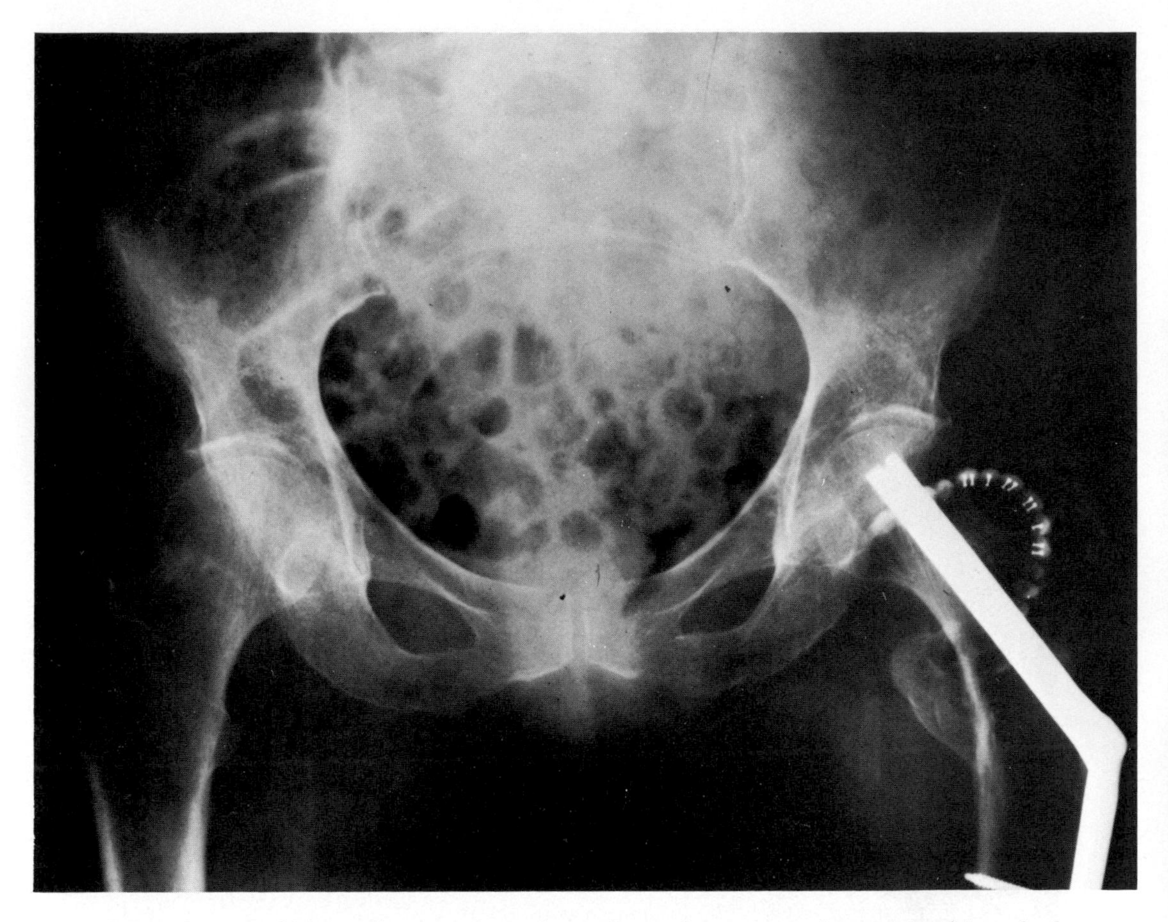

FIG. 7-5. *A biting pain. Somehow the patient's denture was imaged in this recheck examination of the pelvis. The denture was, for some reason, lodged below the left buttock.*

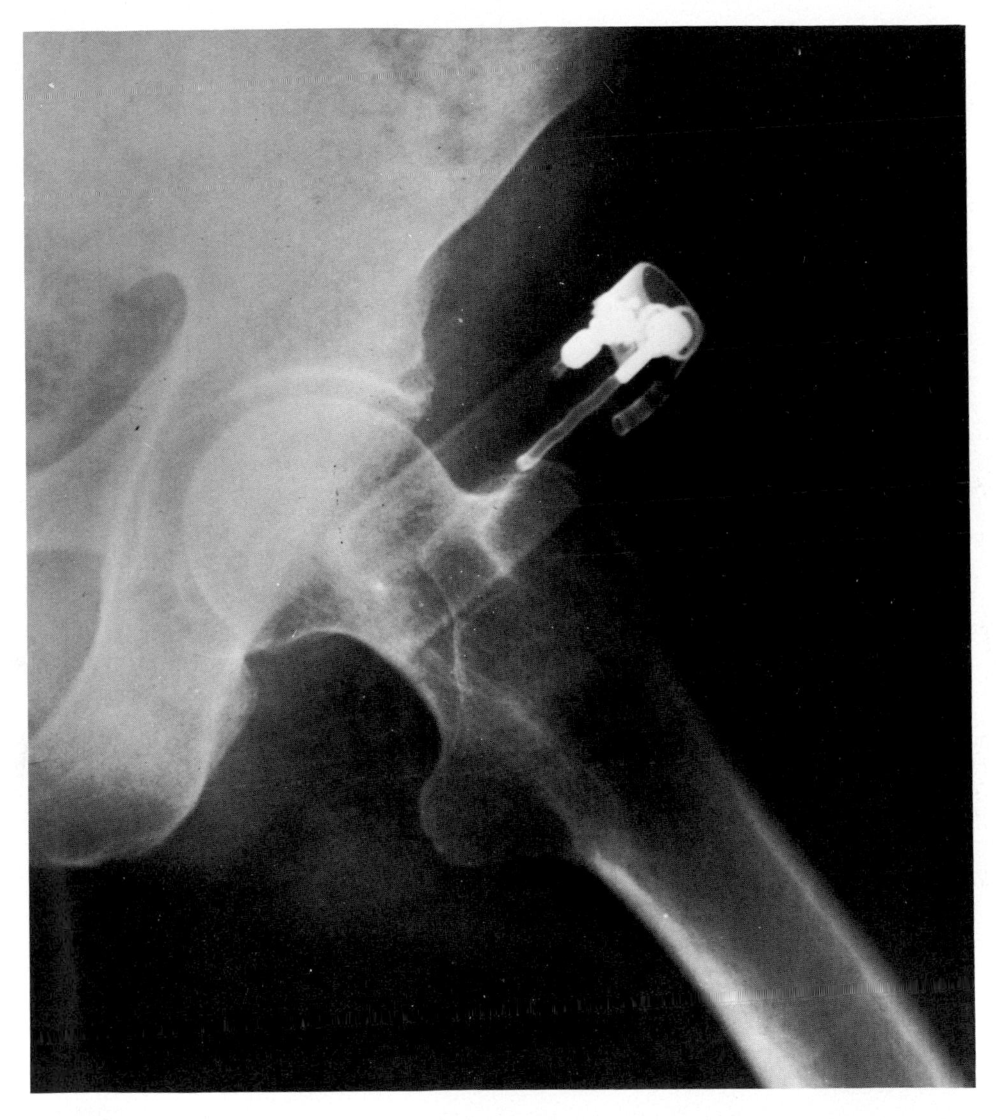

FIG. 7-6. *A disposable cigarette lighter was inadvertently left in this patient's pants pocket.*

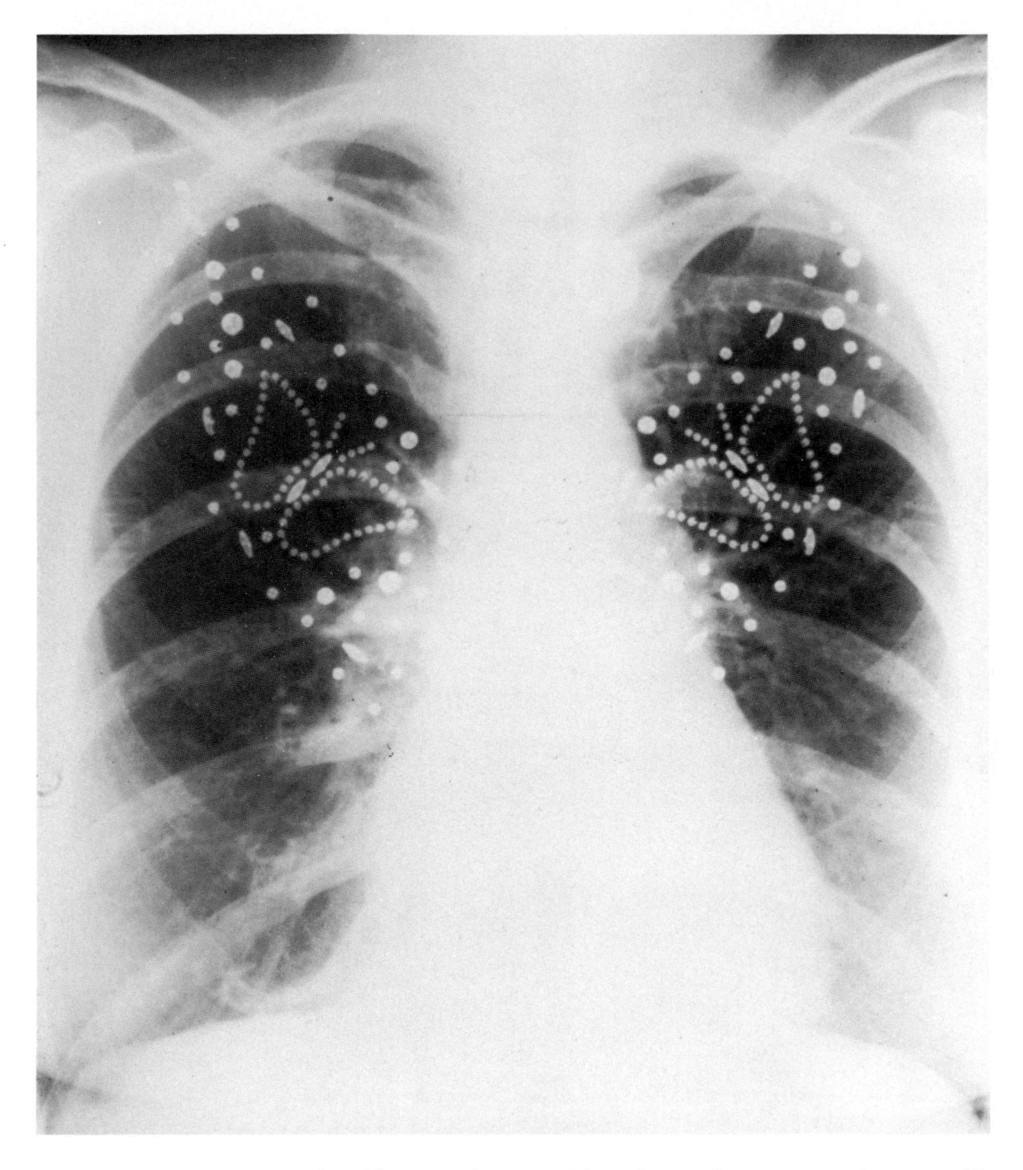

FIG. 7-7. *Some unusual artifacts can be caused by what patients wear. These metallic sequins were embroidered onto the sweater that this patient wore for her radiograph of the chest.*

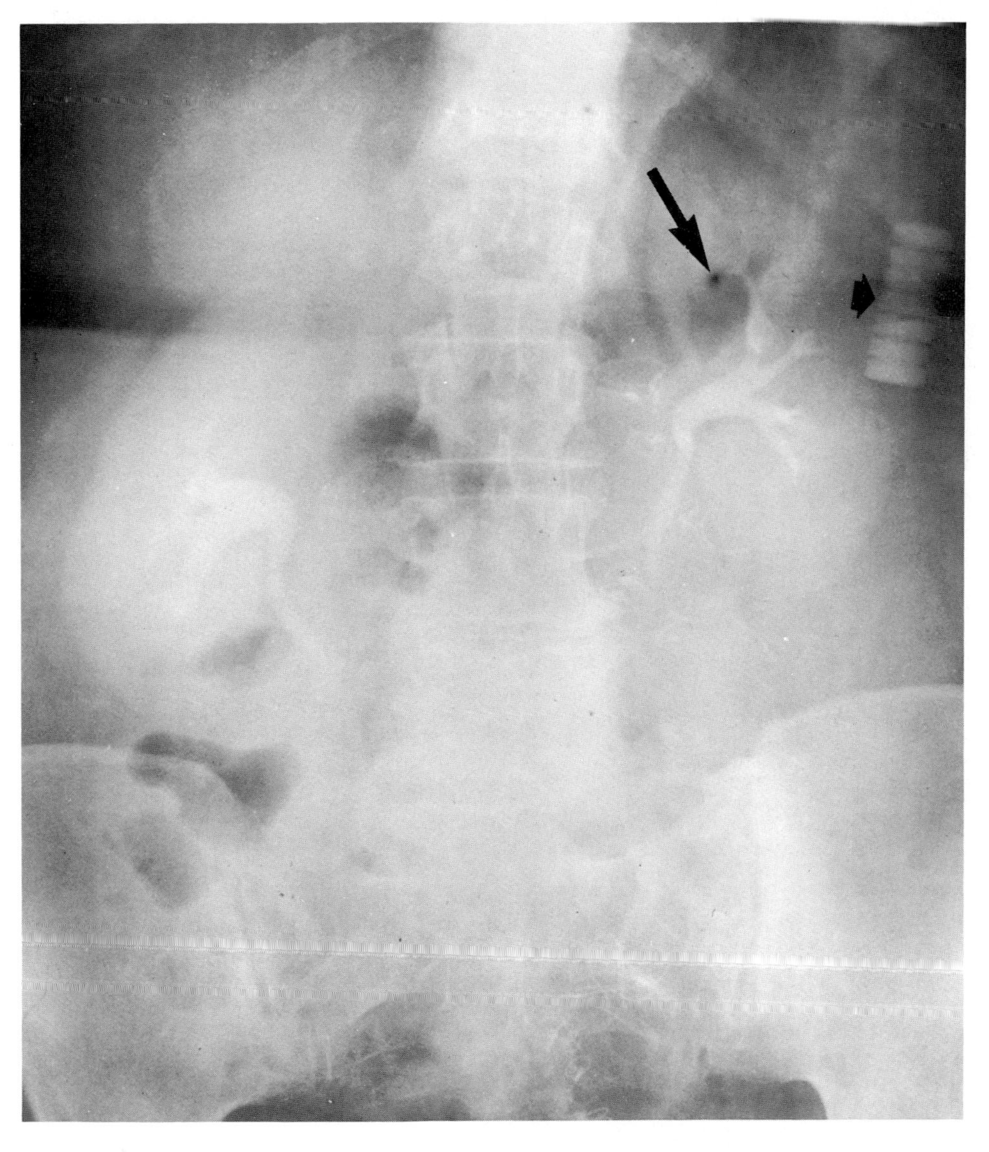

FIG. 7-8. *The artifact indicated by the* large arrow *is a static discharge. The* small arrow *indicates a roll of candy located in the pocket of the patient's pajamas. (Courtesy of Ralph Coates, R.T. [R])*

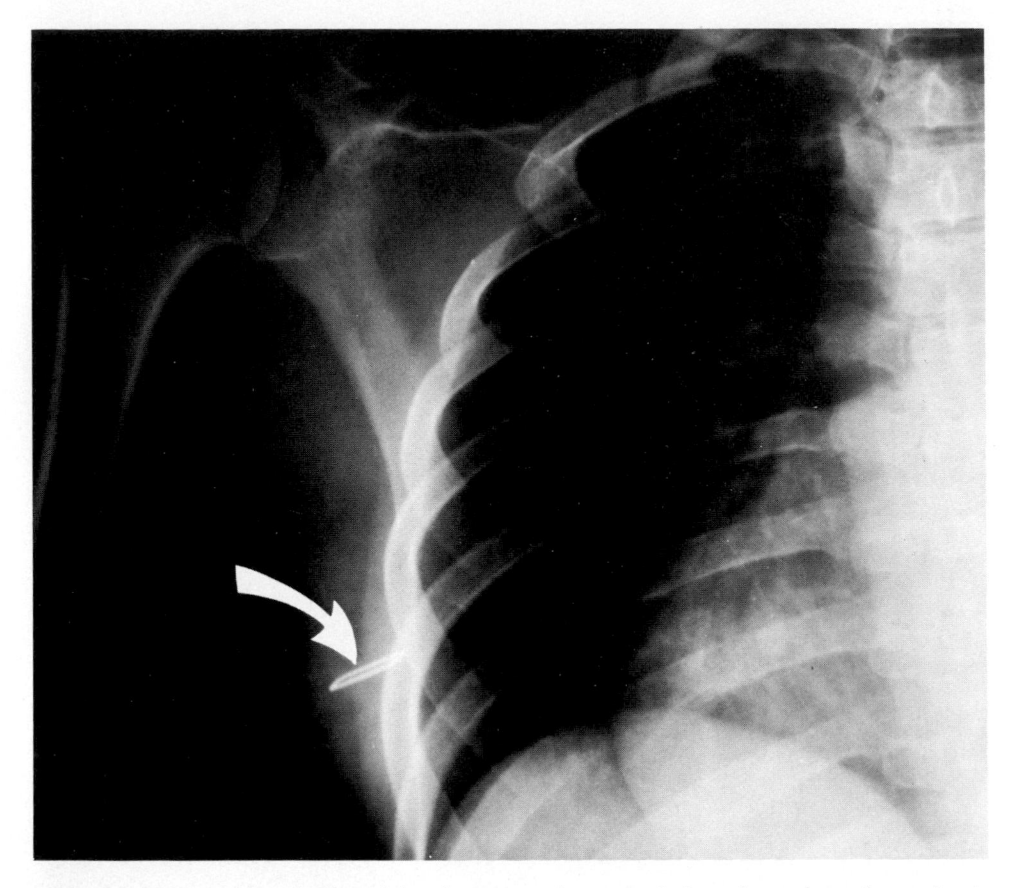

FIG. 7-9. *The object indicated by the* arrow *is a metal clasp located on the patient's undergarment. This type of clasp is used to adjust the length of the straps on bras and slips (Personal experience, senior prom, 1957). If such apparel is not removed, the image of the metal clasp will appear in the radiograph.*

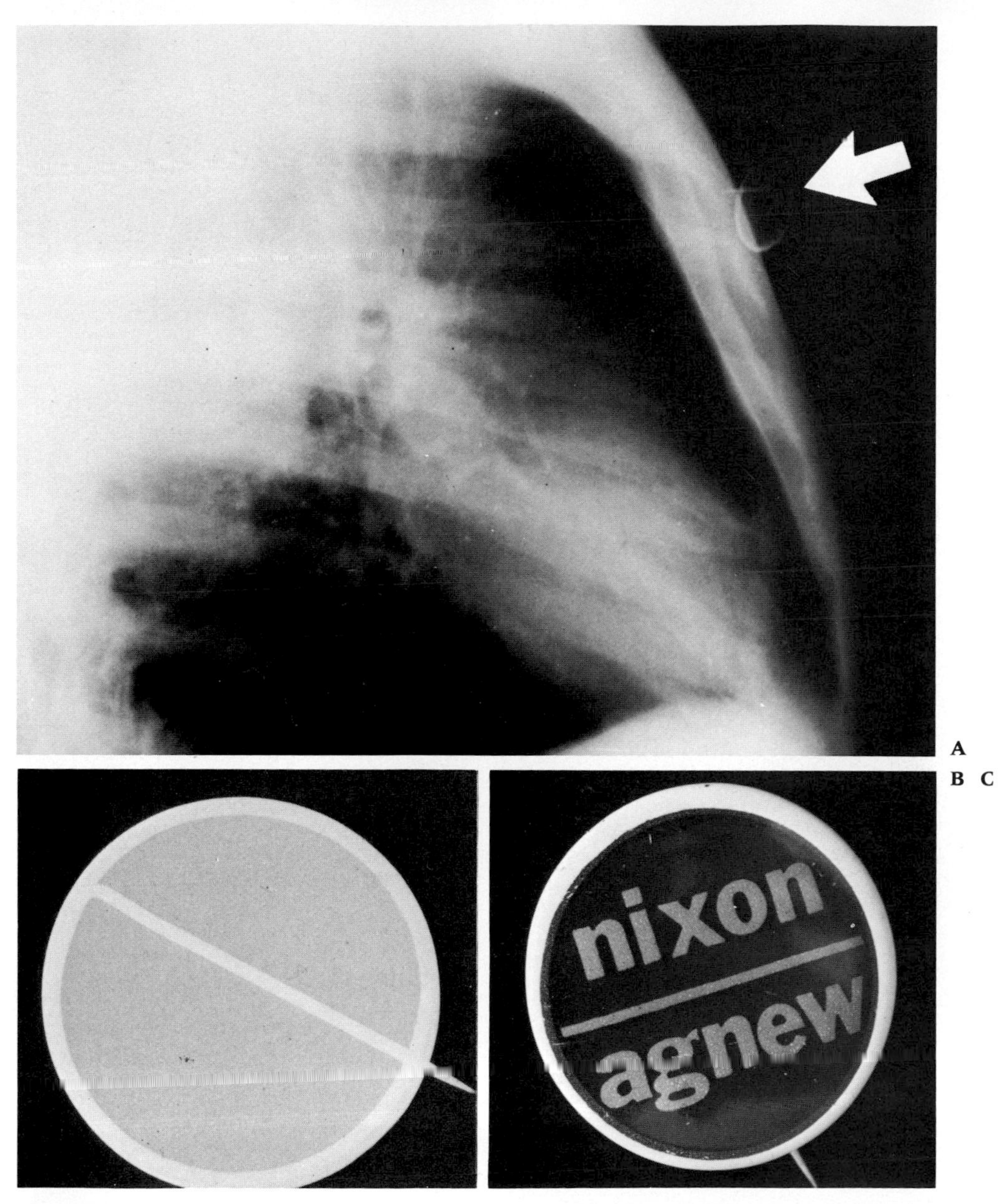

FIG. 7-10. (A) *The artifact* (arrow) *in the upper portion of the chest is the image of a loop of wire inside a button pinned to the front of the patient's gown. Notice the similarity in the appearance of the outline of the artifact in* A *to that of the radiograph* (B) *made of a metal button* (C).

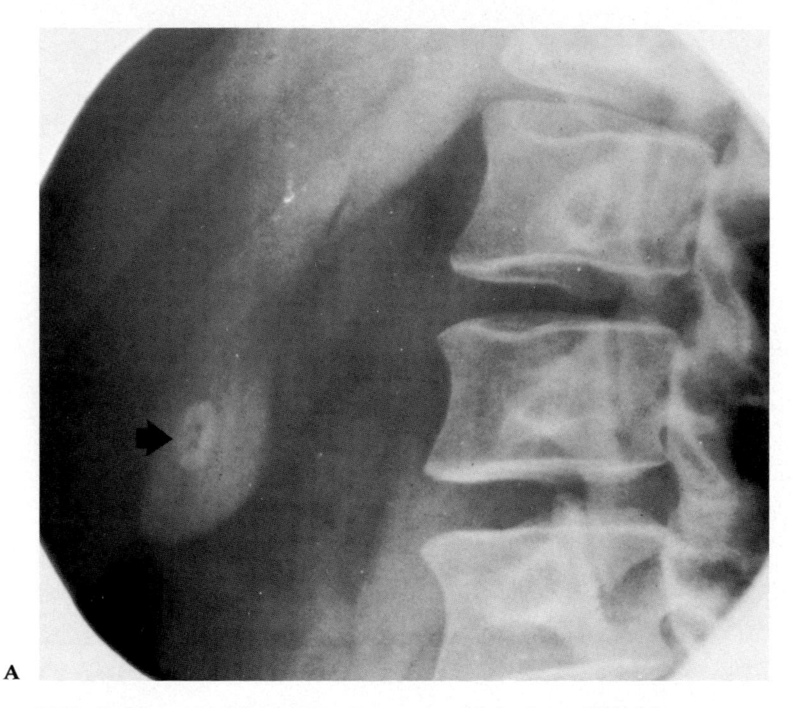

FIG. 7-11. (A) *The image* (arrow) *within the gallbladder appears to be that of a calculus. However, notice the small holes in the central portion of the structure. This is the image of a button located on the patient's robe.* (B) *A radiograph of this button presents the same image as the artifact in* A. (C) *In a repeat radiograph of the gallbladder, the pseudocalculus has disappeared.*

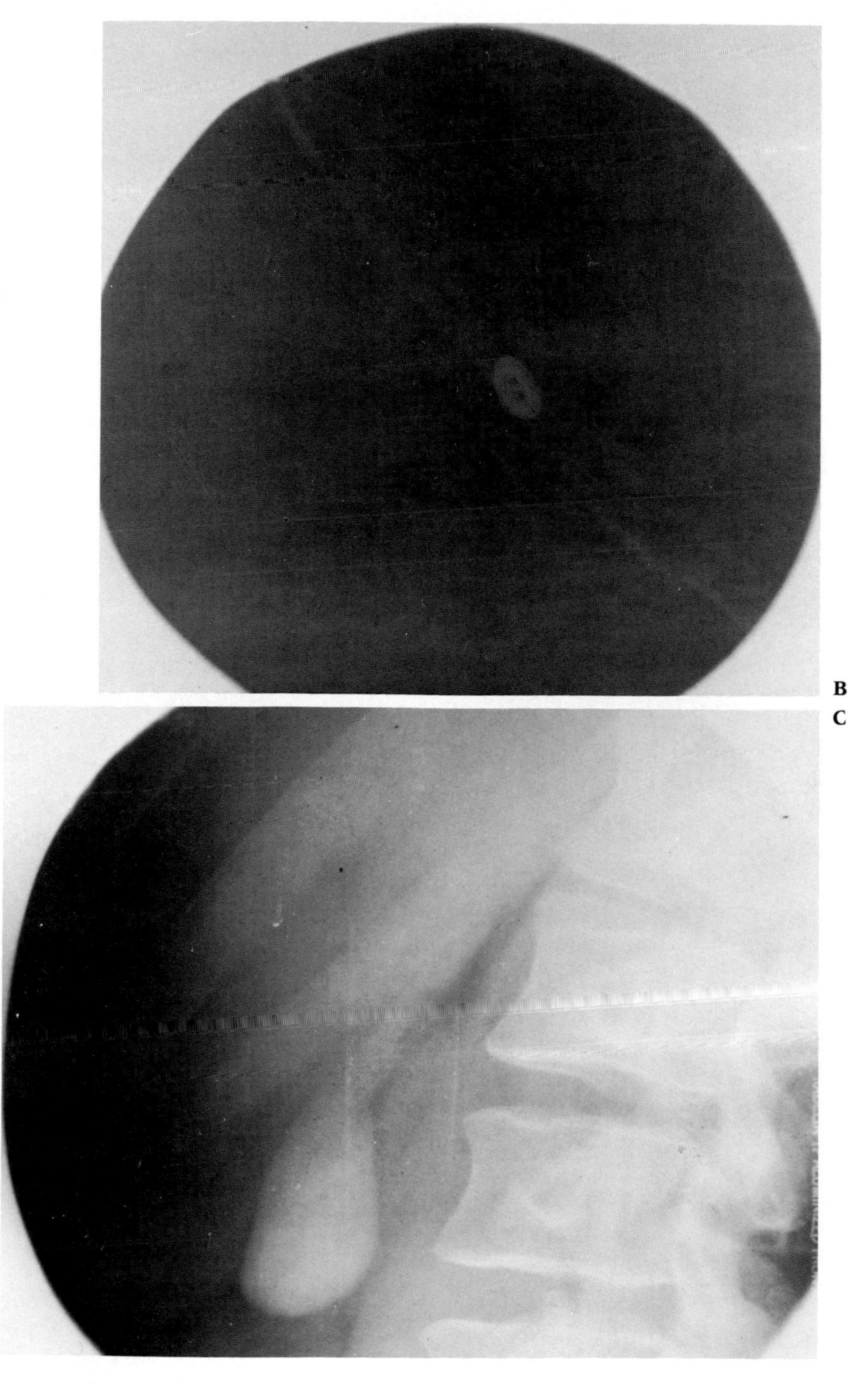

FIG. 7-11. *continued*

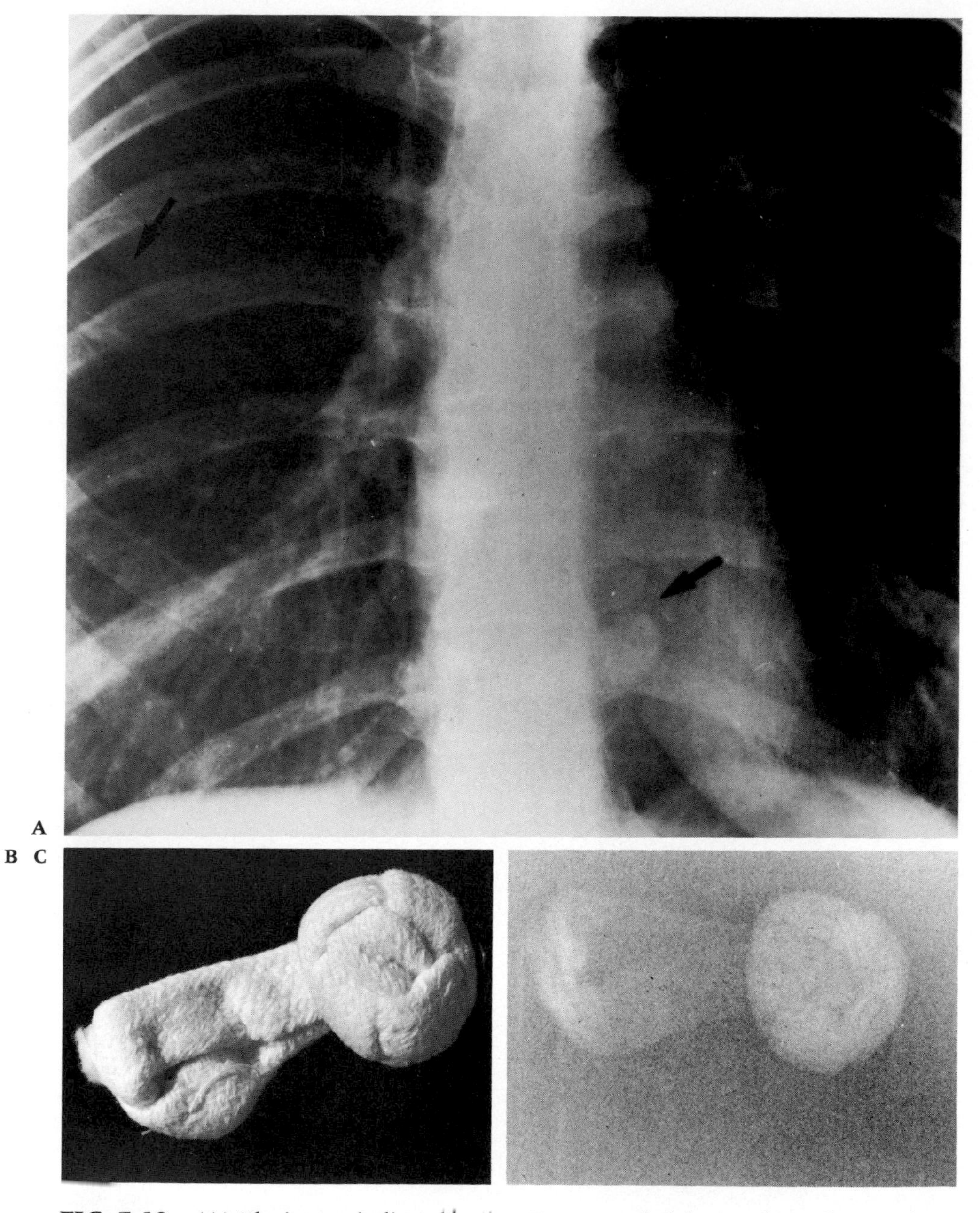

FIG. 7-12. (A) *The images indicated by the* arrows *are cloth buttons located on a pajama top inadvertently left on the patient for a chest radiograph that was obtained prior to performing a mammographic study to evaluate a questionable lump in her breast.* (B) *A photograph of the type of button responsible for the appearance of the artifacts depicted in* A. (C) *A radiograph of the cloth button. Notice the similarity of the image to the artifact in* A.

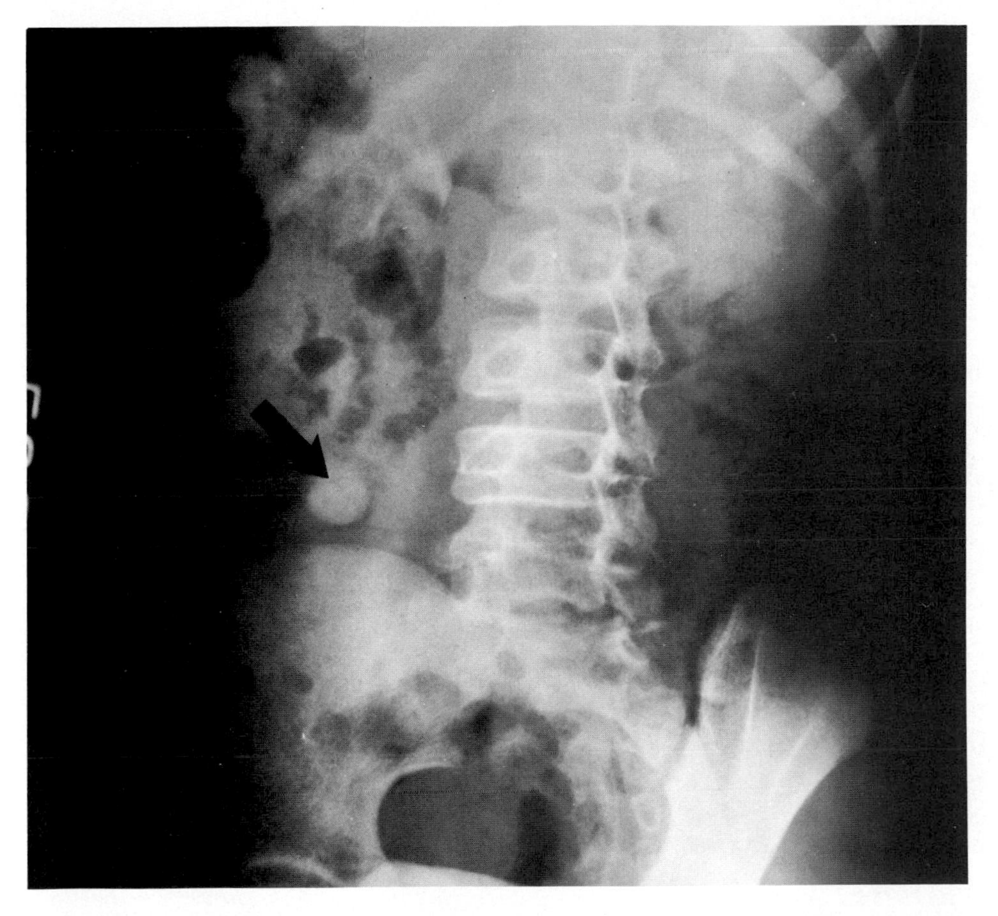

FIG. 7-13. *After looking at the preceding figures, you might conclude that the artifact* (arrow) *in this radiograph is a button on a garment worn by this young patient. Actually, it is the patient's belly button, or umbilicus, which is slightly herniated.*

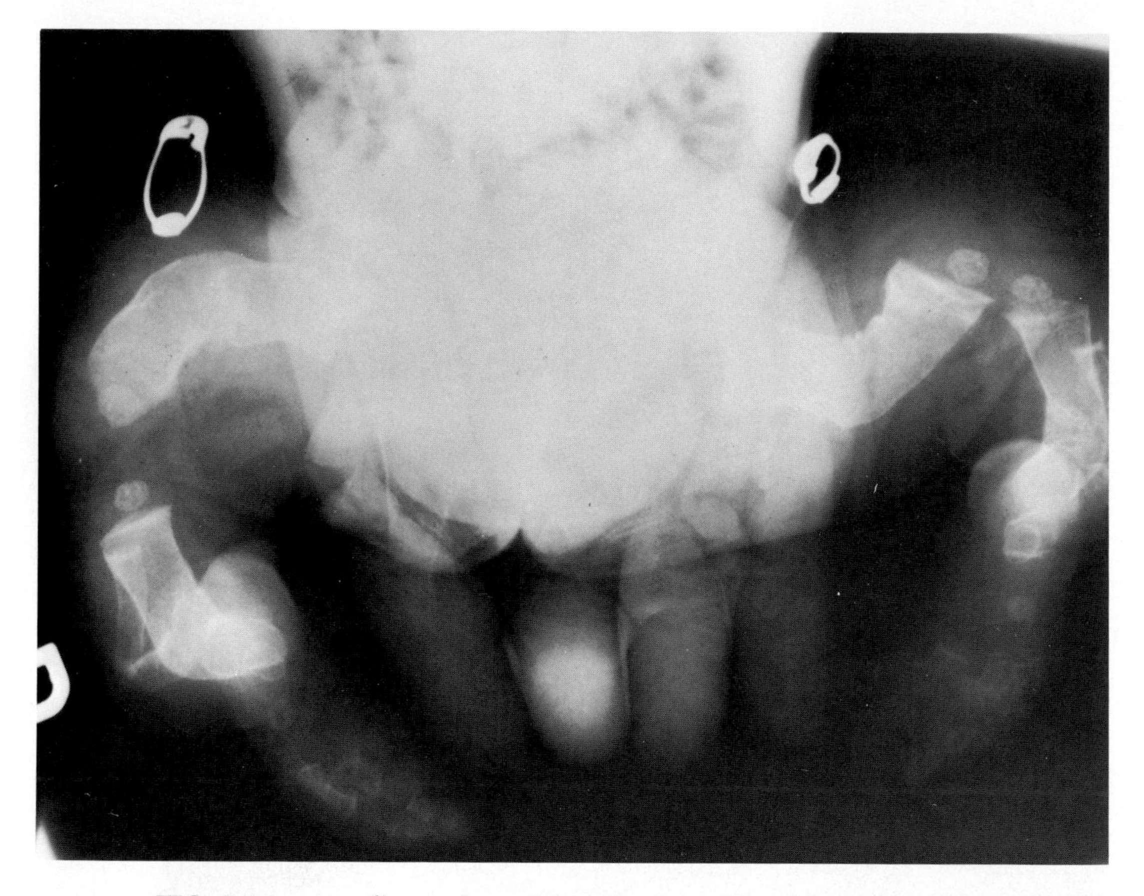

FIG. 7-14. *According to the radiologist's report, this pelvic radiograph of a newborn demonstrates a condition known as osteogenesis imperfecta, in which the bones fracture during uterine life and the child is born with deformities. Aside from these malformations, notice the fingerlike images. This is the image of a surgical glove, which had been filled with water and placed under the child's diaper. The nurses used this technique to support and cushion the pelvis and prevent further injury. Compare the images of the safety pins with those in Figure 1-4, p 6. (Courtesy of Dave Sack, R.T. [R])*

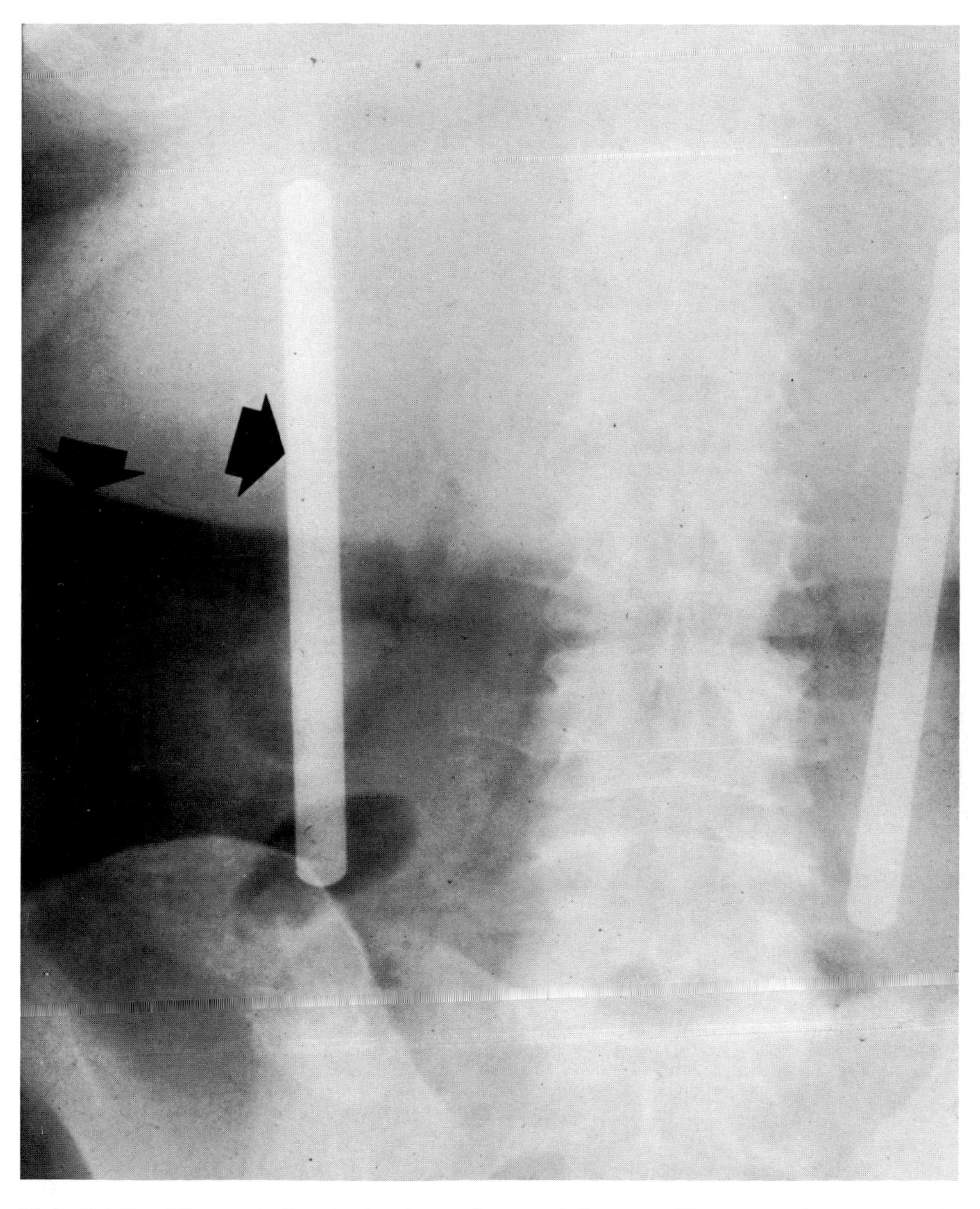

FIG. 7-15. *The vertical strip* (horizontal arrow) *is a metallic support in a corset. The dark band* (vertical arrow) *is air trapped between the rolls of fatty tissue.*

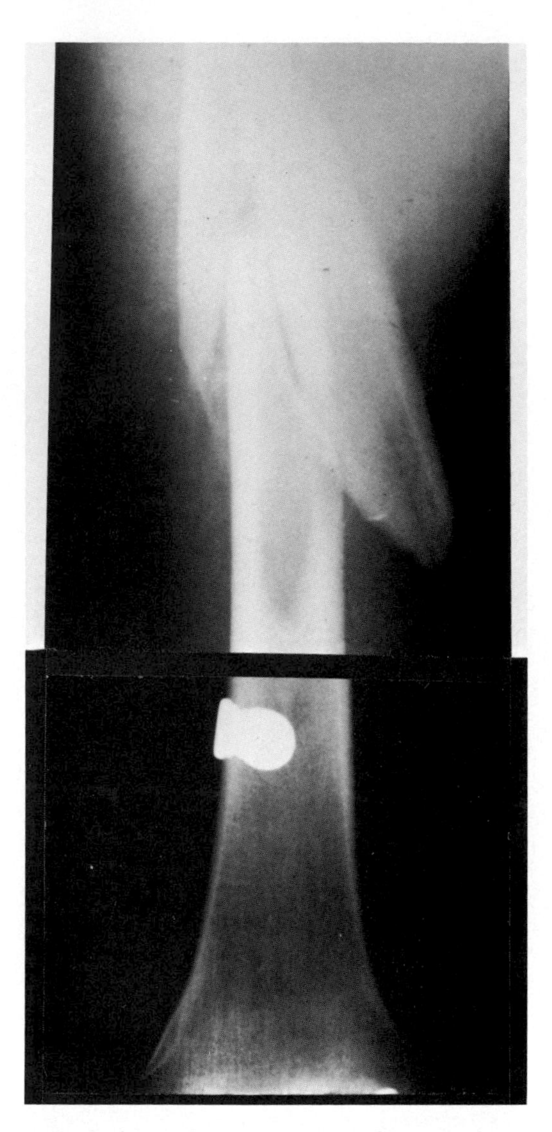

FIG. 7-16. *The patient whose femur is shown in this radiograph had been involved in a motorcycle accident and was hospitalized for numerous injuries, including a fractured femur. The fracture seems to be complicated by the presence of a projectile* (boxed area), *but the artifact is the image of the bowl of a hashish pipe that the patient had concealed in the sling of his traction device. The stem of the pipe is not visible because it was made of bamboo.*

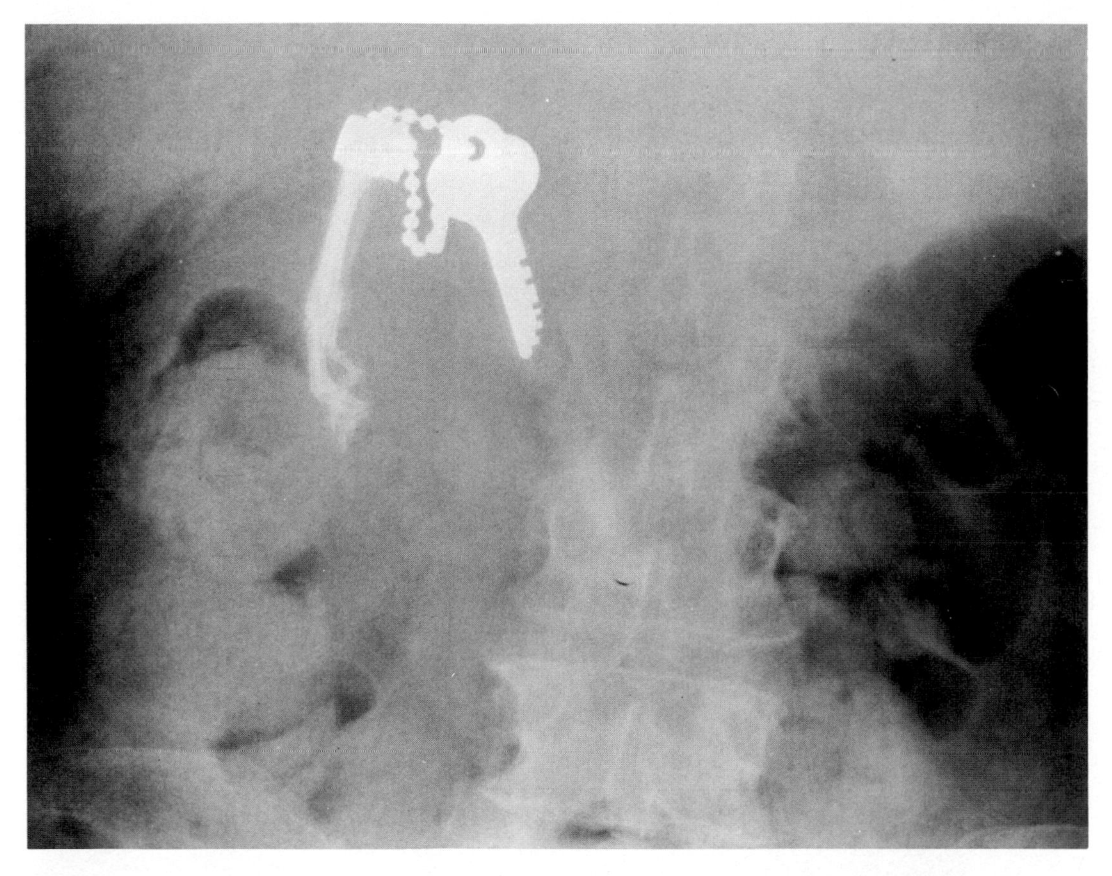

FIG. 7-17. *This keychain was located in the patient's pajama pocket. The image of the bones is from a rabbit's foot.*

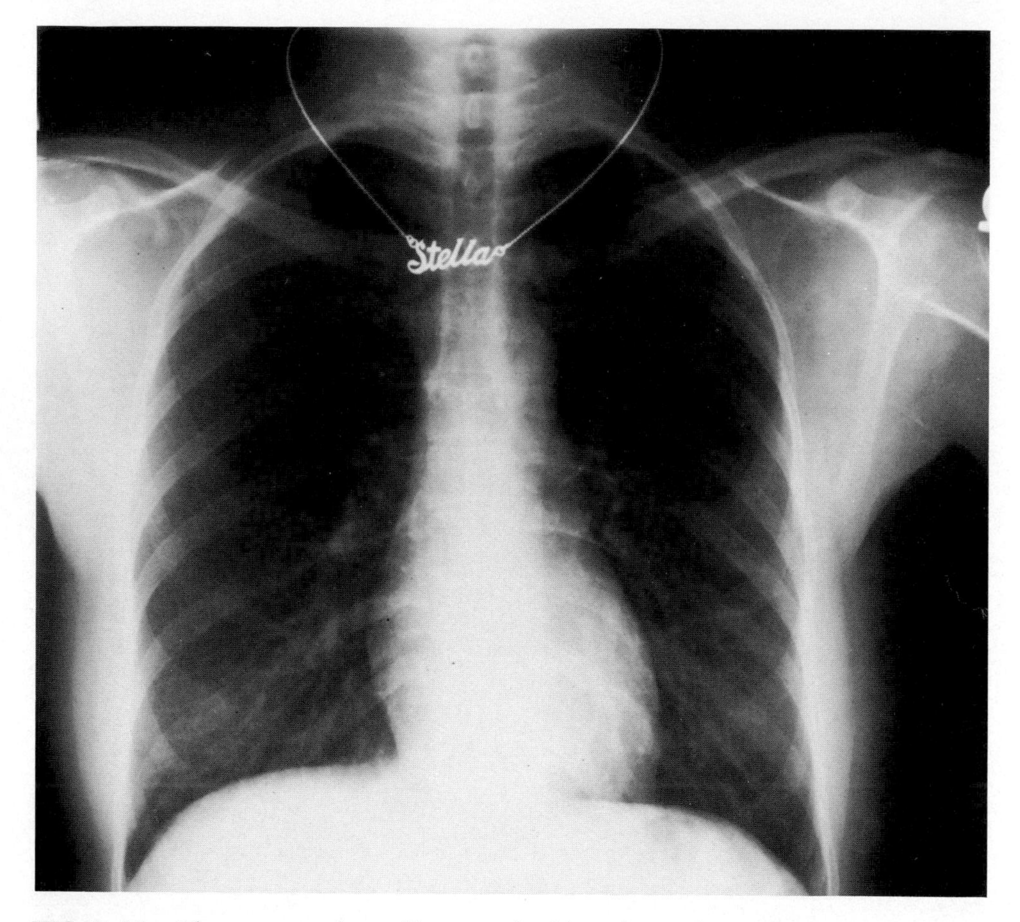

FIG. 7-18. *The name on the necklace seen in this radiograph suggests that the patient's name is Stella. In this instance, however, ''Stella'' means star: the necklace was worn by a male patient named Bruce.*

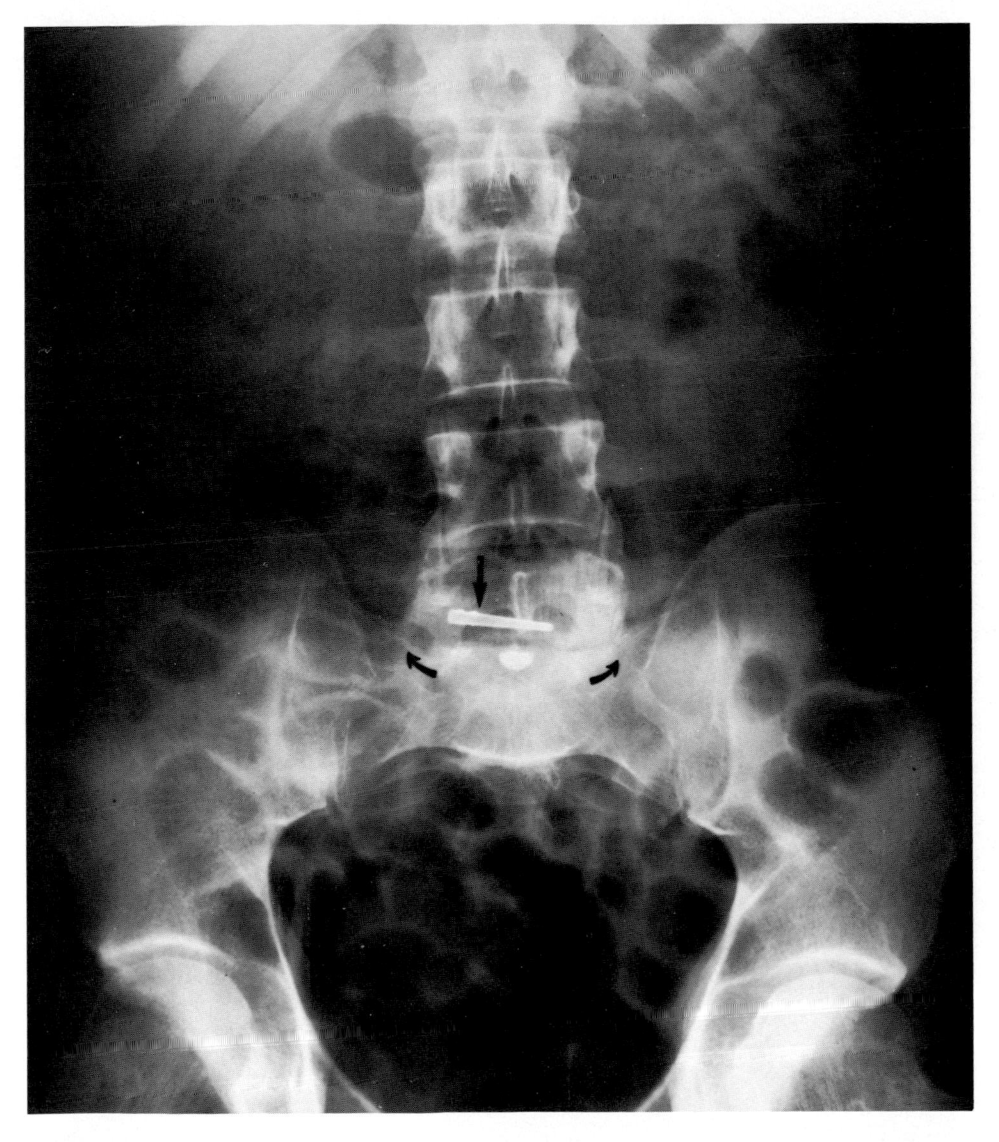

FIG. 7-19. *The outline of the plastic hair comb that was mistakenly left on the patient's abdomen for this radiograph is not visible; however, the metal clip attached to the comb and the patient's pajamas is imaged over the last lumbar vertebra* (straight arrow). *The radiopaque configuration below the clip* (curved arrows) *is residual iophendylate (Pantopaque) from a previous myelogram.*

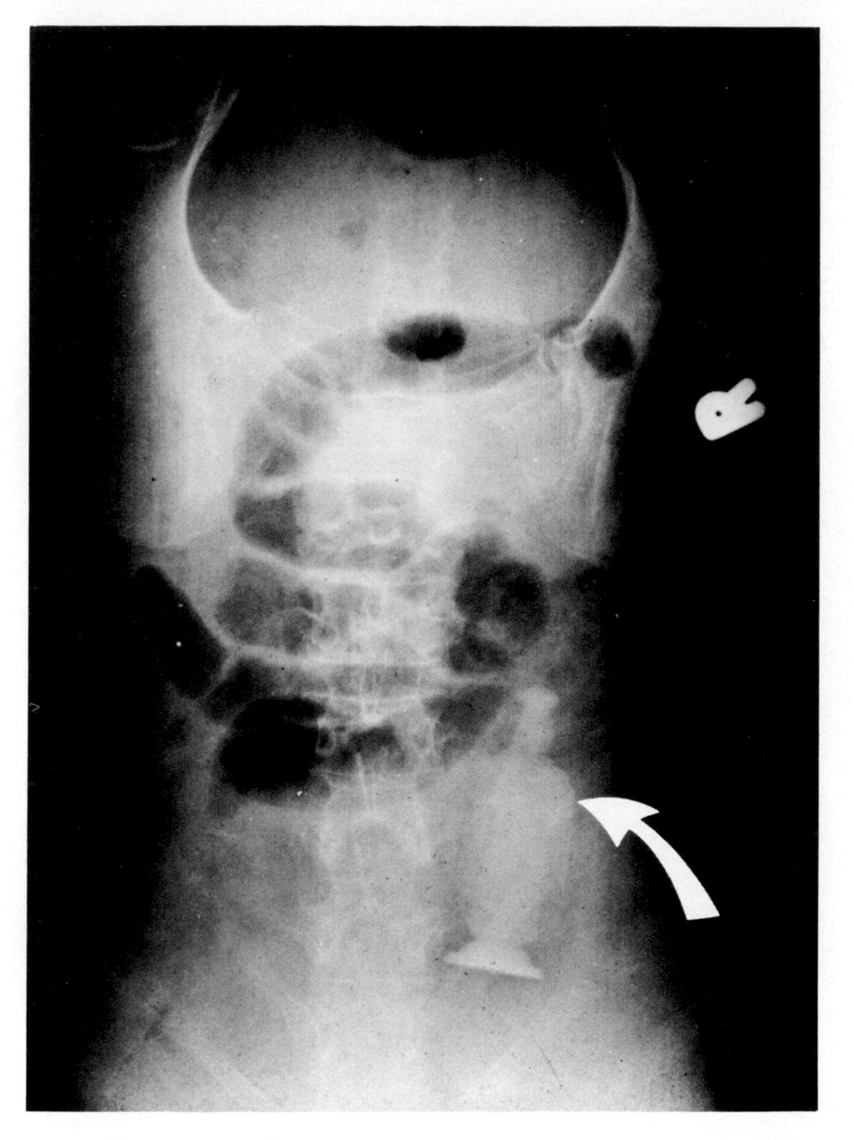

FIG. 7-20. *This radiograph is purposely inverted in order to aid in identification of the artifact indicated by the* arrow. *The image is a statuette of the Infant of Prague located within the patient's pajama pocket.*

FIG. 7-21. *The "coiled spring" configurations encircling the distal portions of the foot in* A *and* B *are metallic heating elements woven into electric socks that the patient wore during the radiographic examination. For the benefit of the reader residing outside the snowbelt, these socks are very popular among sportsmen and those who must work outside in freezing temperatures. The only external difference between these and conventional woolen socks is the wires that lead up the trouser leg to a small battery worn on the belt* (C) *or in some instances attached to the upper portion of the sock* (D). ▷

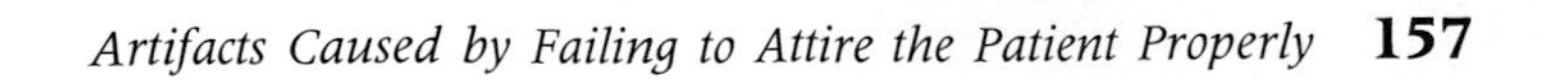

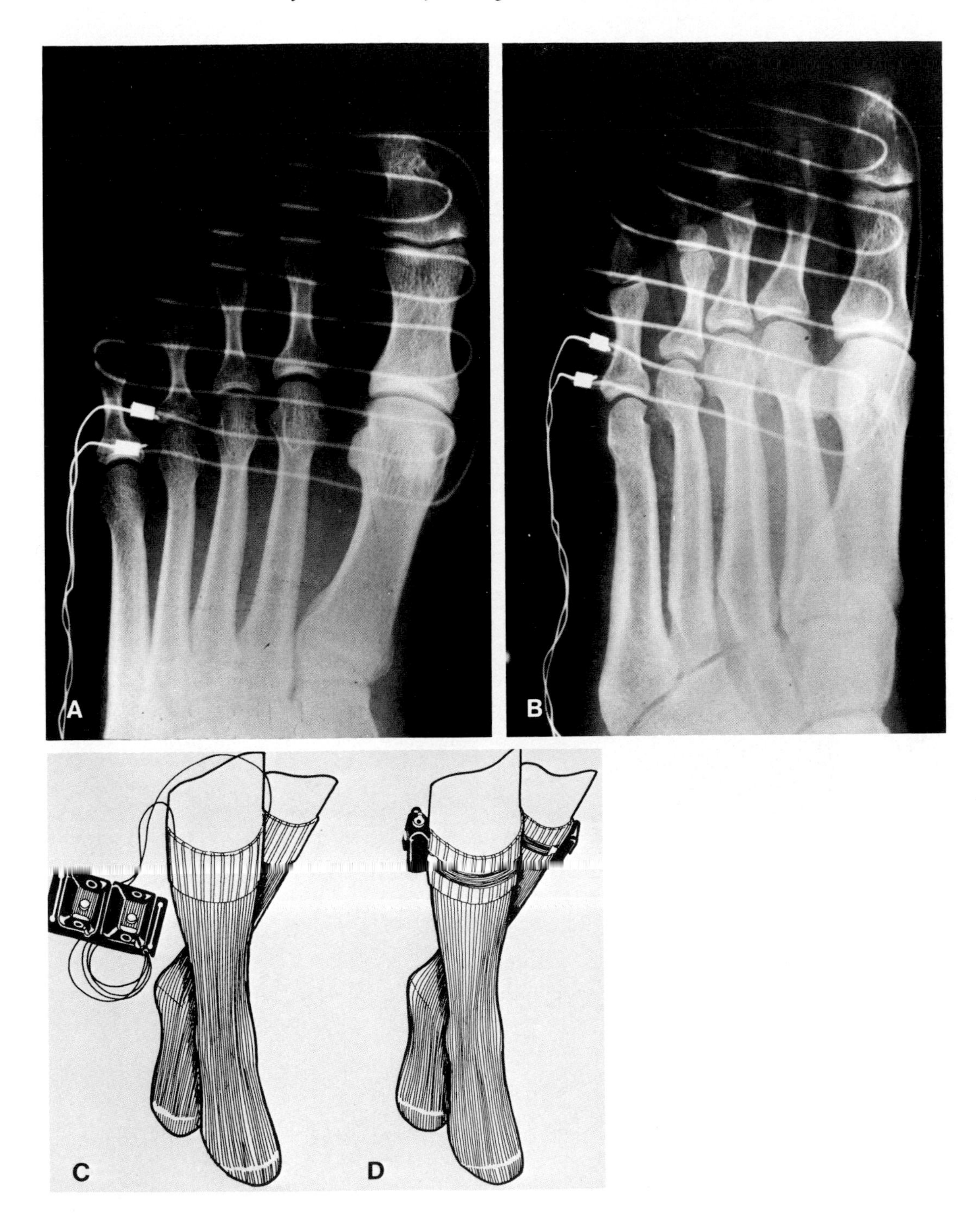
A
B
C
D

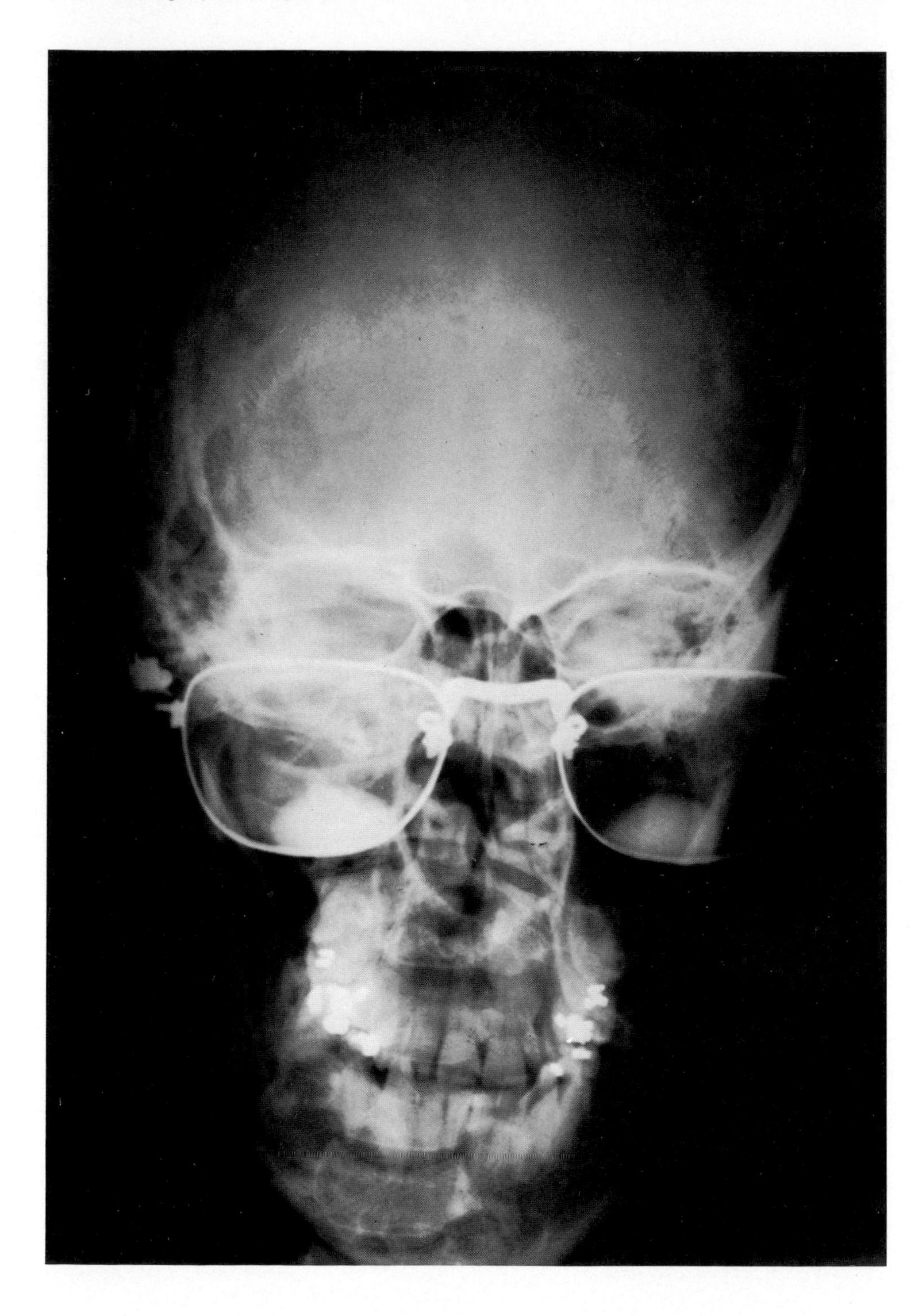

◁ **FIG. 7-22.** *Both the on-call technologist and the patient must have been half asleep when this radiograph of the skull was obtained during the early morning hours. Notice that the patient was wearing glasses for this posteroanterior projection of the skull. (Courtesy of Ralph Coates, R.T. [R])*

FIG. 7-23. *The patient shown in this radiograph of the chest had injured her ribs and right arm playing softball. She insisted on holding her right arm across her abdomen during the radiographic examination. Notice the outline of the arm* (small arrows). *Because of this, both arms were proximally positioned close to the chest, and the soft-tissue outline of the medial aspect of her arms was imaged in the lungs* (large arrows).

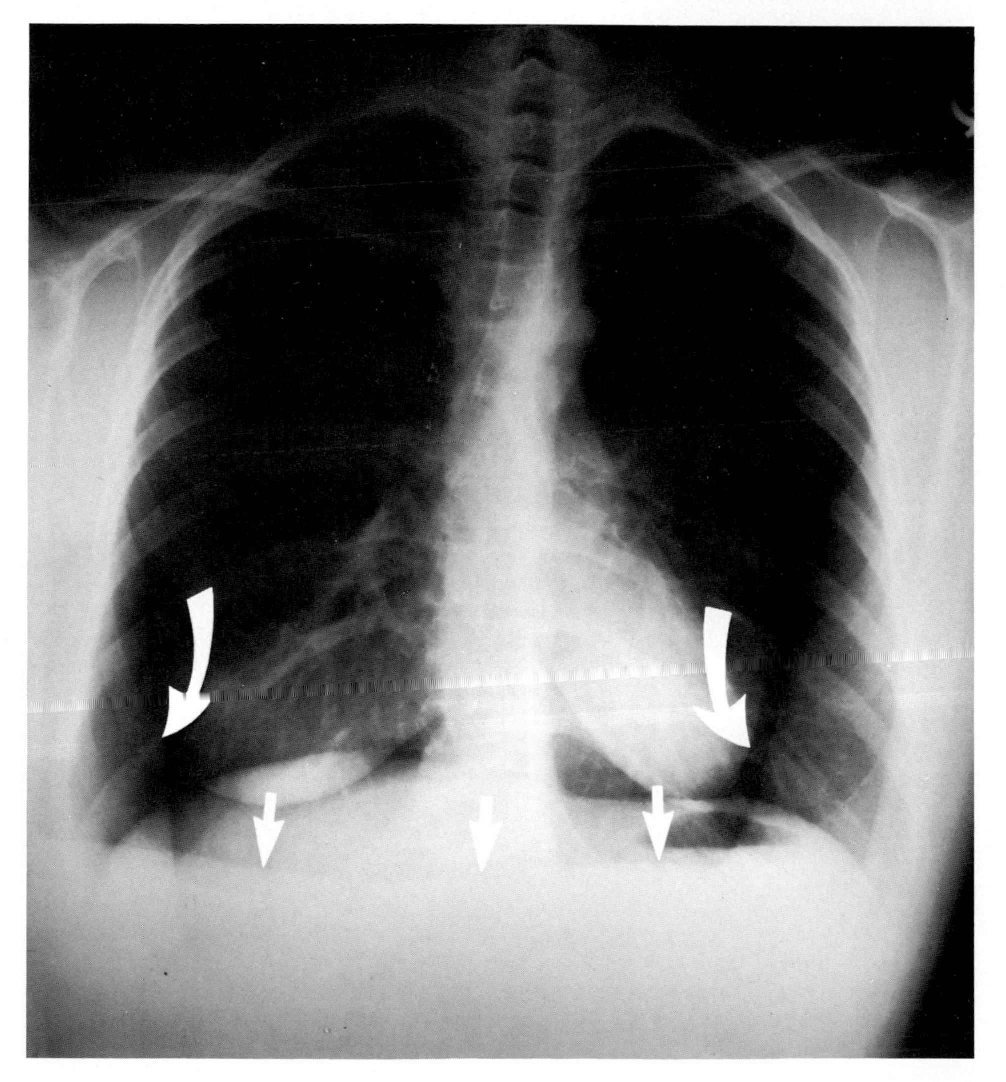

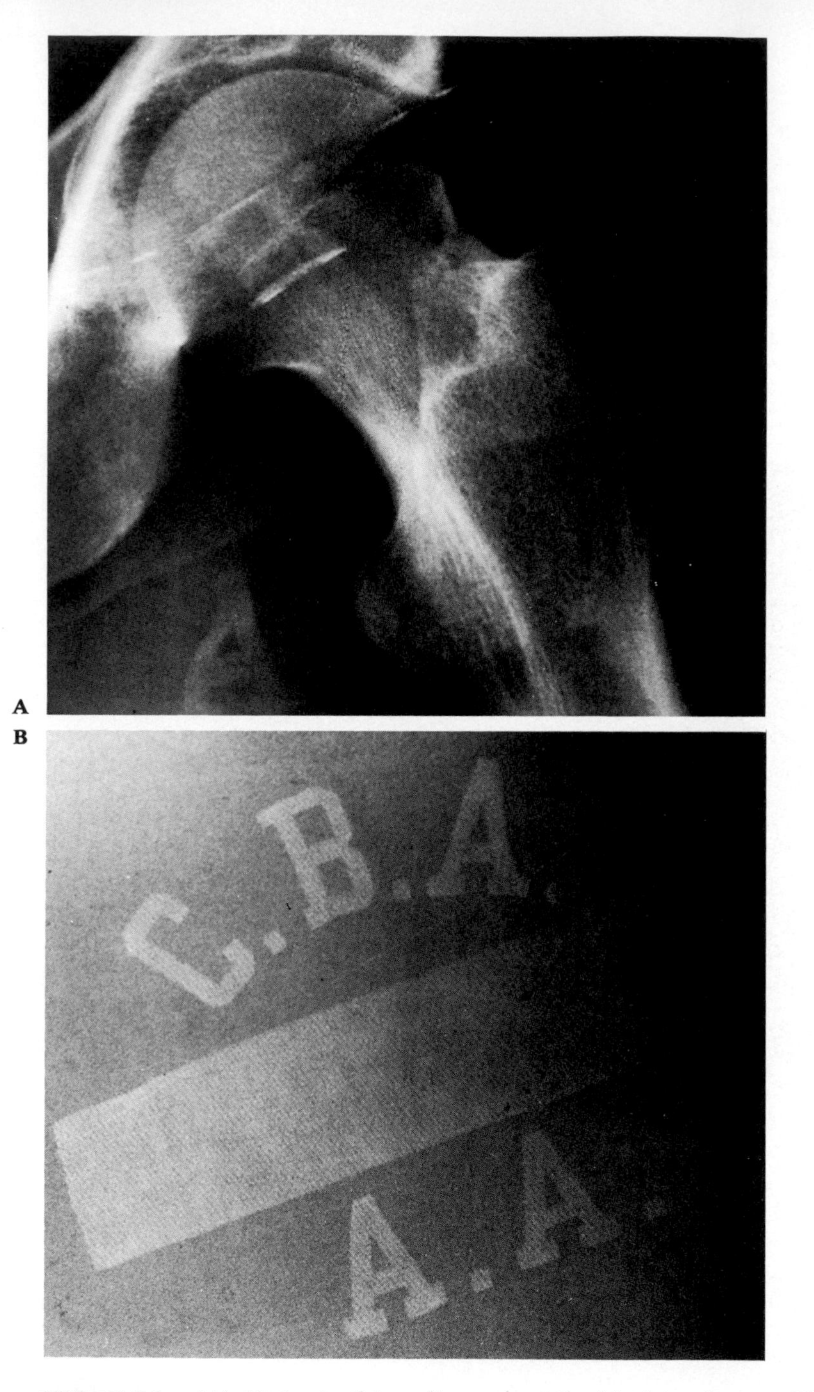

FIG. 7-24. (A) *Notice in this radiograph of the hip joint the streaklike pattern in the region of the femoral head and the appearance of a letter in the soft-tissue area of the thigh. The image is a portion of the labeling imprinted on a pair of gym shorts worn by the patient during the examination.* (B) *A radiograph of the gym shorts responsible for the appearance of the artifacts seen in* A.

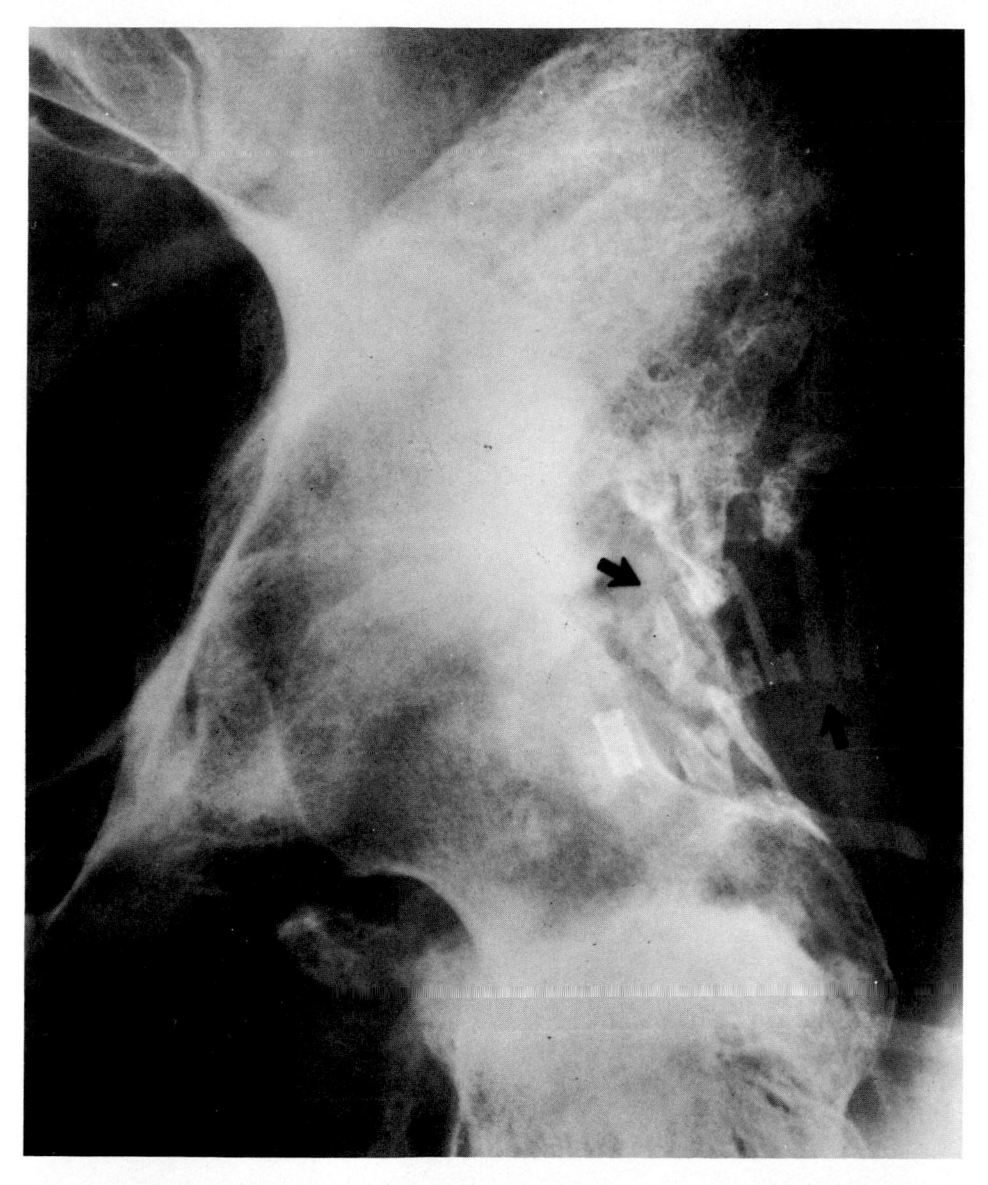

FIG. 7-25. *The medical history of the patient whose hip joint is shown in this radiograph indicated that the soft-tissue calcifications are associated with paraplegia. The appearance of the lettering* (arrows), *however, is again related to an imprint on the patient's shorts.*

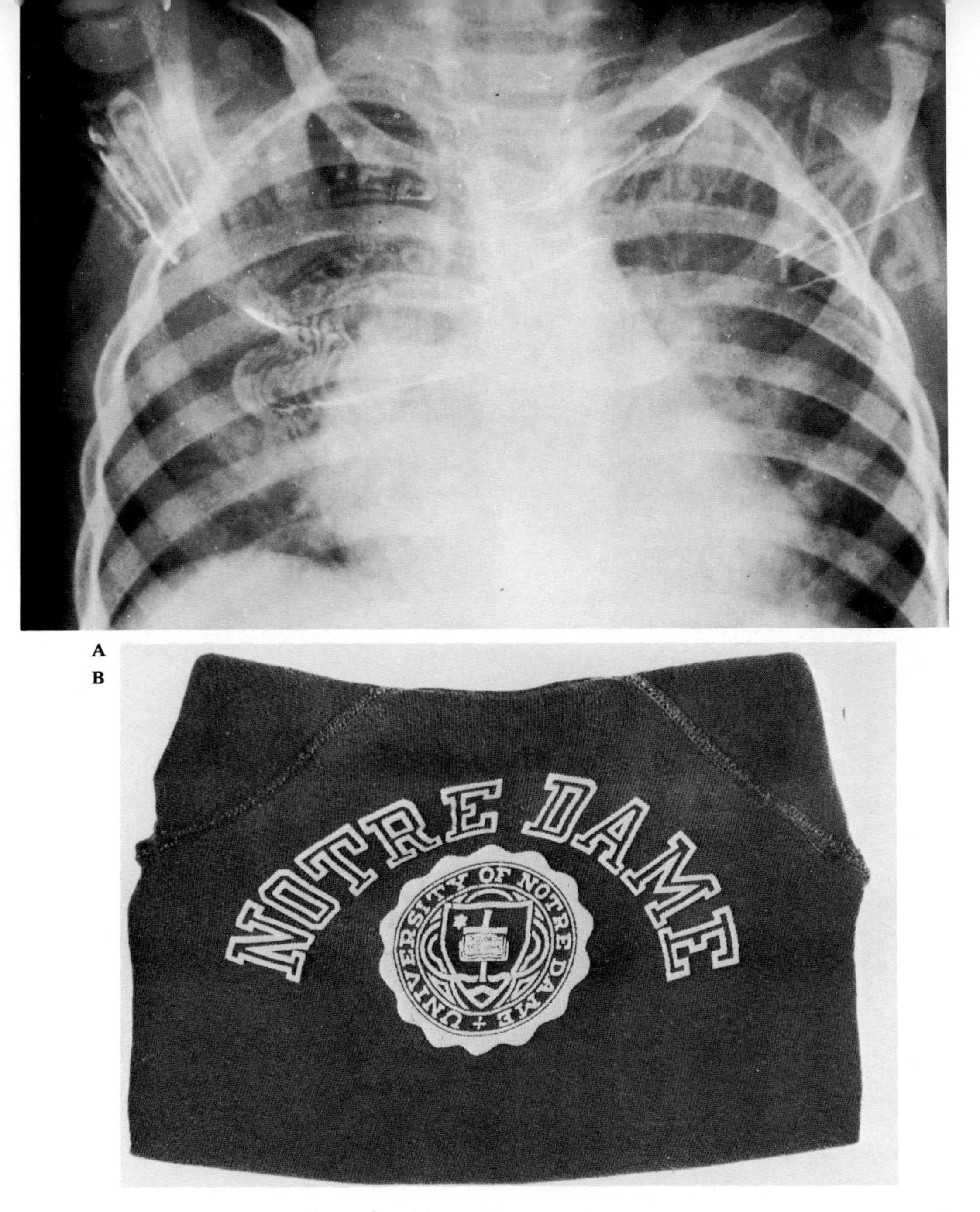

FIG. 7-26. (A) *The rather bizarre image in the lung field in this chest radiograph resulted from the technologist's failure to have the youngster remove a cotton jersey imprinted with lettering and an insignia* (B). *The matted fabric and the radiopaque nature of the lettering were responsible for the appearance of the artifact. (From Sweeney RJ: On the Technical Side. Radiol Technol 46, No 6:462–463, 1975)*

Metallic Content of Consumer Goods

RADIOPAQUE LETTERING

The appearance in radiographs of the lettering on gym shorts and shirts in Figures 7-24 through 7-26 indicates the presence of a substance that is opaque to x-ray. There are specifications regarding the amount of metallic content of such materials used in consumer goods.[1] For example, surface coatings cannot contain lead or lead compounds in excess of 0.06%, and barium compounds cannot be present in excess of 1% of the total content of such coatings.[9] Because of these standards, lettering that is opaque to x-ray should be relatively rare in patient attire. But these standards are not always complied with. Consider the case of Curious George.

> The Curious George doll was imported from South Korea and distributed nationwide by the Knickerbocker Toy Company of Edison, NJ, between March 1977 and April 1978. The paint used to print the name on the sweater of the stuffed animal

FIG. 7-27. (A) *Despite rigorous governmental standards, the metallic content of some consumer goods is dangerously high. A case in point is Curious George, whose sweater was imprinted with lettering high in lead content.* (B) *Notice in this radiograph of the doll the radiopaque appearance of the lettering on the sweater. The metal structure directly below the lettering is a musical component within the toy.*

contained between 12% and 24% lead (Fig. 7-27). As a result, the Consumer Product Safety Commission and the manufacturer recalled a half-million stuffed animals, which, in addition to Curious George, included an Easter chick called Chick-A-Dee and two dogs called Puppy Love and Scooby-Doo.[10]

RADIOPAQUE TATTOOS

In 1955 Pendergrass reported that certain dyes used in tattooing are radiopaque and can be detected in the radiograph if the tattoo is not superimposed over dense anatomic structures (Fig. 7-28).[7] The opacity of the dyes is attributed to the presence of metal salts used in certain substances to outline various features in the tattoo. According to Andrews and Greenbaum, the substances most commonly employed in tattooing are carmine, indigo, vermilion, india ink, and cinnabar. Vermilion and cinnabar are the only ones that contain radiopaque metal salts.[2,5]

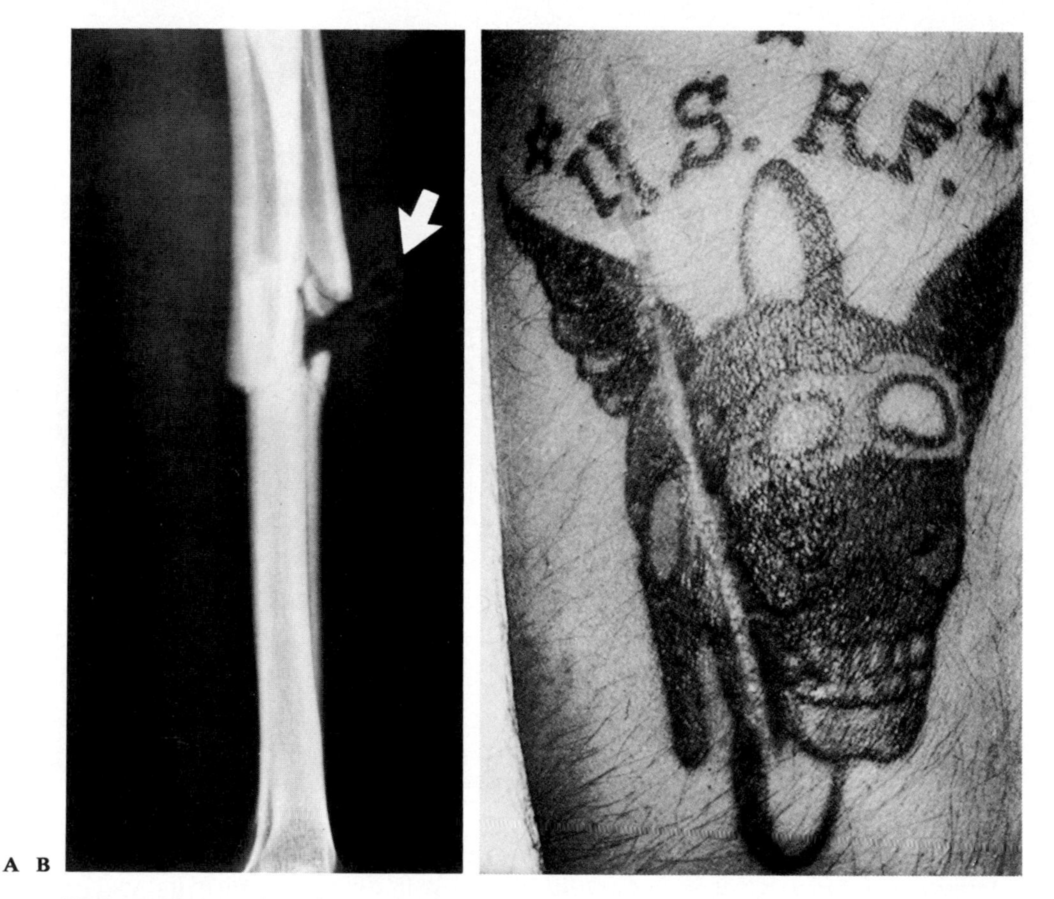

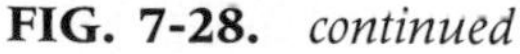

FIG. 7-28. *continued*

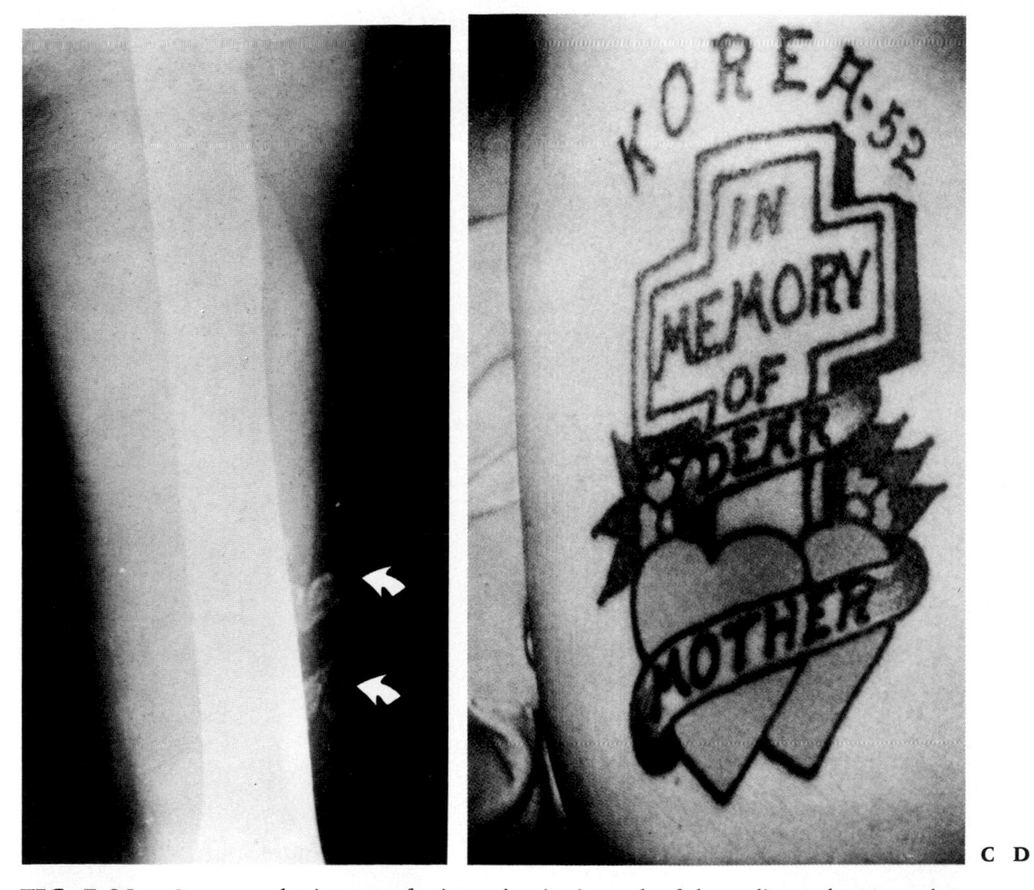

FIG. 7-28. *Compare the images of minus density in each of the radiographs* (A *and* C, arrows) *with the corresponding tattoos* (B *and* D). *The outline of the goggles etched in the tattoo of the aviator's head in* B *was responsible for the figure-eight marking in* A. *The outline of the two hearts and cross in* C *revealed the presence of radiopaque dyes in the tattoo shown in* D. *(From Pendergrass RC: Tattoos in radiographs. Med Radiogr Photogr 31, No. 1:43, 1955. Courtesy of Eastman Kodak. Copyright © by Eastman Kodak Company, Rochester, NY)*

Artifacts Caused by Matted Fabrics

Even if the patient is properly attired for the radiographic examination, his gown, as well as matted sheets, blankets, and the like, may be responsible for the appearance of various striated patterns in the radiograph. The problem arises when the patient is given a Dior-type creation—Max Dior, that is, not Christian. (Max operates a tent and awning business in the lower east side of New York City.) When a large gown is worn by a slender patient, the excess fabric may become matted and produce a characteristic image.

The problem of matting can be eliminated by using disposable rather than fabric-type gowns, or by having a selection of various sized gowns available for the patient. Although the striated pattern of the matted fabric may be easily recognized, in some other cases its origin may not be so easy to determine. Compare the image of the calcified arteries in Figure 7-29, *B*, with the radiograph made of a segment of a patient's gown (Fig. 7-29, *C*).

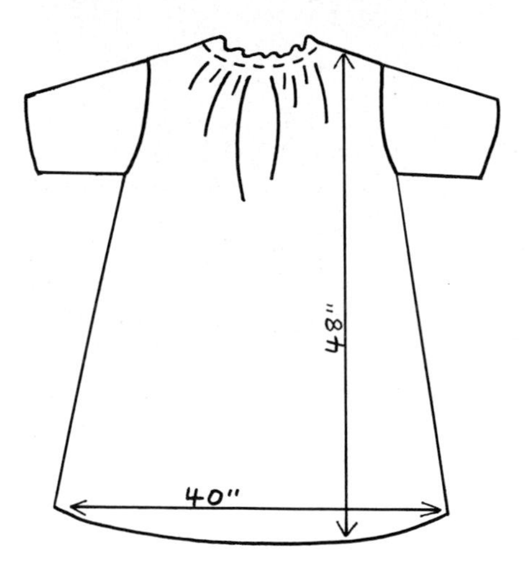

In a study conducted in 1979 at the Indiana University Medical Center, it was determined that when flame-retardant synthetic materials such as those used in patient clothing, sheets, and surgical drapes were laundered, the materials were responsible for the appearance of an unusually high number of unwanted images in radiographs. The study involved the analysis of a number of hospital gowns chosen at random. Each of the gowns radiographed on a pressed-wood phantom produced a blotchy, striated pattern, which, in the opinion of the author, could be responsible for the appearance of various artifacts that often require a repeat radiographic study.[8] The most interesting aspect of this study is that by comparison, when flame-retardant fabric was laundered under certain conditions, the surface properties of the fabric were altered in such a way that the potential for the appearance of such artifacts was increased. Figures 7-29 through 7-32 demonstrate a variety of artifacts that resulted from matted fabrics.

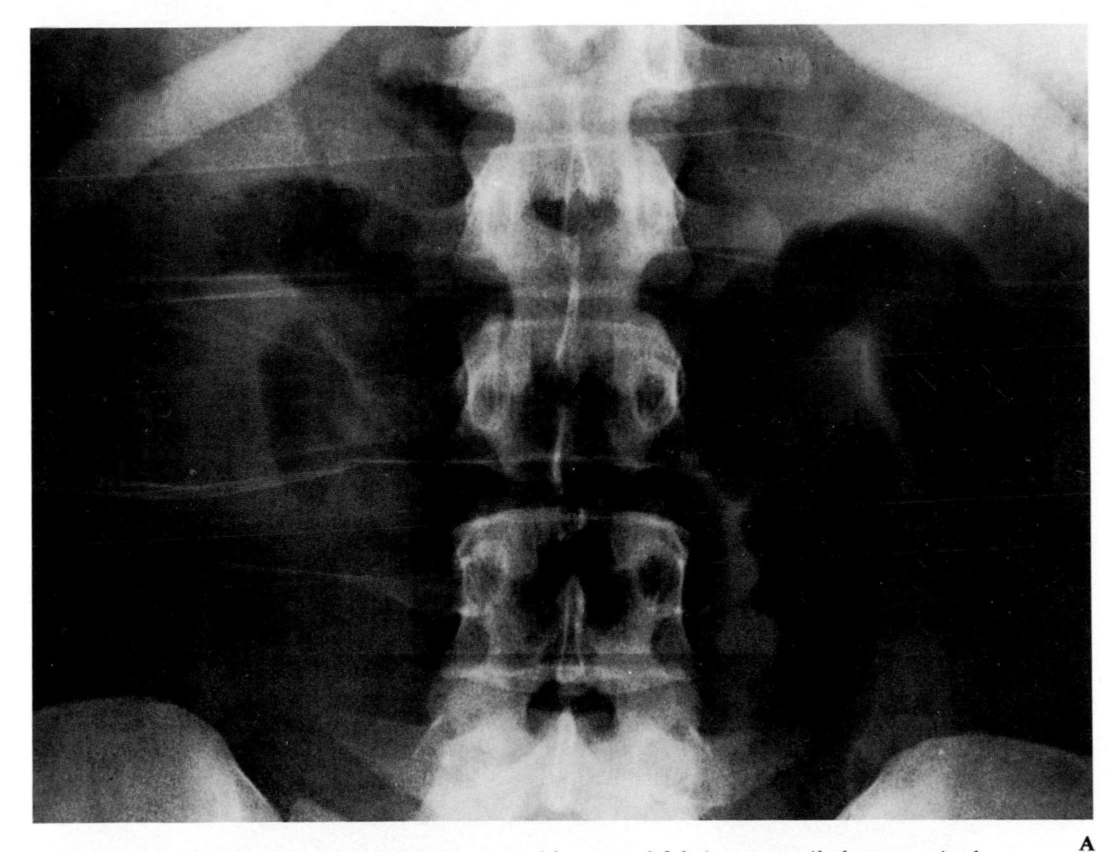

FIG. 7-29. *The characteristic striations caused by matted fabrics can easily be recognized in certain examinations.* (A) *The multiple striations occurred in this radiograph of the abdomen when excess fabric became matted under the patient. In* B, *however, the images of calcified arteries* (arrows) *look very similar to the striations in a radiograph of a patient's gown* (C, arrows).

(Fig. /-29 continues.)

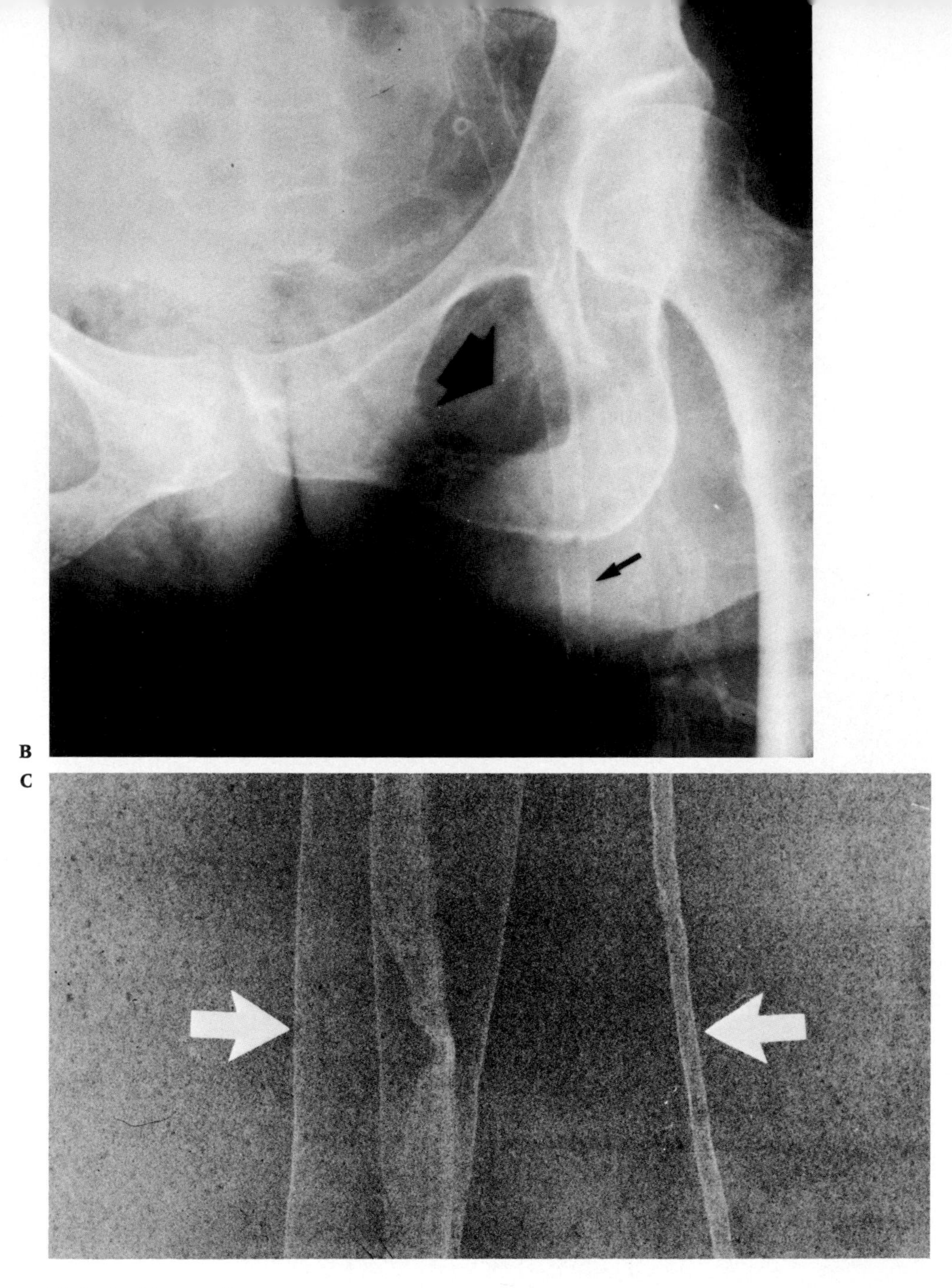

FIG. 7-29. *continued*

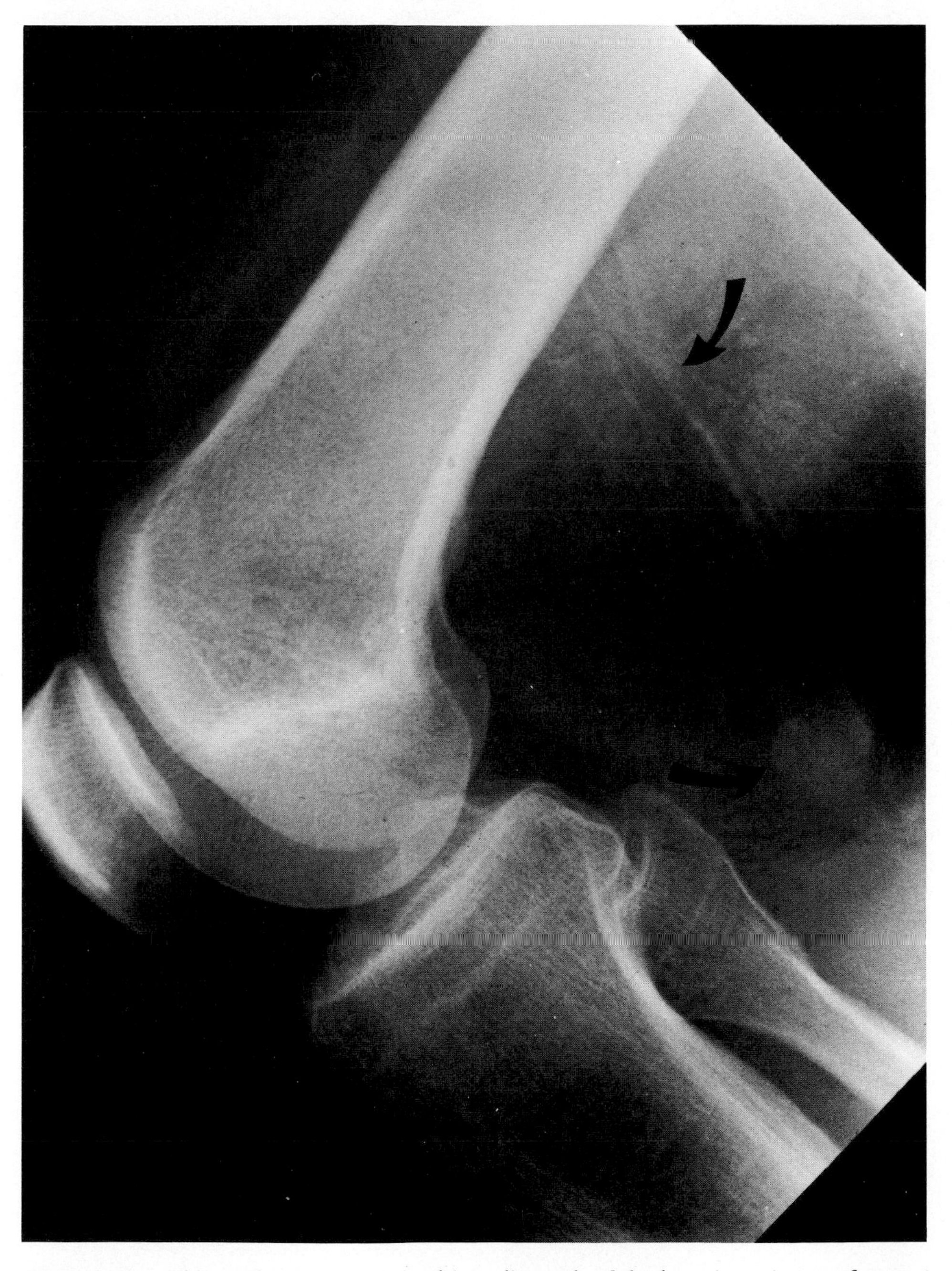

FIG. 7-30. *The artifact* (arrows) *in this radiograph of the knee is an image of matted fabric superimposed in the soft-tissue region of the knee. The artifact occurred because the technologist, rather than have the patient remove his trousers, rolled the fabric above the knee.*

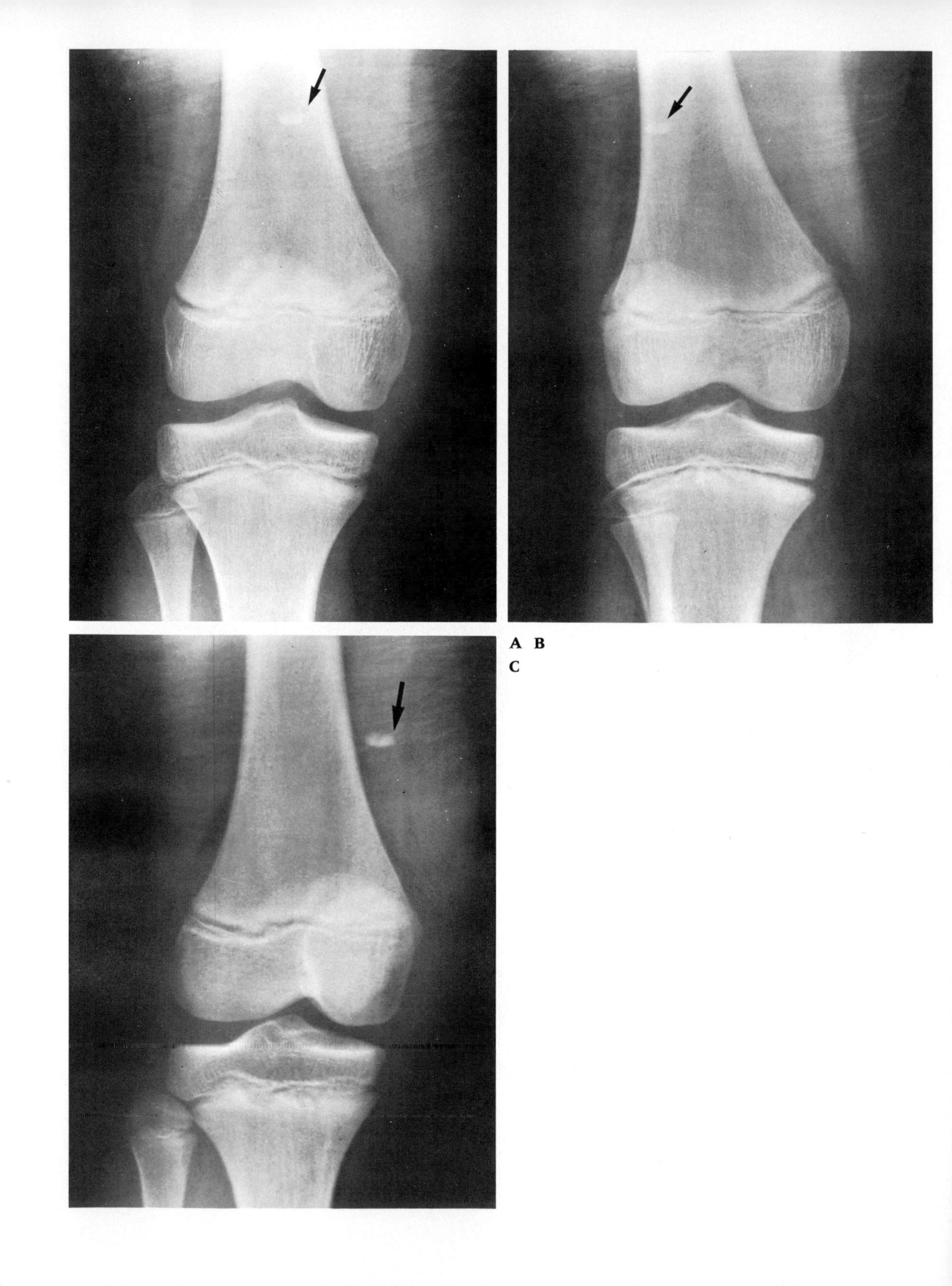

A B
C

◁ **FIG. 7-31.** *This radiographic study of the knee was requested after the young patient had been injured while playing football.* A *and* B *demonstrate an opaque density* (arrows) *over the shaft of the femur. The artifact is a small stone located in the cuff of the patient's trousers. Notice also the streaks of the matted fabric in the soft-tissue region of the distal femur. The stone is particularly clear in the internal oblique projection* (C, arrow). *(From Cullinan JE, Cullinan AM: Illustrated Guide to X-ray Technics, 2nd ed., p 110. Philadelphia, JB Lippincott, 1980)*

FIG. 7-32. *The matted fabric in the upper region of the thorax in this radiograph is the image of a necktie. Rather than remove the patient's tie, the technologist displaced it laterally for this radiograph of the cervical spine. (Courtesy of Ralph Coates, R.T. [R])*

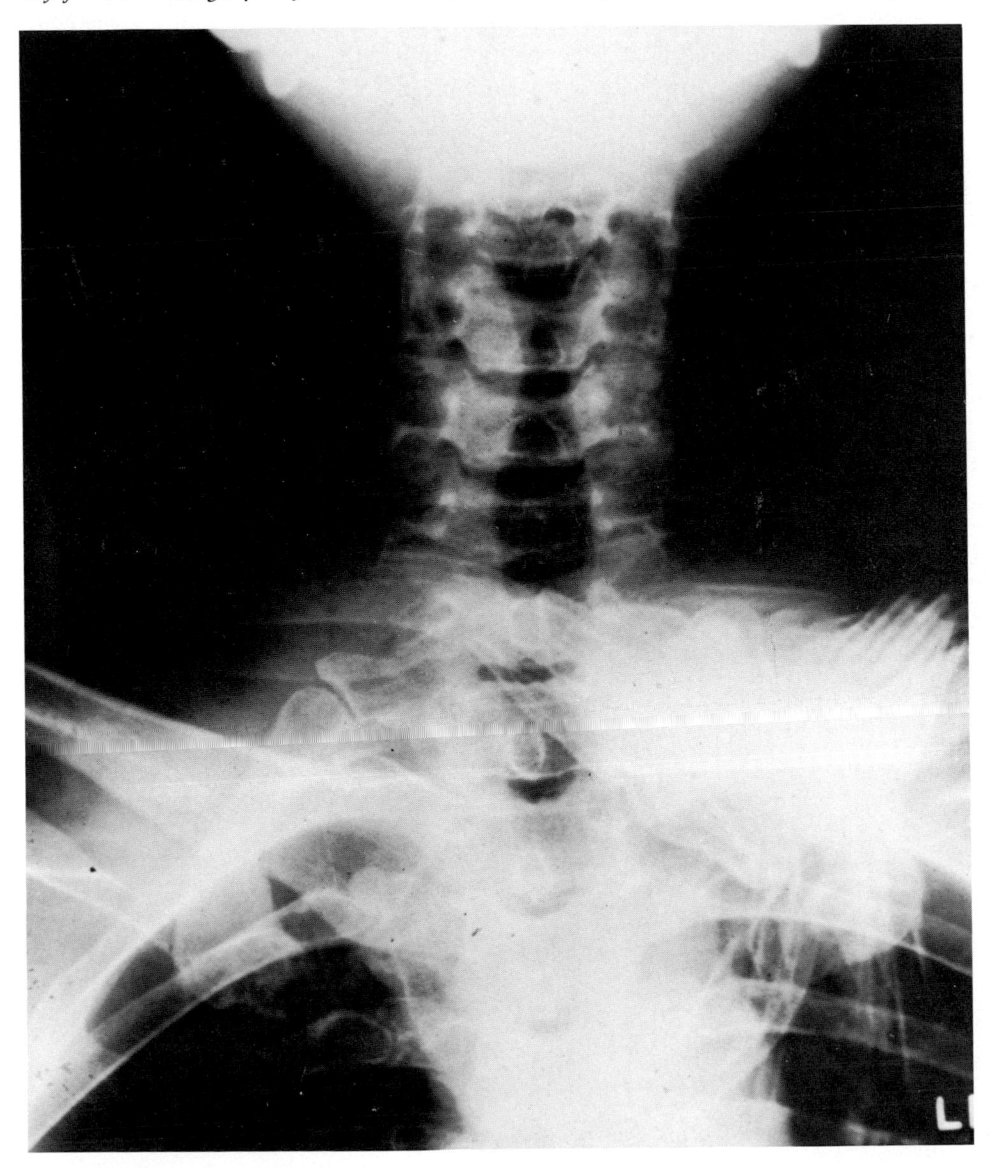

Wigs, Bobby Pins, and Related Artifacts

The failure of the technologist to instruct the patient to remove clothing and jewelry for the radiographic examination may result in obscure shadows that often appear superimposed in the region of interest. Because a patient is more likely to follow instructions correctly if he can see reason in them, the technologist should develop a method of informing the patient not only of what he is to do but of why he is to do it. For example, instruct the patient to remove hair ties, barrettes, and the like before you examine the skull and cervical spine. Although it is reasonable to assume that the patient understands the importance of removing such objects from the hair for a radiographic examination of the skull, the patient may not realize that when various angulations of the x-ray tube are employed for certain projections, these objects may be imaged in the region of the cervical spine and upper thorax. Moreover, both the technologist and the patient may have difficulty in locating such objects in the hair, especially when the hair is matted or styled in certain ways. In such situations a small magnet is a very useful device for locating a number of these objects, which often contain a metallic component (see Fig. 7-33).

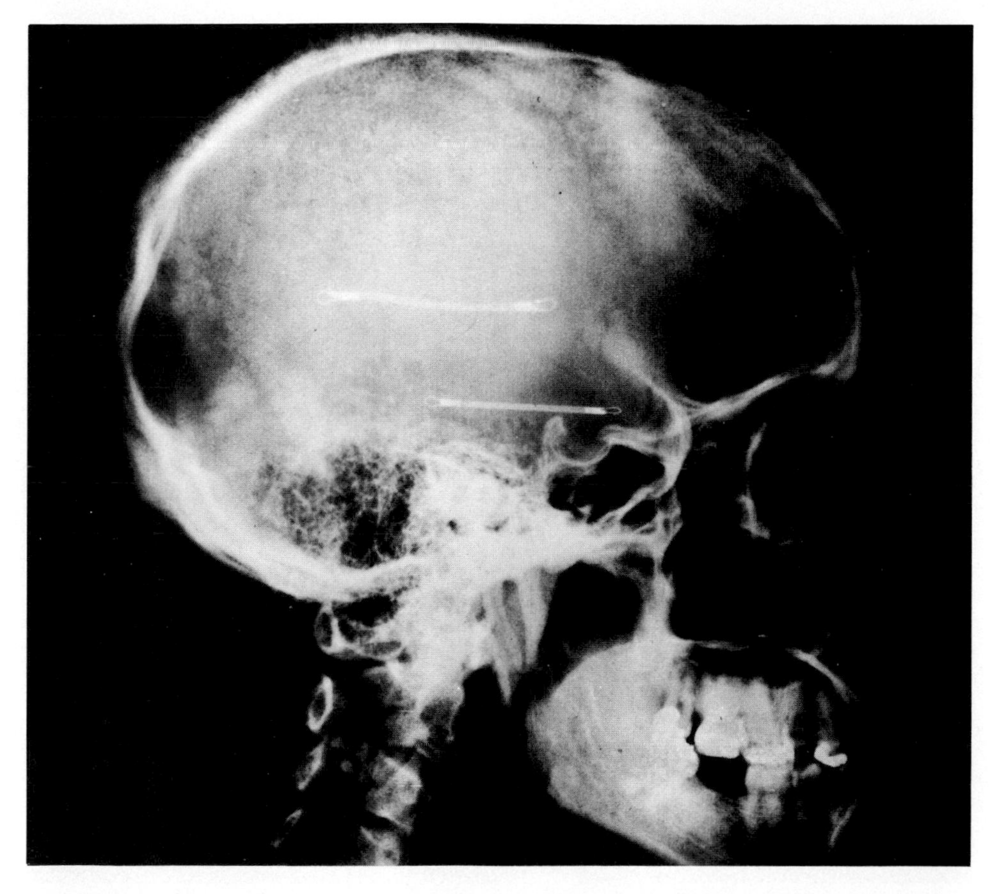

FIG. 7-33. *The two springs visible in this lateral cranial radiograph are located within a wig worn by the patient. It is interesting to note that when asked to remove her wig for the examination, the patient insisted that she was not wearing one.*

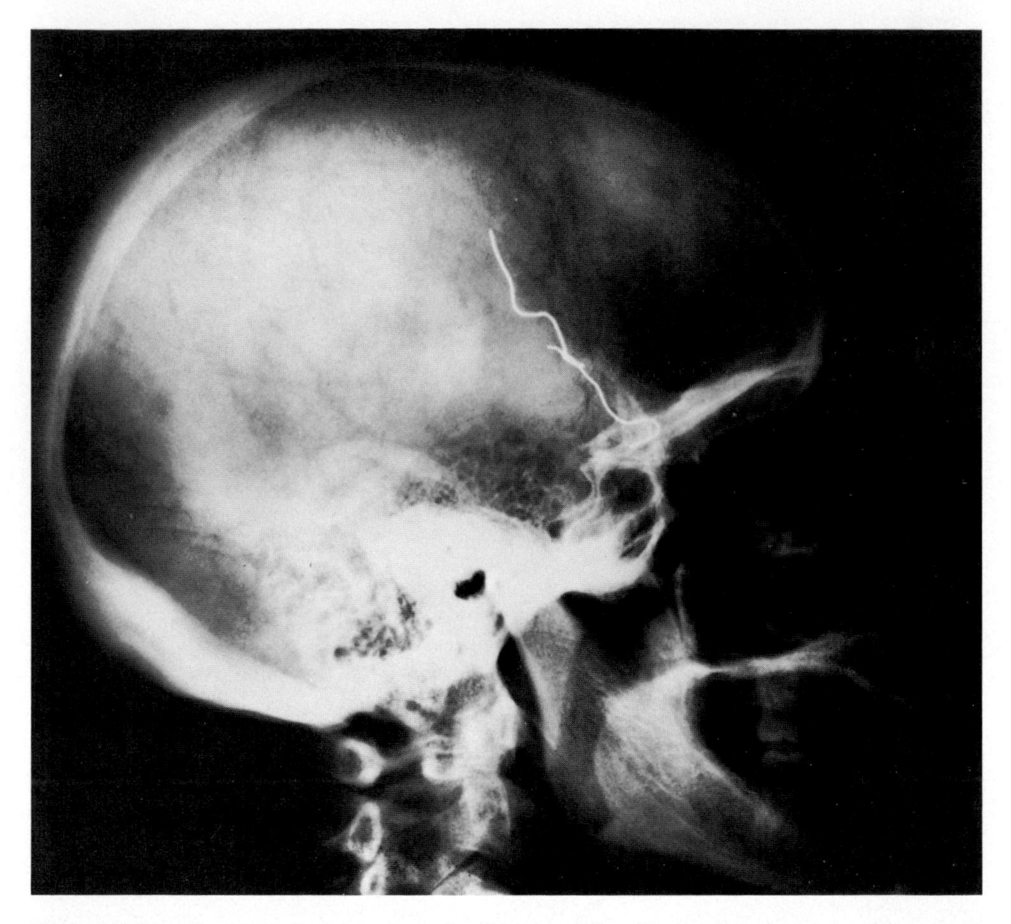

FIG. 7-34. *The strand of wire seen in this cranial radiograph is located within a wig worn by the patient during the examination. (Courtesy of Ralph Coates, R.T. [R])*

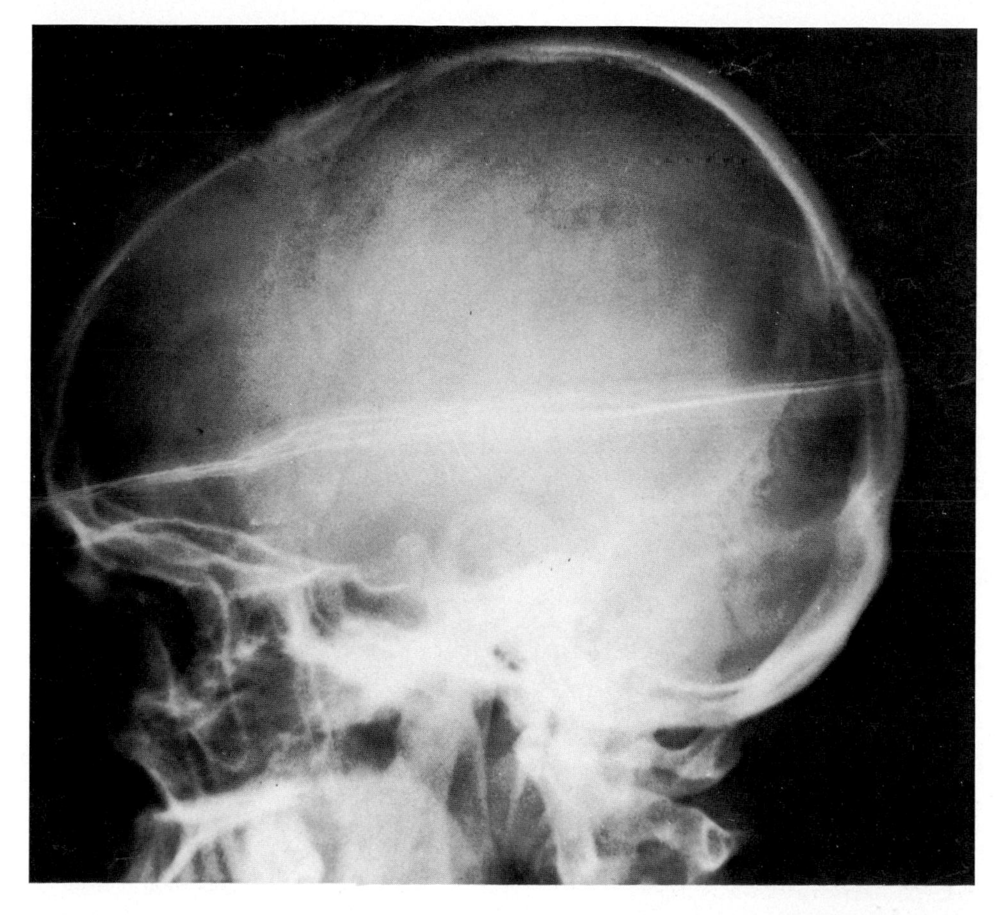

FIG. 7-35. *The artifact in this cranial radiograph is an image of an elastic hair band inadvertently left on the patient for a radiographic examination of the skull.*

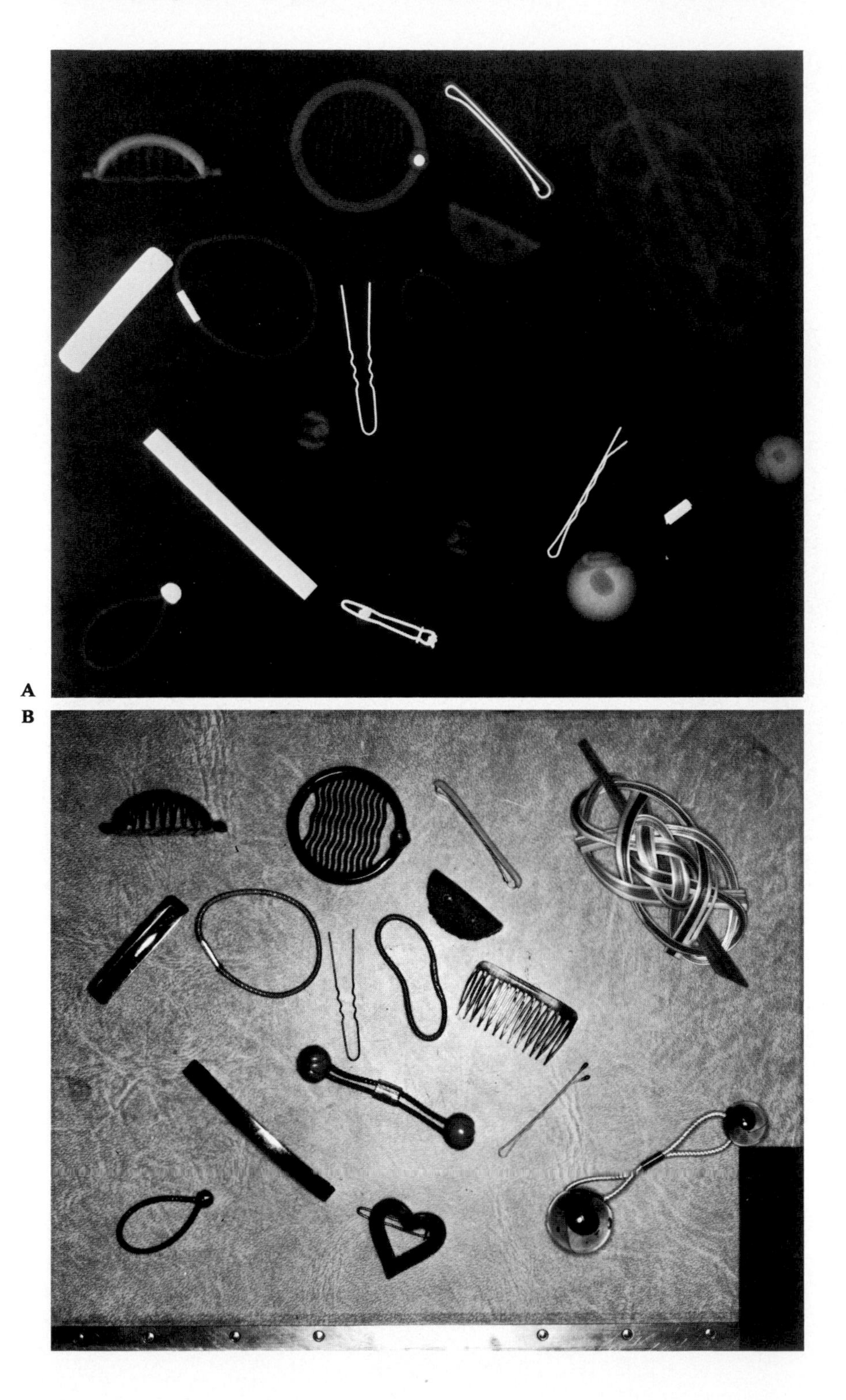
A
B

◁ **FIG. 7-36.** (A) *A radiograph and* (B) *a photograph of a variety of hair ties, barrettes, and the like, which may often appear in radiographs of the skull, cervical spine, and upper thorax. Although the lack of radiodensity and the position of such objects are the main reasons why they escape detection in the radiograph, notice that many contain metal components or other dense materials that should be easy to identify.*

FIG. 7-37. *The artifact in this radiograph of the skull is the image of an elastic tie used to hold a ponytail in place. Compare the appearance of the metal component within the elastic tie* (arrow) *with the images depicted in Figure 7-36.*

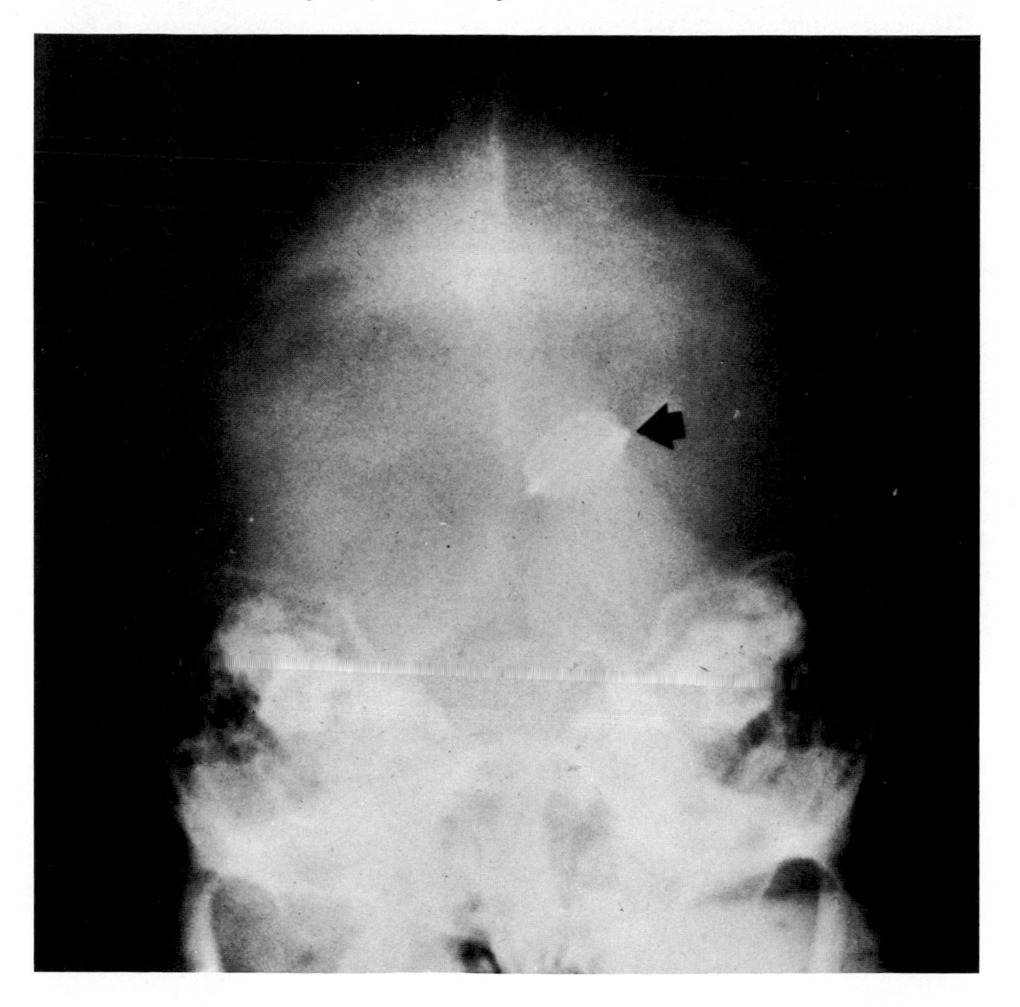

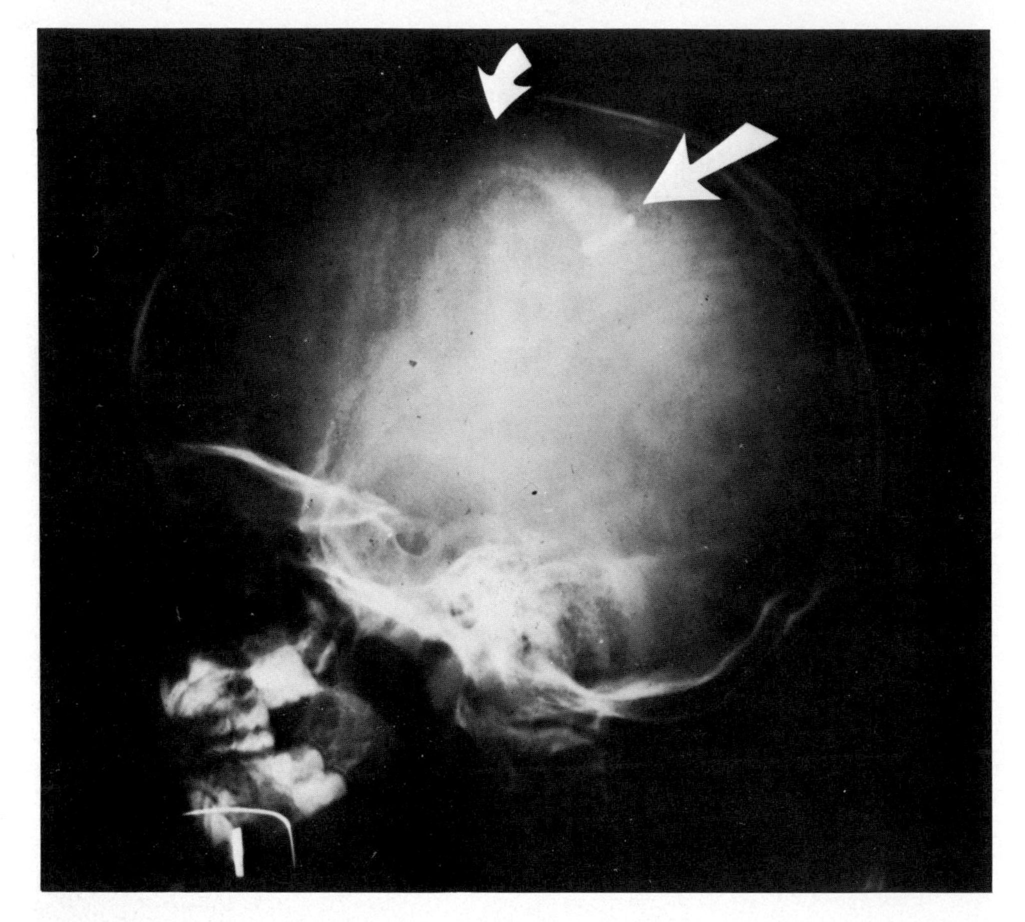

FIG. 7-38. *This lateral radiograph of the skull demonstrates a small tuft of hair tightly bound with a rubber band* (arrows).

FIG. 7-39. (A) *The radiopaque configurations imaged in the apices of the lungs are ringlets of hair* (B). *When you encounter a patient with a long braid of hair, or with hair worn in the fashion depicted, you should instruct the patient to pin up her hair in order to eliminate the problem. (Incidentally, the patient informed me that her hairstyle is artificial and that the hairpiece is simply pinned under her scarf or bandanna.) (From Sweeney RJ: On the Technical Side. Radiol Technol 50, No. 6:728–729, 1979)* ▷

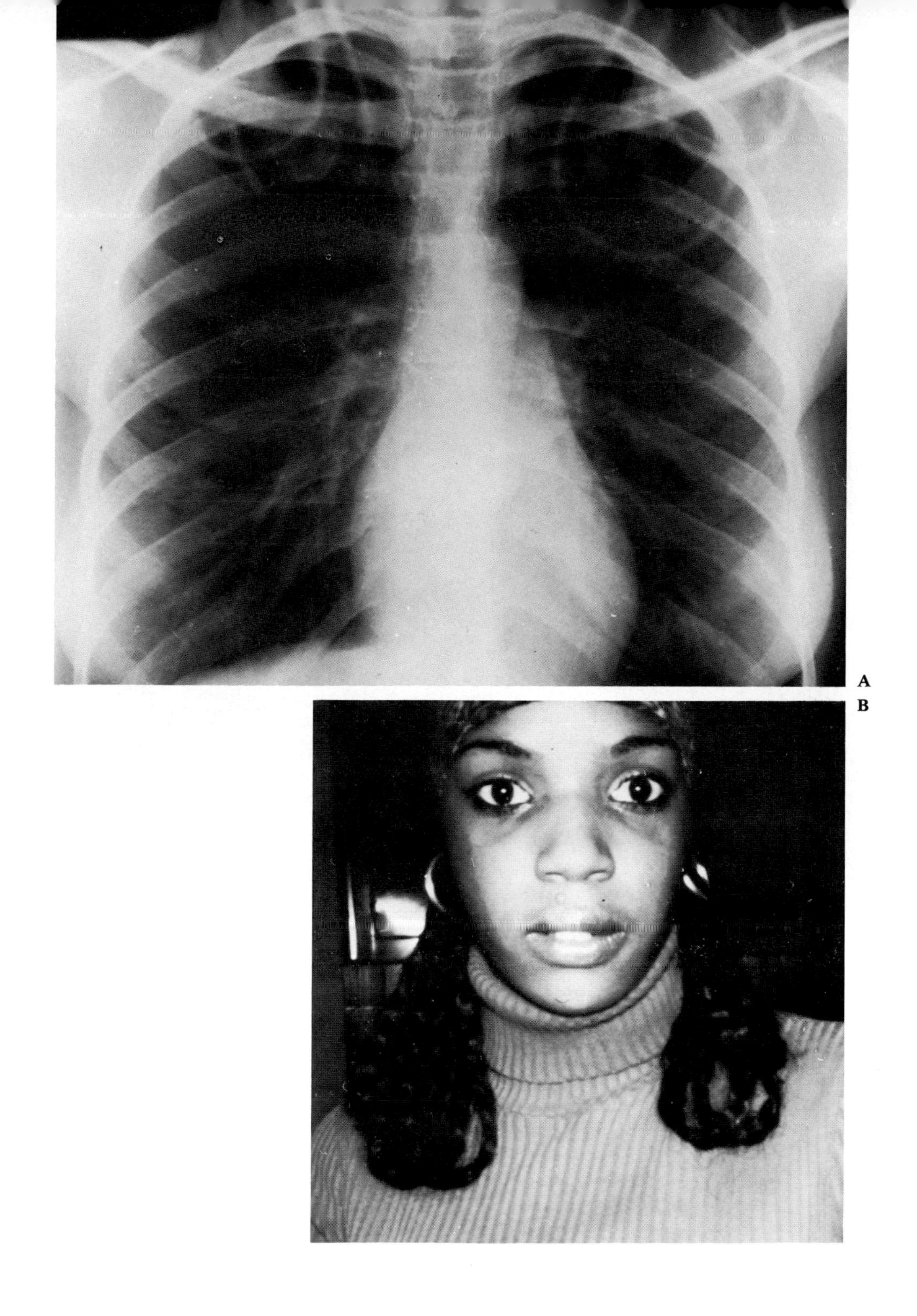
A
B

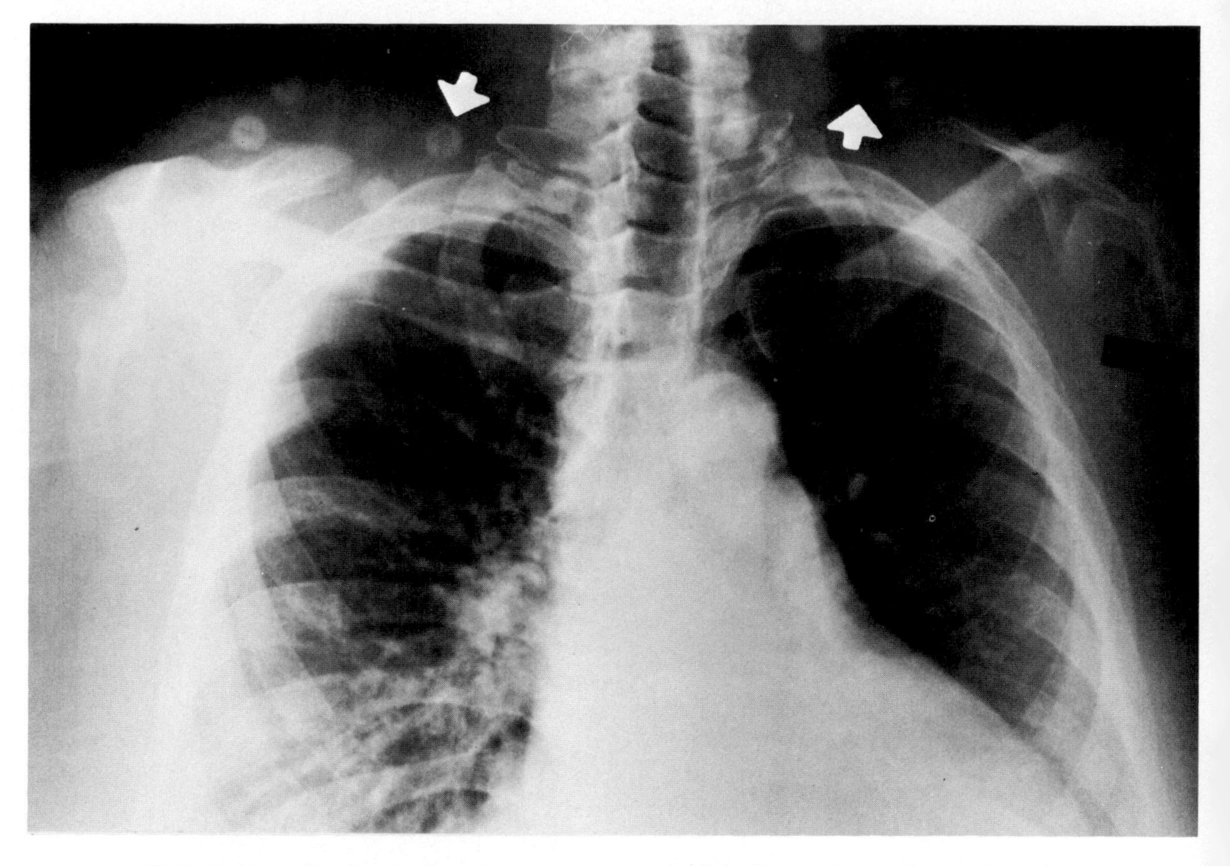

FIG. 7-40. *The circular images* (arrows) *seen in this chest radiograph are plastic beads that were attached to the patient's braided hair at the time of the examination. (Courtesy of Ralph Coates, R.T. [R])*

FIG. 7-41. (A) *The streaklike artifact of minus density* (arrow) *is a long braid of hair imaged over the region of the chest* (B). *In order to eliminate such problems, the technologist should instruct the patient before doing the examination to pin up the braided hair.* (B, *courtesy of Gunta Avens)* ▷

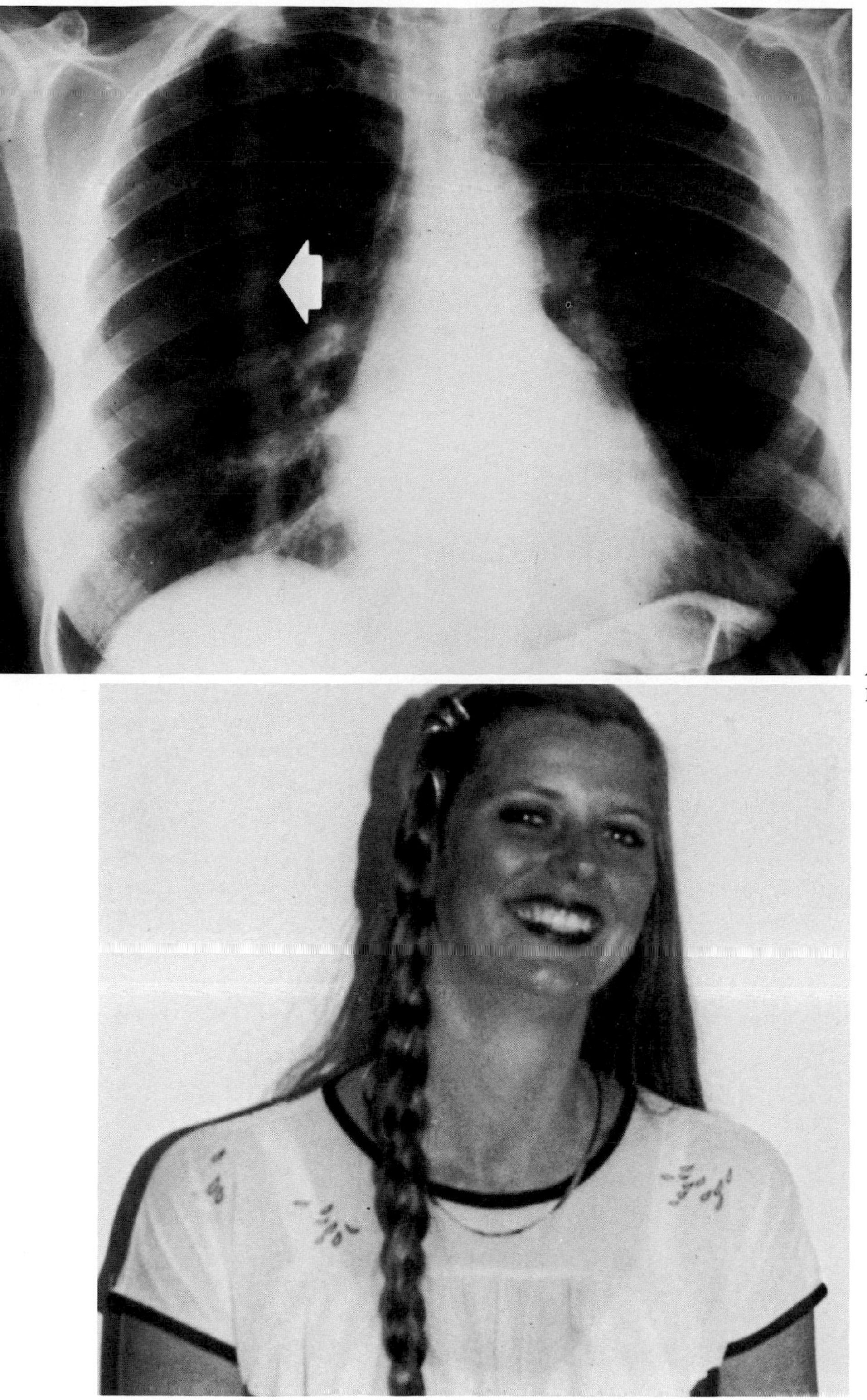
A
B

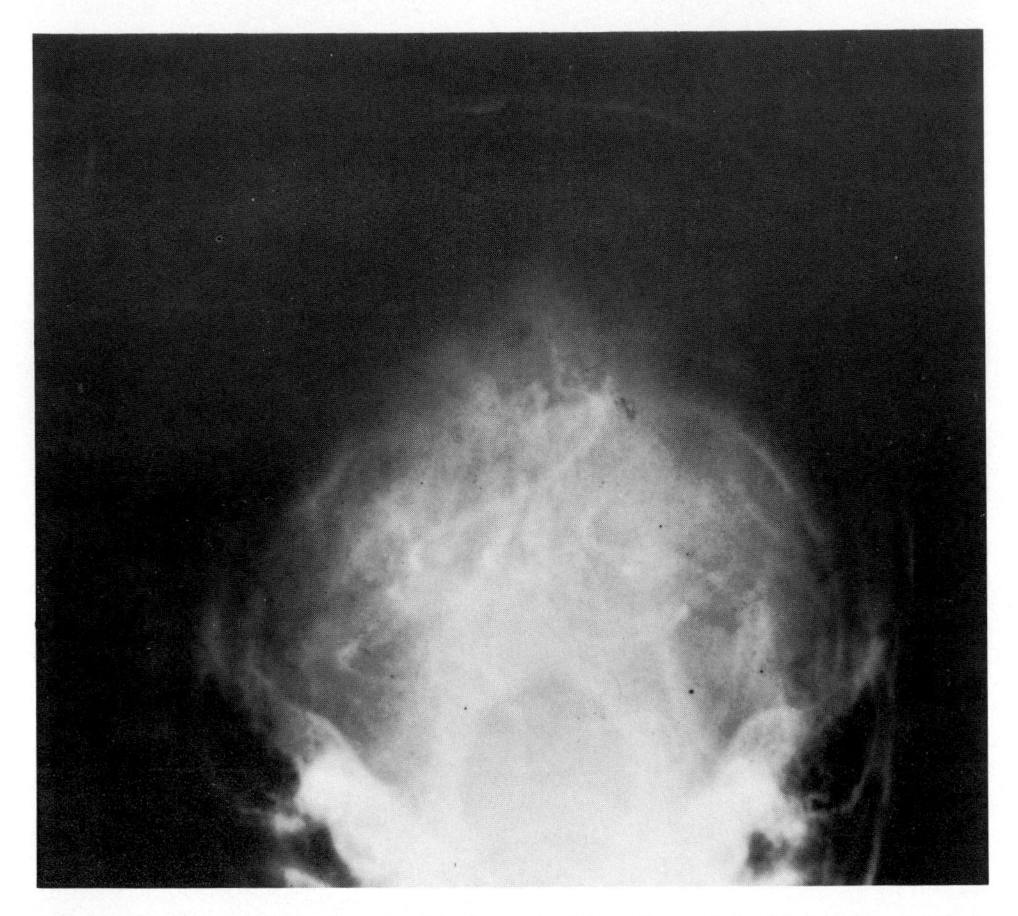

FIG. 7-42. *The patient whose skull is shown in this radiograph had fallen from a bicycle and sustained a laceration of her scalp. A skull series was requested prior to débridement and suturing of the wound. The striations in the radiograph resulted from the matting of the patient's hair with mud, which occurred when she fell onto the pavement.*

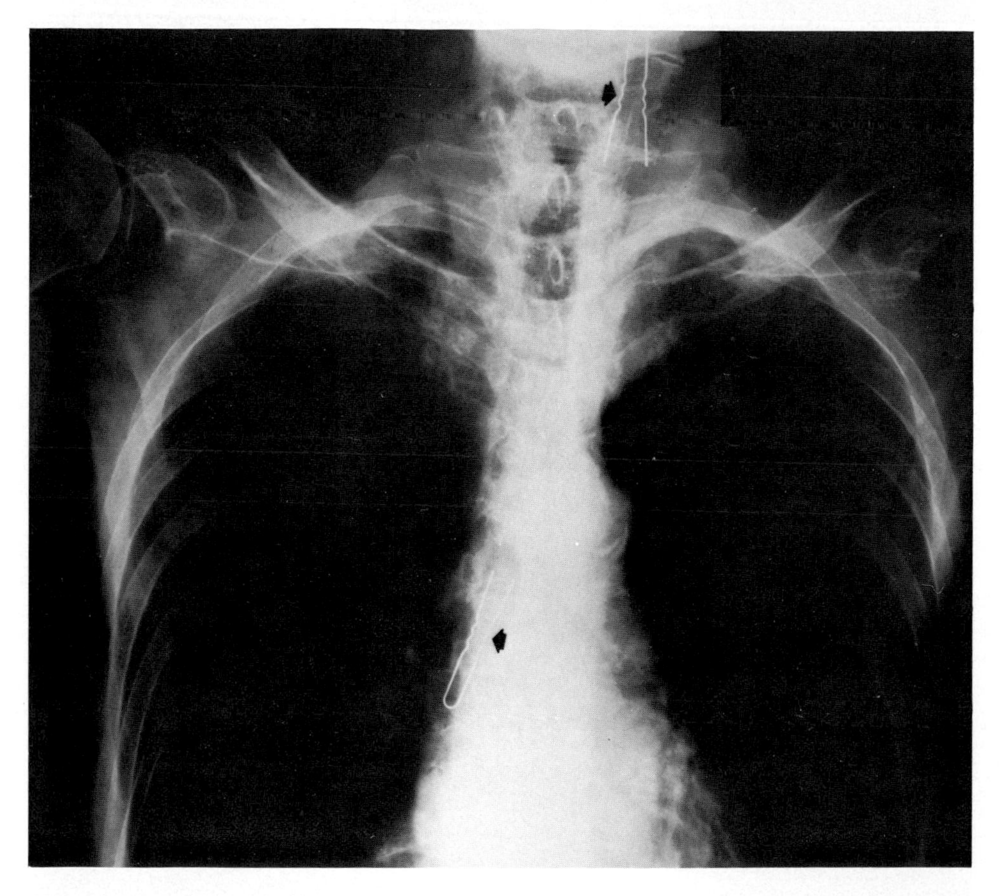

FIG. 7-43. *The two bobby pins* (arrows) *seen in this overexposed radiograph of the chest were used to secure cloth bows to a long braid of hair.*

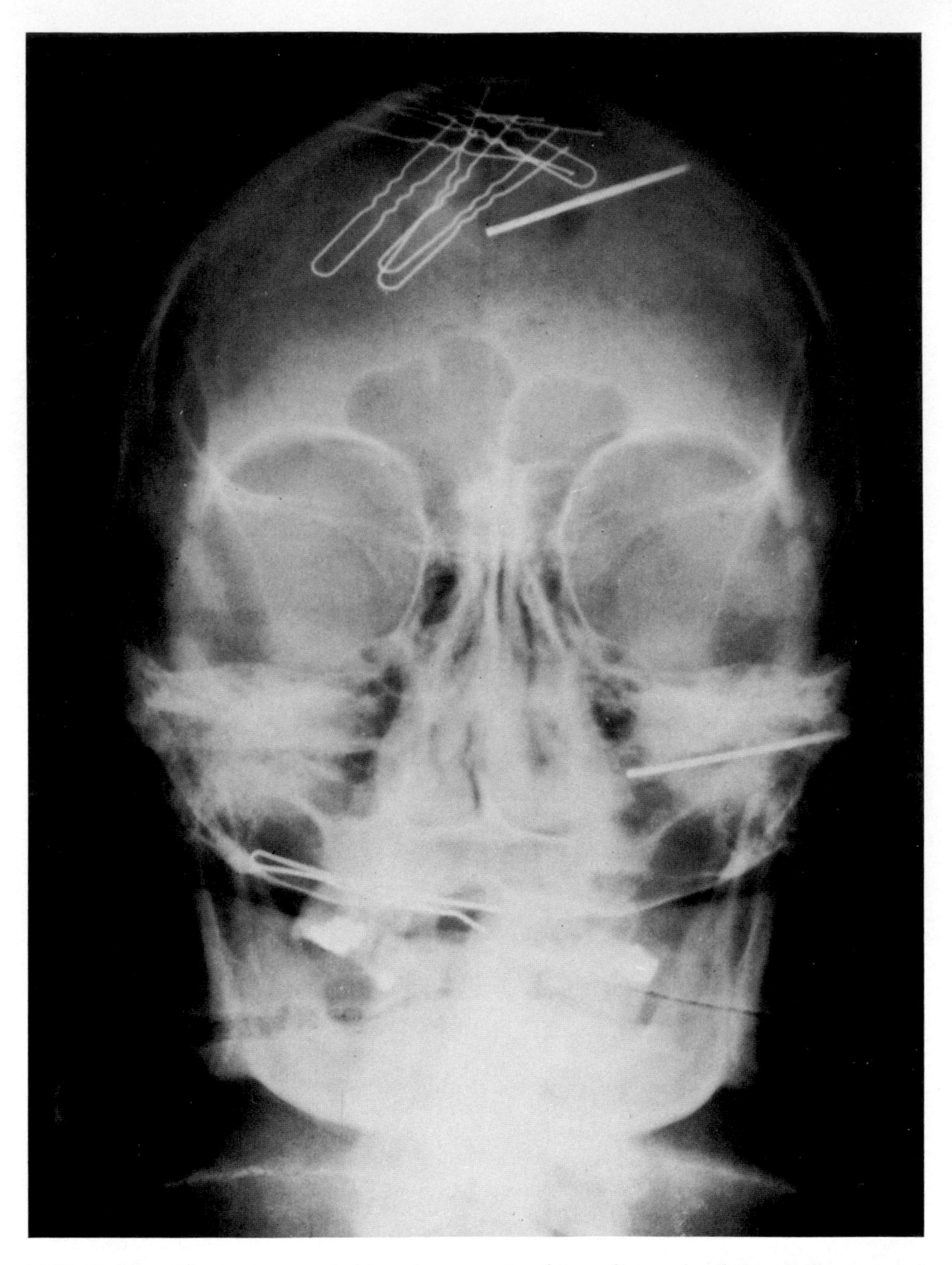

FIG. 7-44. *The numerous bobby pins seen in this radiograph of the skull were not removed from the hair because they were beneath a compression bandage applied by a physician in the emergency room. However, when permitted, you should always attempt to remove any objects located in the hair. A small magnet is a very useful device for locating metallic objects when the matted condition of the hair, as well as the varied location of the hairpins, presents a problem.*

References

1. American Standards Association Standard Z66.1: To Minimize Hazards to Children From Residual Surface Coating Materials, 1964
2. Andrews GC: Diseases of the Skin. Philadelphia, WB Saunders, 1939
3. Dewing SB: Modern Radiology in a Historical Perspective, p 29. Springfield, IL, Charles C Thomas, 1962
4. Eastman Kodak, Med Radiogr Photogr 46, No. 3, 1970
5. Greenbaum SS: Dermatology, Diagnosis and Treatment. Philadelphia, FA Davis, 1949
6. Klickstein HS: Wilhelm Conrad Röntgen on a New Kind of Rays: A Bibliographic Study, Vol 1, p 24. Mallinckrodt Classics of Radiology. St Louis, Mallinckrodt Chemical Works, 1966
7. Pendergrass RC: Tattos in radiographs. Med Radiogr Photogr 31, No. 1:41–43, 1955
8. Smith WL: Radiographic artifacts caused by the laundering of flame retardant materials. Radiology 123:625–626, 1979
9. U.S., Federal Register, Title 16, Part 1303: Ban of Lead Containing Paint and Certain Consumer Products Bearing Lead Containing Paint, 1977
10. Worsham J: "Her croup leads to a recall." *Boston Globe,* August 25, 1978

Suggested Reading

EFFECTIVE COMMUNICATION

Bell ME: Patient–technologist interpersonal relationship and how it can be improved. Radiol Technol 50:41–44, 1978

Brogdon BC: The eternal triangle. Radiol Technol 35:181–184, 1963

Calabrese R: Interpersonal communication skills for technologists. Radiol Technol 49:759–764, 1978

Carey B: Human relations for radiologic technologists. Radiol Technol 38:278–283, 1967

Carroll QB: Improving patient cooperation. Radiol Technol 51:68–71, 1979

Fischer HW: Radiology Departments: Planning, Operation and Management, pp 369–382. Ann Arbor, Edwards Brothers, 1982

Foster FM: Effective personnel communications in radiology department management. Radiol Technol 43:209–213, 1972

Gurley LT, Mays PS: Patient education in diagnostic radiology. Appl Radiol 5, No. 3:66–67, 1976

Hunter WK: Preparation is more than a laxative. Radiol Technol 52:513–515, 1981

Ireland SJ, Hansen EU: Brief encounter: Origin of patient communication. Radiol Technol 50:33–36, 1978

Kahn PA: Radiologic technologists can communicate clearly. Radiol Technol 49:637–641, 1978

Laws PW: How patients view the efficient use of diagnostic radiation. Radiol Technol 47:245–249, 1976

Neuhaus B: Our professional image: As others see us. Radiol Technol 49:485–489, 1978

Petrello J: Your patients hear you, but do they understand? RN 39, No. 2:37–39 1976

Sweeney RJ: A system designed to improve communication between the patient and technologist. Radiol Technol 47:295–297, 1976

Wedel CS: Patient communication: The final step towards professionalism. Radiol Technol 50:27–31, 1978

ARTIFACTS CAUSED BY MATTED FABRICS

Caffey J: Pediatric X-ray Diagnosis, Vol 2, pp 1573–1590. Chicago, Year Book Medical Publishers, 1973

Keats TE: An Atlas of Normal Roentgen Variants That May Simulate Disease. Chicago, Year Book Medical Publishers, 1975

ARTIFACTS CAUSED BY BRAIDED HAIR

Keats TE: An Atlas of Normal Roentgen Variants That May Simulate Disease, p 5. Chicago, Year Book Medical Publishers, 1975

8 ARTIFACTS RELATED TO FOREIGN BODIES

Basic Techniques for the Investigation of Foreign Bodies

When you suspect the presence of a swallowed foreign body, preliminary radiographs should include right-angle projections of the entire alimentary tract (Fig. 8-1). Cullinan suggests that when radiographs are requested of infants and young children suspected of swallowing foreign objects, you should employ a moderate-kilovoltage, non-Bucky technique for the radiographic studies.[1] He recommends that a preliminary radiograph be made of the entire alimentary tract by positioning the child diagonally on a 14″- × -17″ cassette. The child's head should be turned to one side to demonstrate the nasopharynx and the soft-tissue structures of the neck. A technique employing high milliamperage and short exposure time is recommended in order to avoid blurring of linear-positioned metallic objects.[1] For larger children you may need to use multiple films in order to demonstrate the neck, chest, and abdominopelvic regions.

Whenever there is a possibility of there being a foreign object in the nasopharynx, a lateral view of the neck to include the nasopharynx is mandatory.[3] The importance of such a procedure is demonstrated in Figure 8-2.

*Detecting Foreign Bodies in the Extremities and Other Soft-Tissue Regions: An Alternate Approach**

Foreign objects can enter the body under many different circumstances and can be of a variety of types. They may often present certain imaging problems to the technologist. A few of the reasons for these problems are as follows:

The lack of radiodensity of certain foreign substances

The low contrast inherent in some of the imaging methods used for recording the extremities (*e.g.*, direct exposure without intensifying screens)

Loss of contrast due to the problems associated with radiation scatter in certain anatomic regions

Obviously, restricting the x-ray beam and employing a grid will control the last of these factors, but in the case of a grid exposure, an intensifying screen must be employed. Otherwise the exposure factors have to be adjusted beyond an acceptable range to compensate for the lack of sensitivity of the screenless exposure holder and the radiation absorption of the grid.

Although many departments have substituted tabletop exposures with high-definition screens for traditional screenless exposures, there is still a great deal of interest in finding better ways to enhance the quality of the image. An alternative method consists in employing a vacuum-type holder with a single-emulsion film and an intensifying screen. Although designed principally for mammography, this imaging system is superior for the following reasons:

The inherent high contrast of the single film and intensifying screen enhances the visualization of structures.

The vacuum feature improves screen–film contact, thereby enhancing visualization of structural detail.

The single intensifying screen–film combination eliminates the problem of light crossover (parallax).

I have used this technique successfully to demonstrate foreign objects in the soft-tissue areas of the extremities, as well as in the ocular region (see Fig. 8-10). Regardless of where you use it, however, be sure to follow these guidelines:

---

*The material in this section appeared in a slightly different form in Sweeney RJ: Some technical considerations for demonstrating foreign bodies in the eye. On the Technical Side. Radiol Technol 50, No. 3, 1979.

Because a single intensifying screen is employed, it should be cleaned prior to use. This, of course, eliminates the problems associated with intensifying-screen artifacts.

To limit the amount of radiation scatter, employ acute beam restriction. Because the exposures are conducted on the tabletop, place lead masking material over the portion of the exposure holder outside the field of exposure. This will absorb a considerable degree of radiation scatter, thereby enhancing the contrast of the image.

Expose a phantom in the various radiographic positions commonly employed in the department in order to develop a technique chart for the examination of an adult.

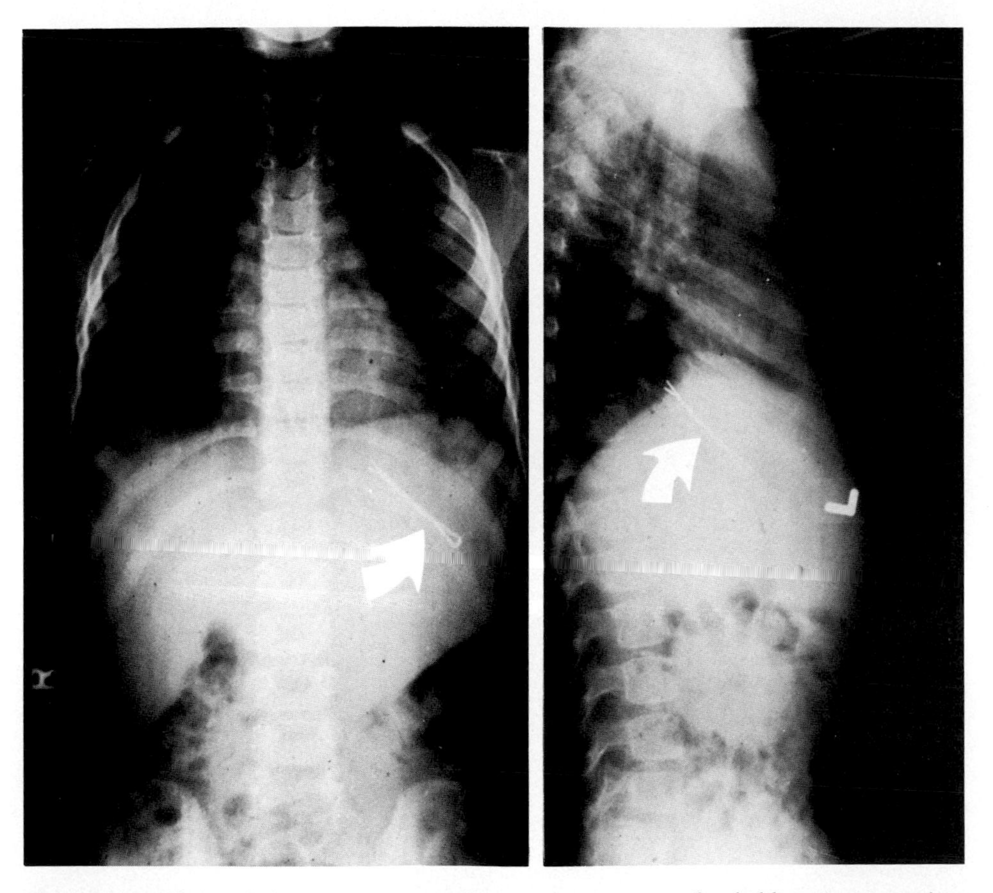

A B

FIG. 8-1. *The radiologist's report confirmed the presence of a bobby pin* (A *and* B, arrows) *in this patient's stomach. Although the identification of such a radiopaque object does not usually present a problem, it is very important that right-angle projections be obtained in order to determine the location of the foreign object.* (A) *Anteroposterior projection.* (B) *Lateral projection.*

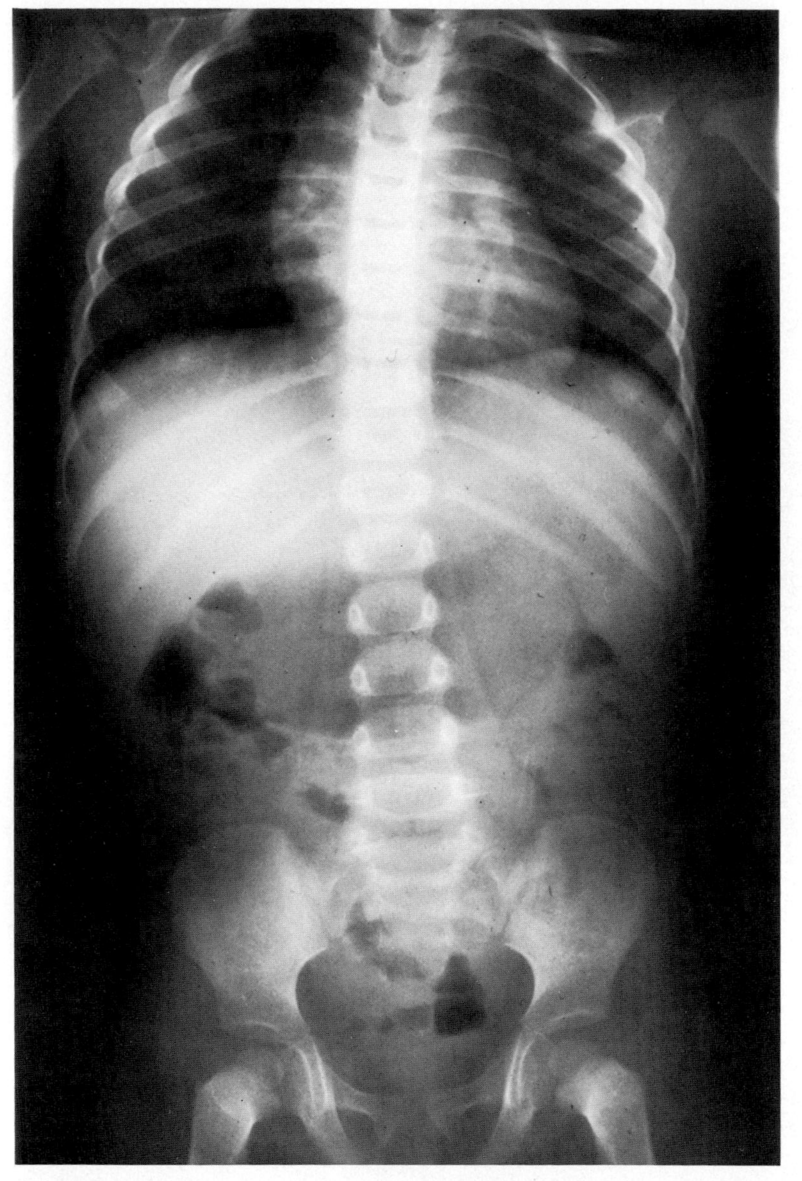

FIG. 8-2 *The babysitter who accompanied this child to the emergency room told the doctor that the child had swallowed a coin. The doctor requested that radiographs be obtained of the child's chest and abdomen* (A) *in order to rule out the presence of a coin there. While in the process of obtaining the preliminary radiograph, the technologist was informed by the babysitter that she did not tell the doctor in the emergency room that the child had been bleeding from the nose after swallowing the coin. After the technologist notified the doctor of this, radiographs of the neck and nasopharynx were requested, and these projections demonstrated that the coin was lodged in the child's nasopharynx.* (B).

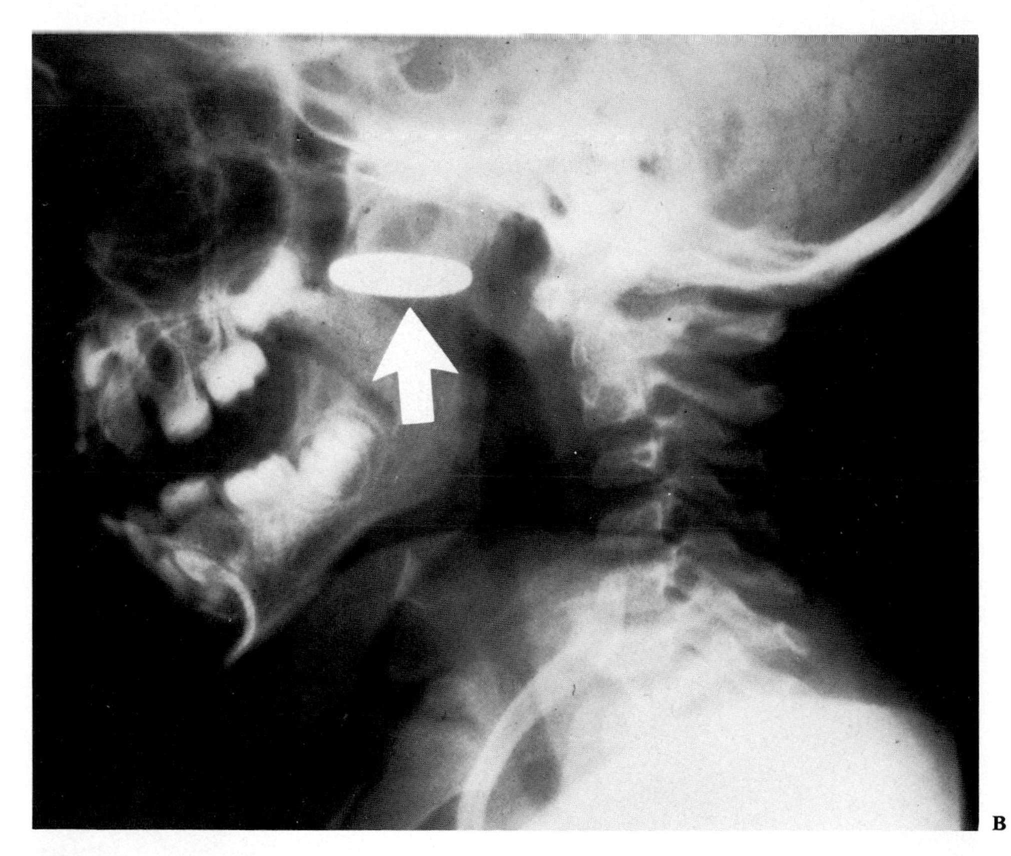

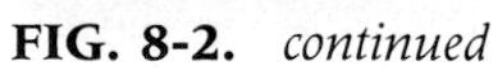
FIG. 8-2. *continued*

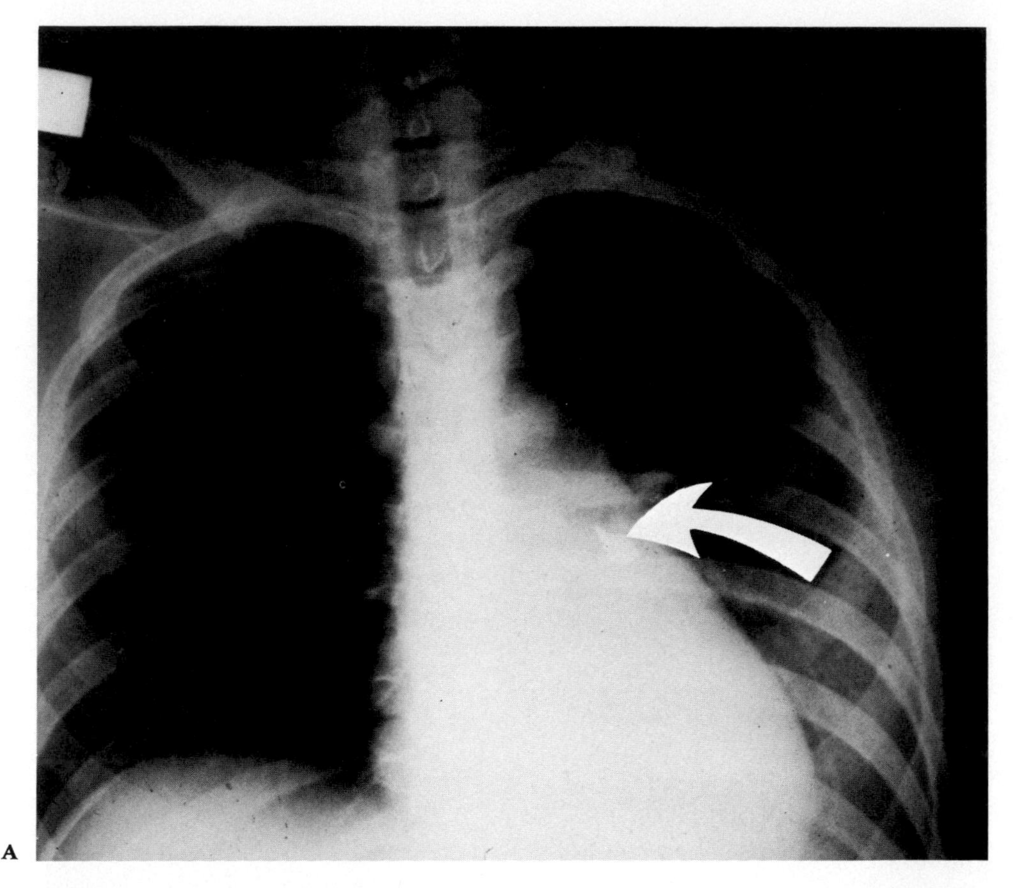

A

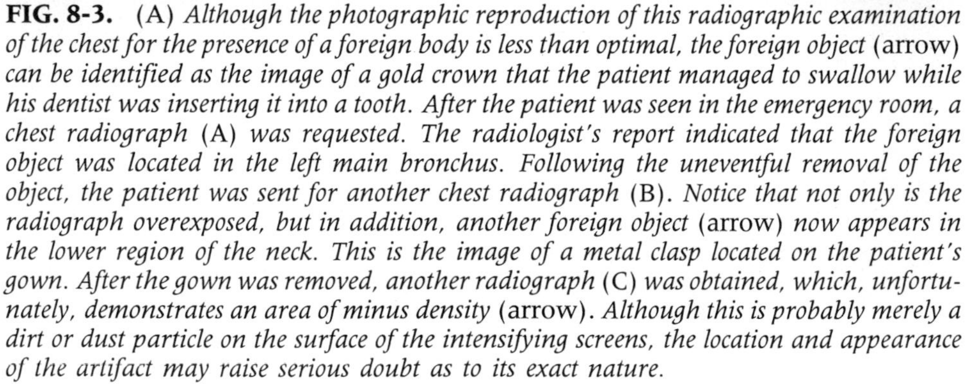

FIG. 8-3. (A) *Although the photographic reproduction of this radiographic examination of the chest for the presence of a foreign body is less than optimal, the foreign object* (arrow) *can be identified as the image of a gold crown that the patient managed to swallow while his dentist was inserting it into a tooth. After the patient was seen in the emergency room, a chest radiograph* (A) *was requested. The radiologist's report indicated that the foreign object was located in the left main bronchus. Following the uneventful removal of the object, the patient was sent for another chest radiograph* (B). *Notice that not only is the radiograph overexposed, but in addition, another foreign object* (arrow) *now appears in the lower region of the neck. This is the image of a metal clasp located on the patient's gown. After the gown was removed, another radiograph* (C) *was obtained, which, unfortunately, demonstrates an area of minus density* (arrow). *Although this is probably merely a dirt or dust particle on the surface of the intensifying screens, the location and appearance of the artifact may raise serious doubt as to its exact nature.*

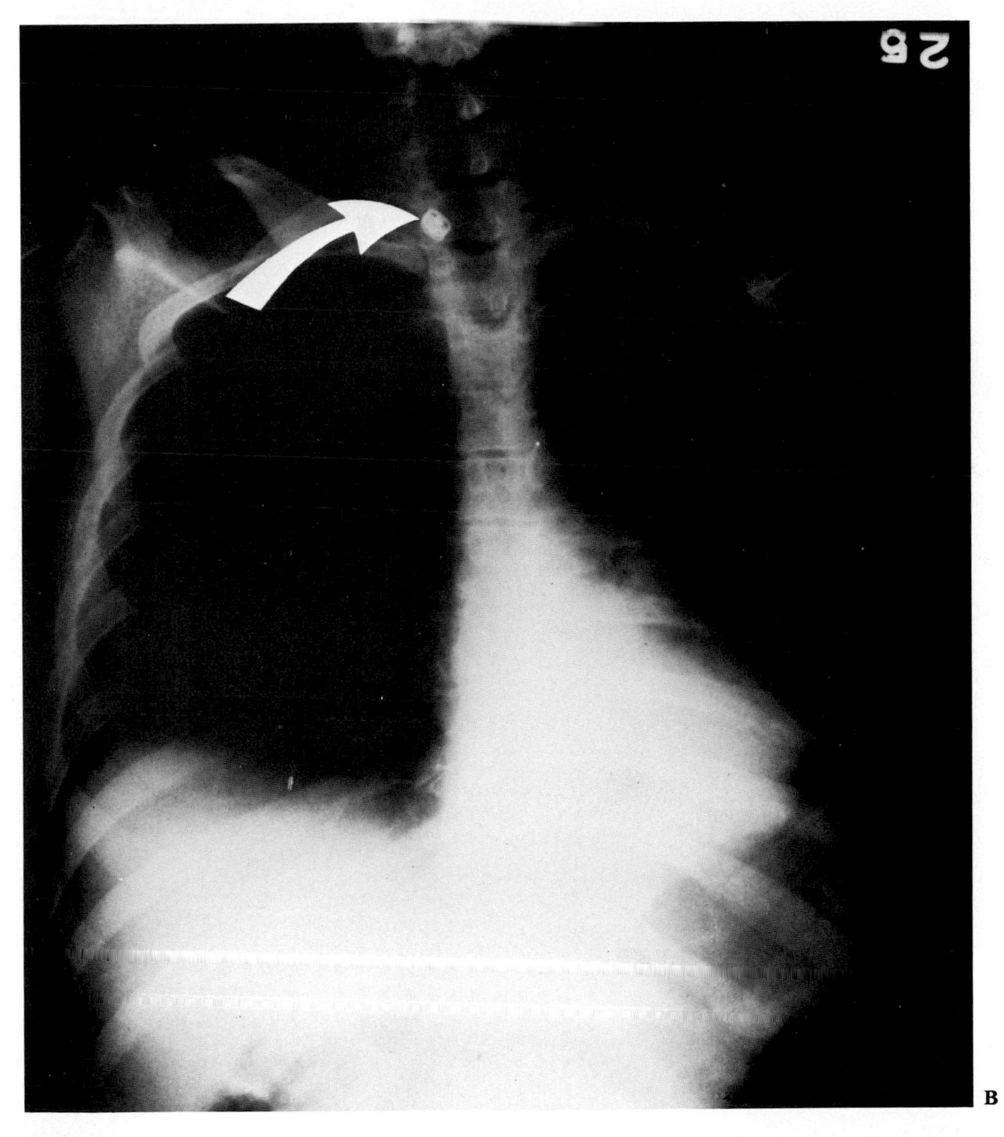

FIG. 8-3. *continued*

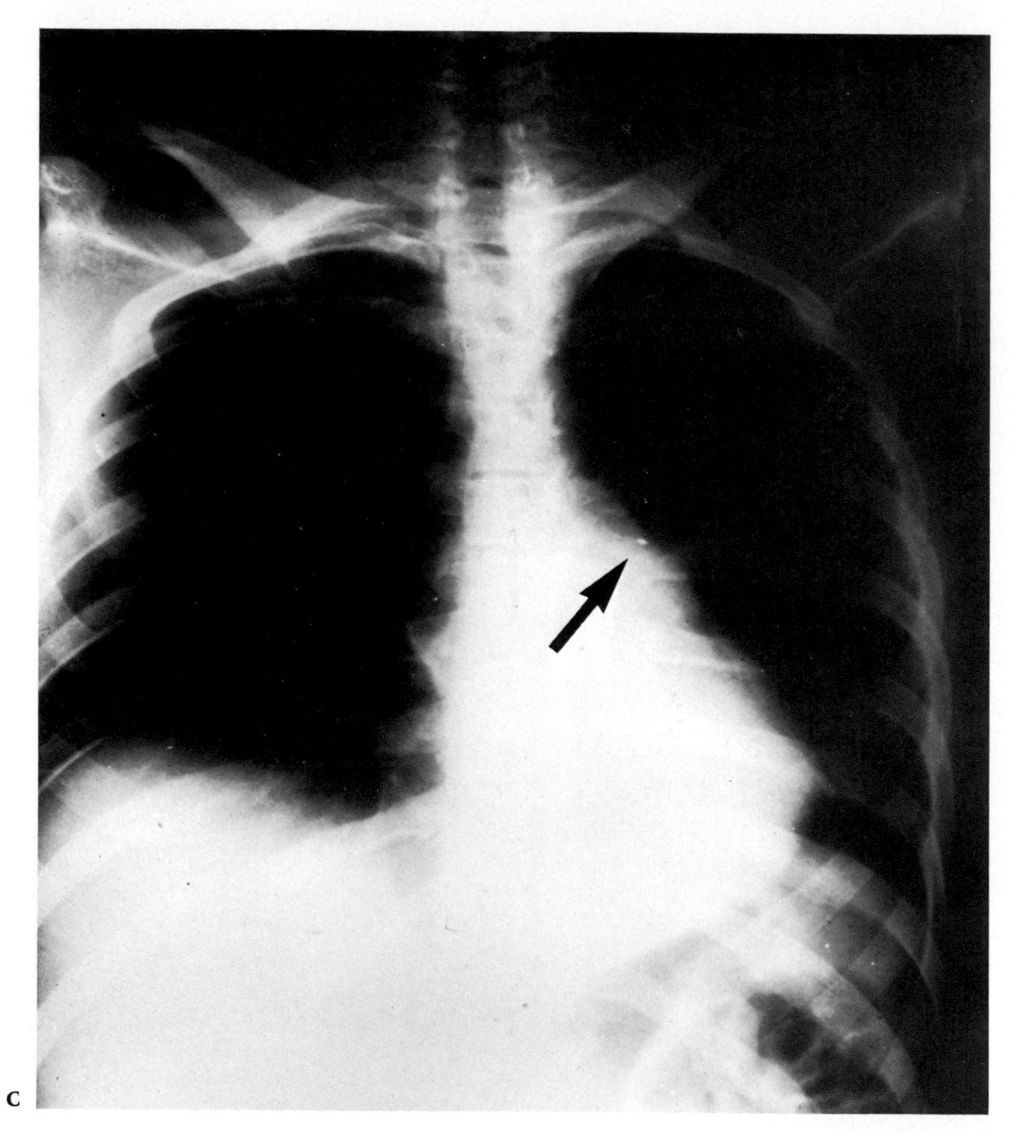

C

FIG. 8-3. *continued*

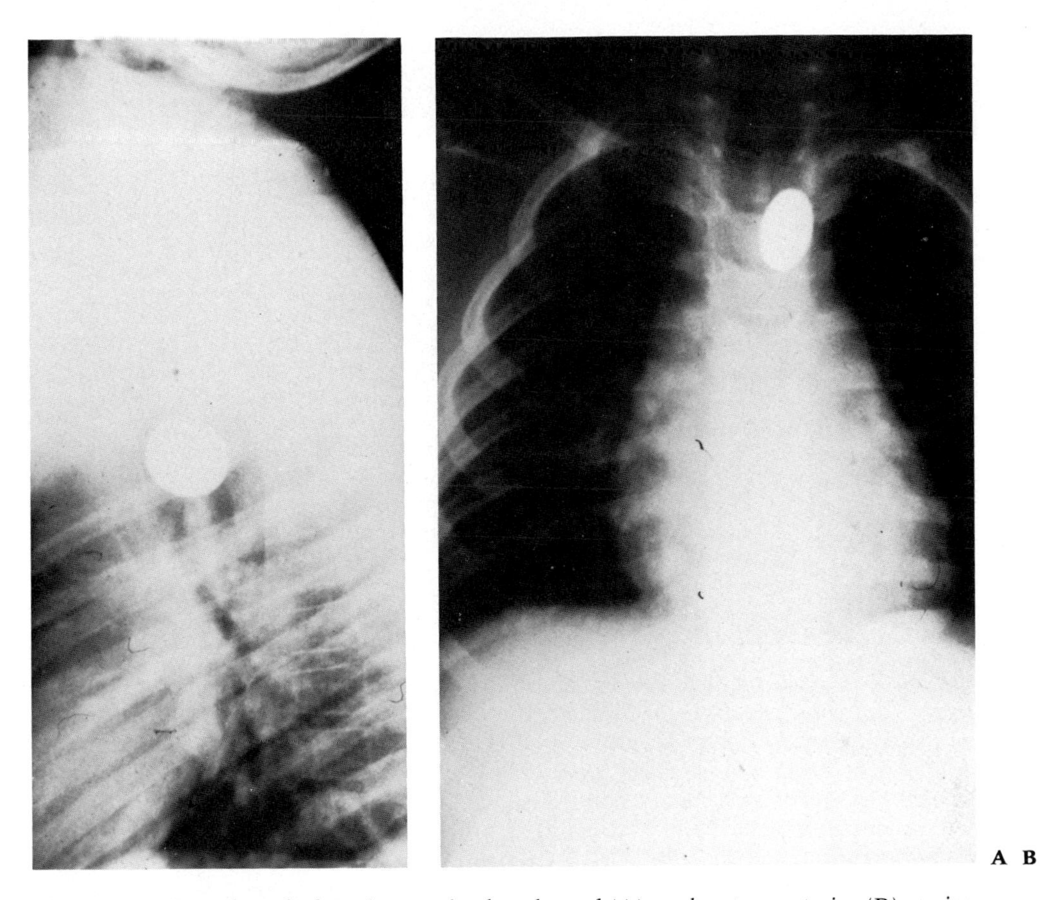

FIG. 8-4. *Even though the coin seen in these lateral* (A) *and anteroposterior* (B) *projections of the thorax was minted in the United States, it is still considered a foreign body. Given the opportunity, children will swallow a variety of objects. Whenever there is a reason to suspect the presence of such objects in the airway or intestinal tract, two views at right angles should be obtained to include these regions.*

A

FIG. 8-5. *During the early morning hours, a middle-aged woman came to the emergency room complaining of pain in the upper region of her chest and lower neck. The patient gave no history of injury, and following a brief examination, the physician requested radiographs of her shoulder and chest. After the physician studied the radiographs and confronted the patient with his findings, she admitted that weeks before, she had inserted several sewing needles* (A *and* B, arrows) *into her shoulder after reading about the benefits of acupuncture in the relief of pain.* (B) *This portion of one of the chest radiographs demonstrates the presence of sewing needles in the soft-tissue region of the shoulder.*

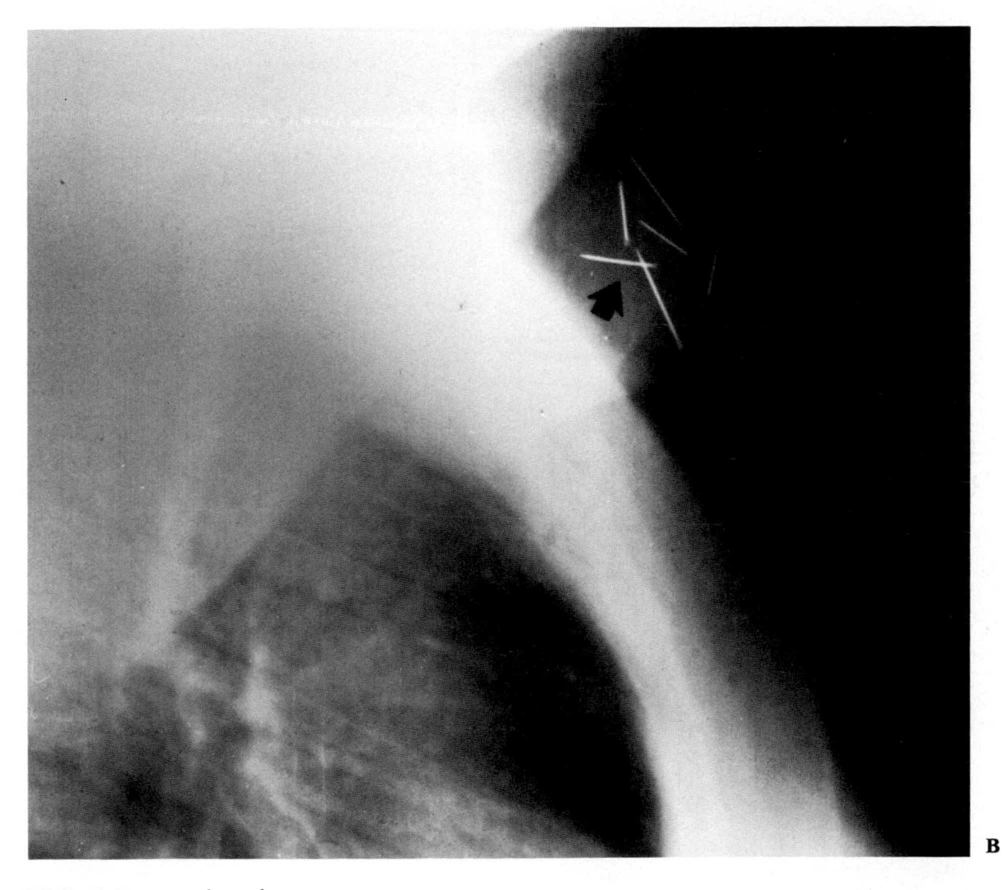

FIG. 8-5. *continued*

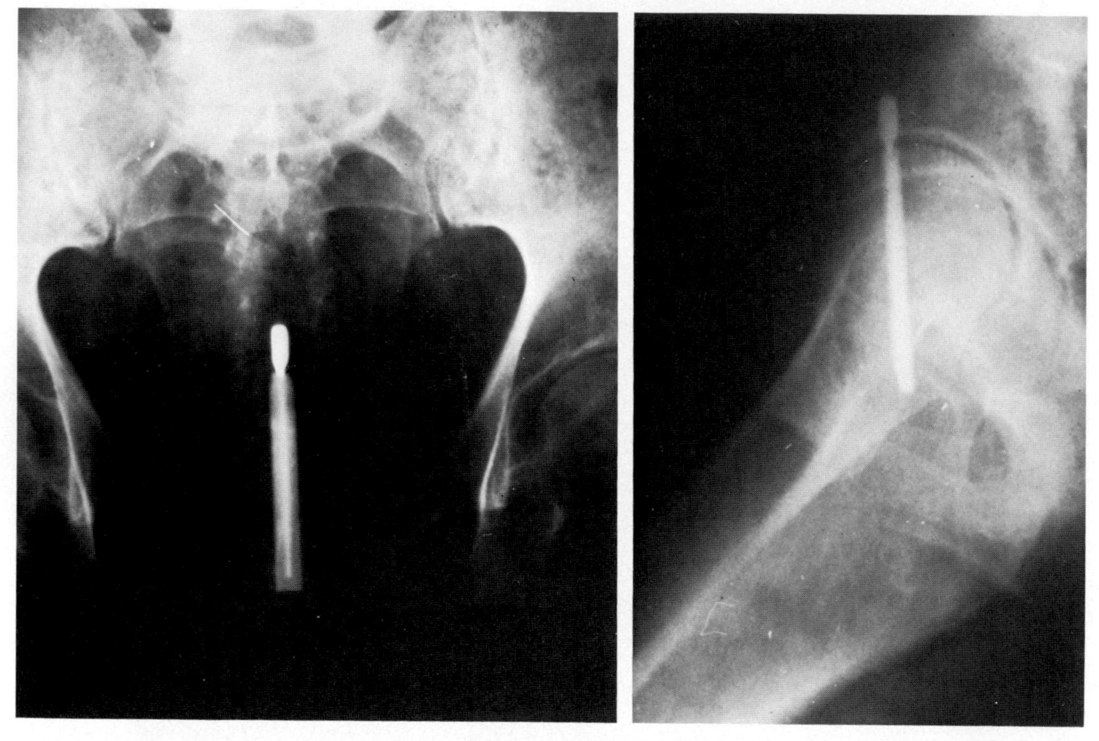

A B

FIG. 8-6. *It is not often that the radiologist has the opportunity of not only making a diagnosis, but also determining the patient's temperature. The radiologist's report confirmed what was suspected: the patient had inserted a thermometer into his bladder.* (A) *Anteroposterior projection.* (B) *Lateral projection.*

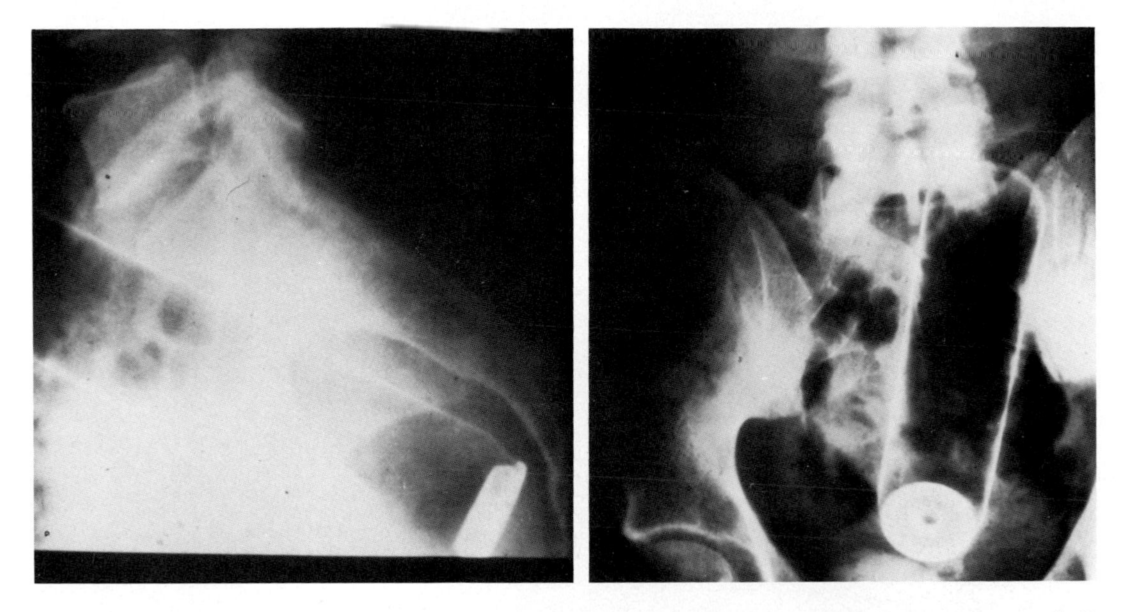

A B

FIG. 8-7. *The cylindrical object in these abdominal radiographs is the image of a can of butane fluid used to refill cigarette lighters. Notice that the lateral projection* (A) *confirms that the object is located within the rectosigmoid region of the large intestine.* (B) *Anteroposterior projection.*

Detecting Glass Artifacts

Glass is radiopaque and can be demonstrated radiographically if the projection is free of overlying dense structures (see Fig. 8-8), a finding that was proven in a study conducted by Felman and Fisher in which 50 pieces of glass found at random in the street were radiographed through 4 cm of water.[2] Figure 8-9 demonstrates the radiodensity of four common particles of glass.

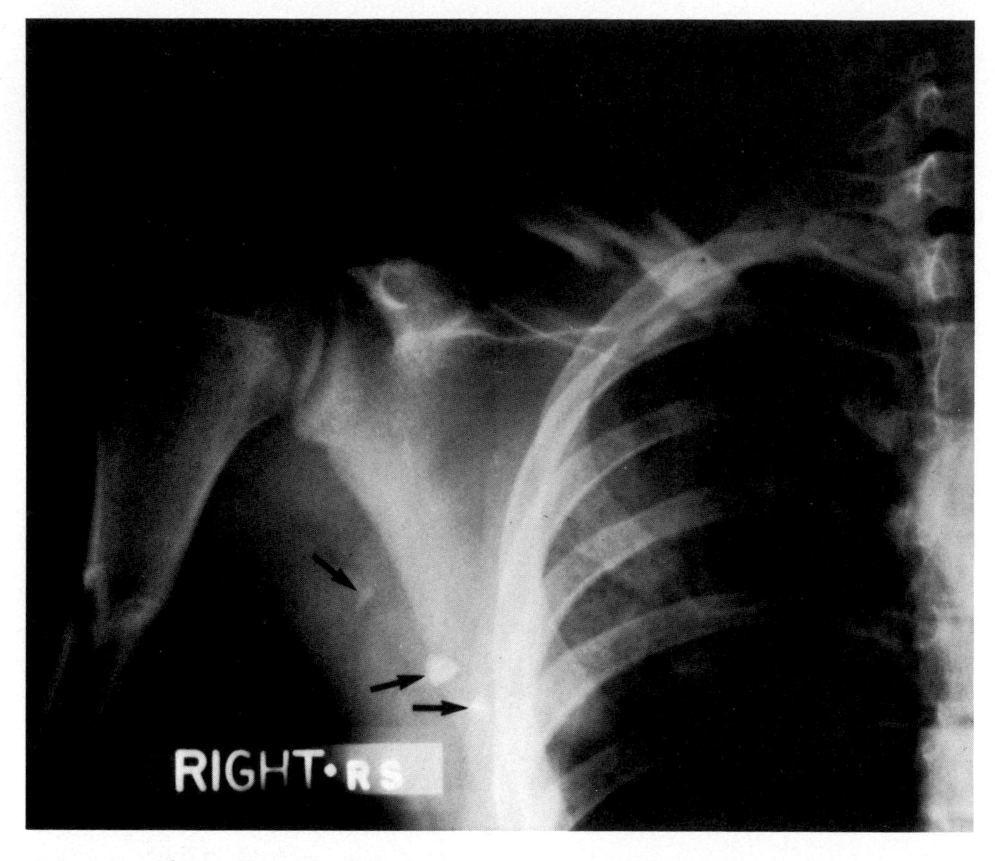

FIG. 8-8. *This radiograph of the shoulder was made of a patient involved in an automobile accident. The artifactual images* (arrows) *are of particles of glass located on a blanket that had been placed under the patient at the scene of the accident.*

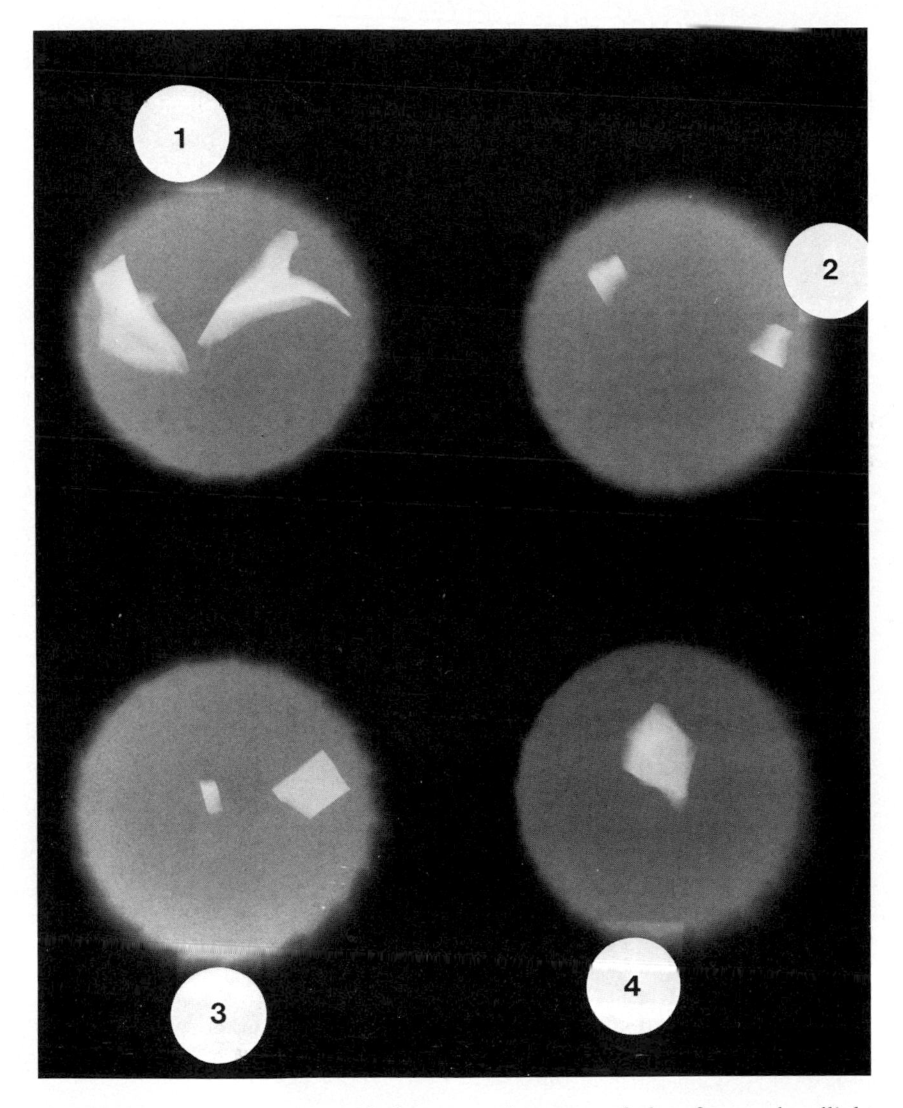

FIG. 8-9. *Radiodensity of four types of glass.* 1, *a fragment of glass from a headlight assembly;* 2, *a piece of windshield;* 3, *a piece of windowpane;* 4, *a piece of wine bottle.*

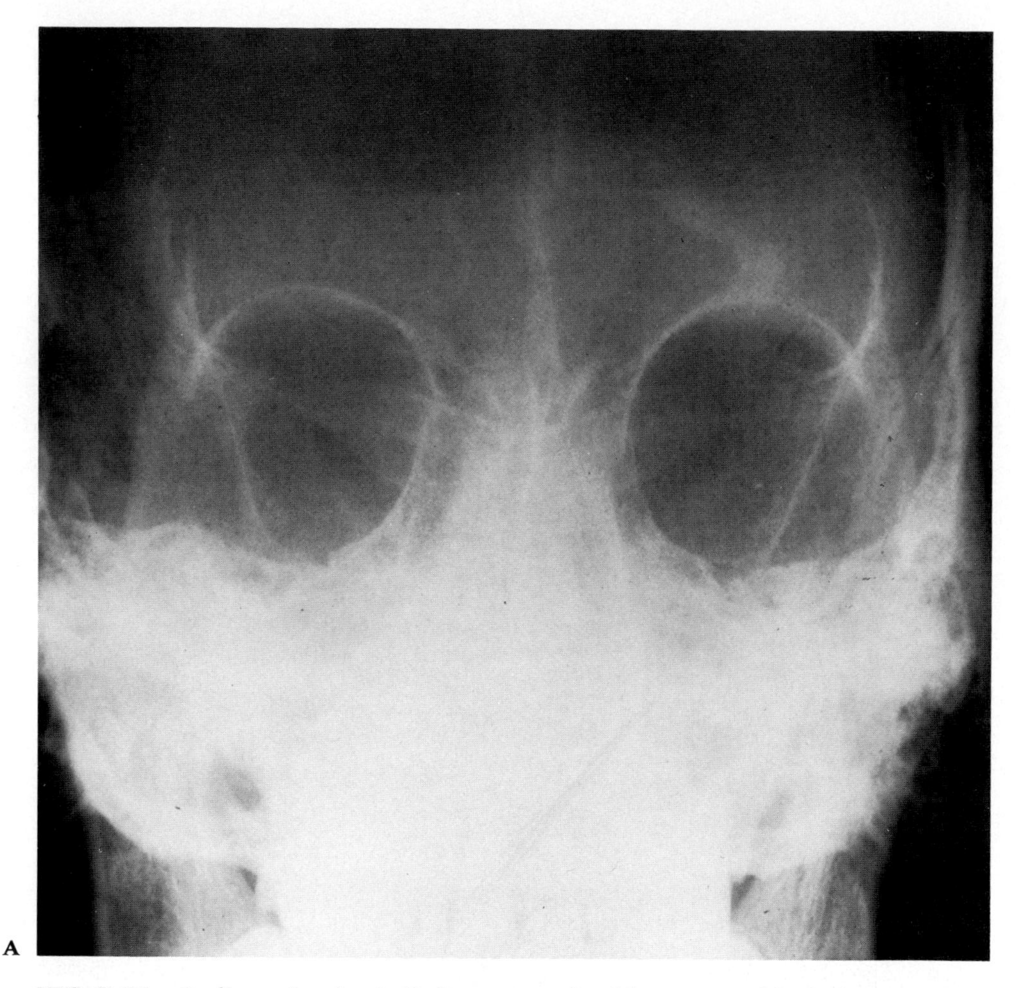

FIG. 8-10. *Radiographs of a skull phantom made with a vacuum film holder containing a single emulsion film and intensifying screen can be employed in order to establish techniques for the radiographic investigation of foreign objects in the body.* (A) *Anteroposterior projection.* (B) *Lateral projection.*

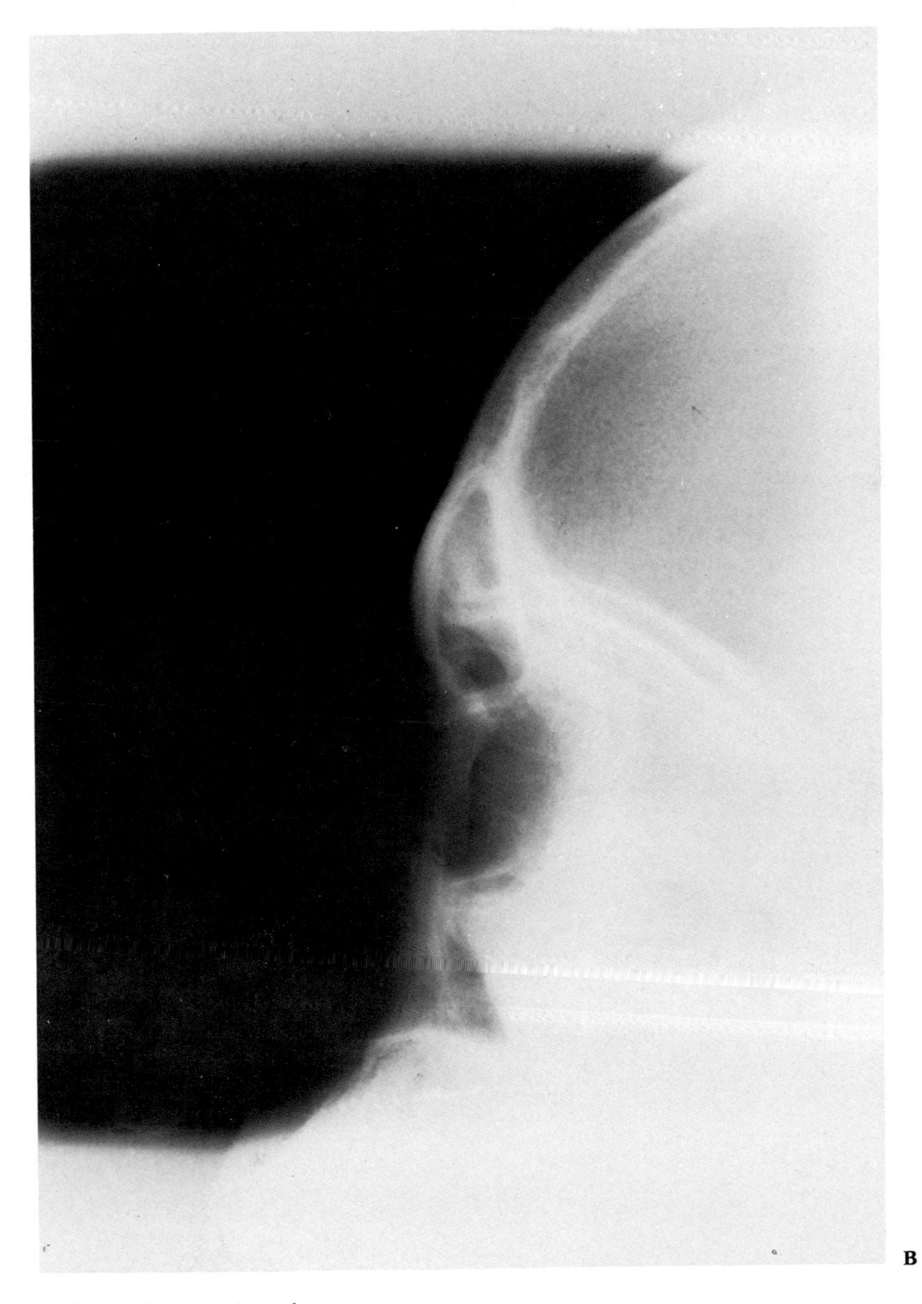

FIG. 8-10. *continued*

References

1. Cullinan JE: Illustrated Guide to X-ray Technics, pp 123–124. Philadelphia, JB Lippincott, 1972
2. Felman AH, Fisher MS: The Radiographic Detection of Glass in Soft Tissue. Radiology 92:1529–1531, 1969
3. Merrill V: Atlas of Roentgenographic Positions and Standard Radiologic Procedures, 4th ed, Vol 1, pp 316–331. St Louis, CV Mosby, 1975

Suggested Reading

TECHNICAL CONSIDERATIONS

Crabb G, Macht SH: Chicken-bone bezoar in rectum. Med Radiogr Photogr 27, No. 3:88–89, 1951

Emerson EB: Bizarre pulmonary foreign body. Med Radiogr Photogr 33, No. 4:112, 1957

Funke T, Gianturco C: Radiographic demonstration of aluminum play coin in nasal cavity. Med Radiogr Photogr 33, No. 4:110, 1957

Griffin FR: Buckshot in the appendix. Med Radiogr Photogr 32, No. 2:62, 1956

Macht SH: Foreign body (bottle) in rectum. Radiology 42:500–501, 1944

McGillivary JE: Foreign bodies in the stomach. Med Radiogr Photogr 30, No. 1:27, 1954

Salb RL: Metallic bezoar. Med Radiogr Photogr 32, No. 1:32, 1956

9 ARTIFACTS CAUSED BY IMPROPER USE OF EQUIPMENT AND ACCESSORIES

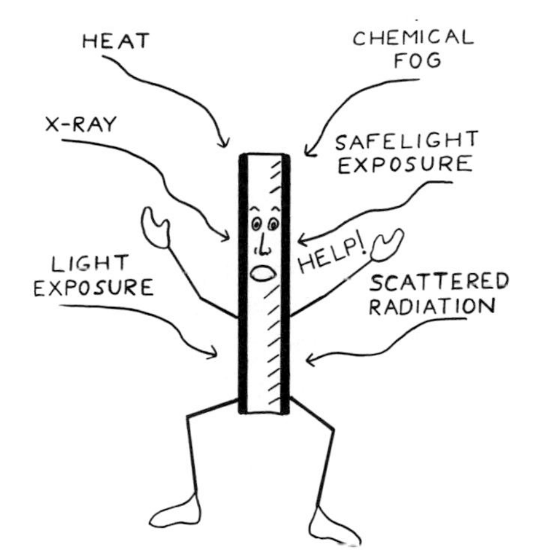

The Influence of Radiant Energy

The most common conditions responsible for the appearance of bizarre images in radiographs result from presensitization or postsensitization of the radiographic film to various forms of radiant energy. All of the artifacts depicted in this chapter could have been avoided if steps had been taken to protect the film during the various stages of handling, exposure, and processing. Many of these artifacts were caused by the direct action of x-ray or by radiation scatter.

The influence of radiation scatter is more difficult to isolate than it at first appears. "We get a more or less fogged film" is by no means a clear definition of the problem, much less a solution. A sufficient solution to the problem of scatter can be attained only when both the essential character and the causes and control of this "fogging effect" have been looked into.

To aid in our understanding of radiation scatter, it will be necessary to define two terms, *fog* and *secondary density effect.*

Fog is undesired density that is independent of the x-ray exposure. It can be caused by the quality of the film, preliminary exposure, after-exposure development, storage, and so forth.

A secondary density effect is an undesired density caused by stray radiation that does not coincide with the primary radiation. This undesired stray radiation causes a decrease in contrast and, by its projective action, a decrease in sharpness.

What, you might ask, is the difference between a decrease in contrast and a decrease in sharpness? These terms are highly controversial and, needless to say, often confusing.

If you were to examine a radiograph of the abdomen made without a grid (it is understood, of course, that radiography of the abdomen requires a grid), you would immediately recognize the problem involved in trying to decide how much of the poor quality is due to fogging with loss of contrast and how much is due to loss of definition (Fig. 9-1). Although in either case no one would deny the "poor visibility of detail," this phrase, though occasionally useful, is very nebulous and imprecise. It implies only that a sharp image may be present but poorly seen, or even that the image is completely invisible. One can only guess how this could be proven experimentally. On the other hand, the effect of scattered radiation on contrast and image sharpness is subject to radiographic proof.

In order to demonstrate the effect of scattered radiation, a cylindrical column of water approximately 6 inches (15 cm) in diameter was irradiated by a narrow, almost cylindrical primary beam (Fig. 9-2, *A*). A lead diaphragm was placed opposite the tube port to control off-focus radiation. A brass cylinder cone was attached to the tube head to restrict the primary field to approximately 2½ inches (6.35 cm) in diameter. The primary rays could not, therefore, escape through the wall of the water vessel. They could reach the bottom only in the form of a small cylindrical field.

In Figure 9-2, *B*, we see the density produced on a film placed under the water column. The black circle in the center marks the primary irradiated field. The blackening outside the circle is caused exclusively by stray radiation generated in the water.

To demonstrate the effect of this scatter radiation on image sharpness, the experiment was carried one step further (Fig. 9-3, *A*). Lead numbers were placed on top of the film holder and arranged clockwise around the base of the water-filled container. Once again the water phantom was irradiated, and the resulting radiograph is seen in Figure 9-3, *B*. The lead figures do not stand out very sharply because the projected stray radiation does not originate from a single point source but from the entire volume of water.

In actual practice, as the human body is traversed by radiation it responds in much the same manner (Fig. 9-4). Unlike the primary rays, the scattered ray produces a shadow that is not truly image forming. Hence, scattered rays that happen to pass the structure tangentially produce a blurred image superimposed on the primary image in the same manner in which a penumbra is formed by geometric factors. How could such a phenomenon be interpreted as other than a loss of definition?

The question now arises as to what percentage of stray radiation is permissible in a given exposure. At present, unfortunately, constants for permissible stray radiation with given exposure factors cannot be formulated because of contradicting values between the minimum quantity of scattered radiation and the minimum dose for the patient.

Many radiologists who, in the course of time, have learned to accept for diagnosis radiographs with a density representing at times more than 50% of scatter radiation will ask themselves whether it is really true that a brilliant exposure with only very slight scatter density reveals more detail. This question often goes unanswered because the radiographic end product is always a compromise arrived at by consideration of the source of radiation, the recording medium, and the subject. Most radiographs can be improved upon to the satisfaction of the observer, provided that the original values and related techniques are known and manipulated accordingly. However, if the original radiograph is acceptable, you should avoid doing repeat exposures in consideration for protecting the patient from excess radiation.[4]

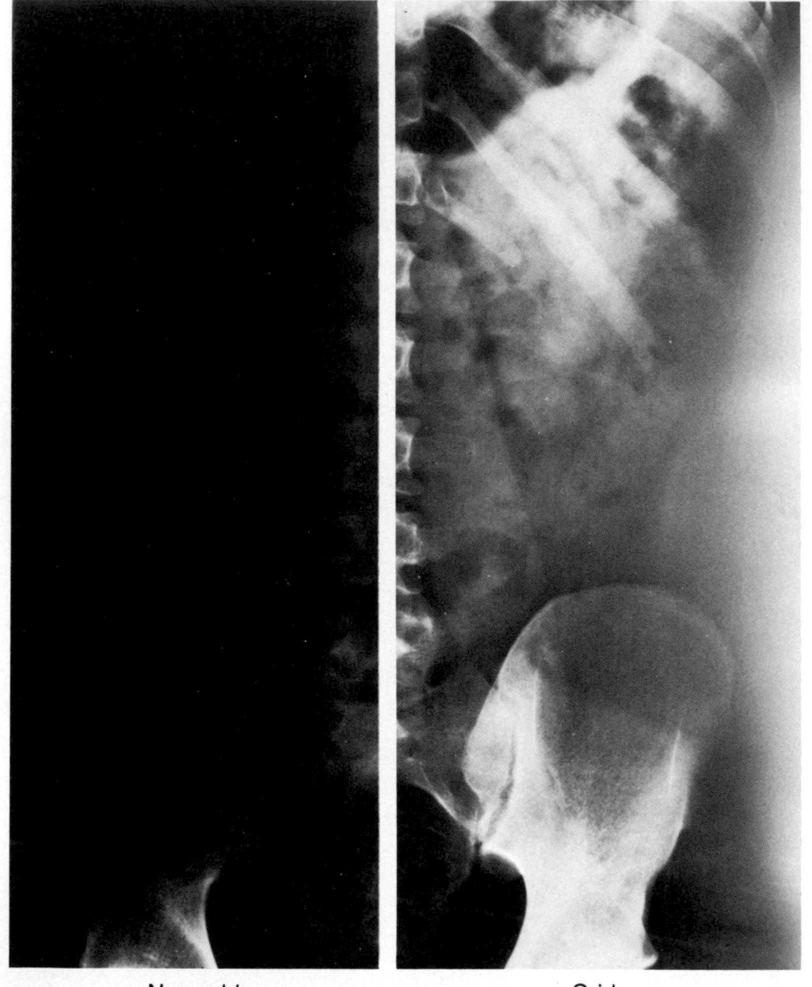

FIG. 9-1. *The difference in image quality between a radiograph made with a grid and one made without is obvious. But how much of the difference is a result of changes in definition, and how much is a result of changes in contrast?*

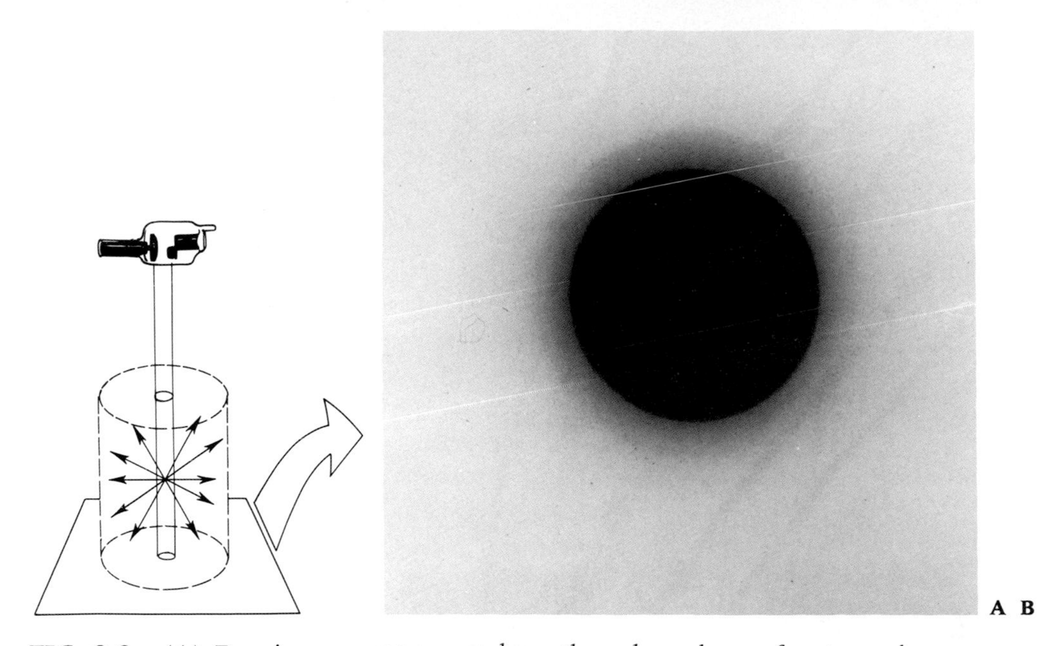

FIG. 9-2. (A) *Focusing a narrow x-ray beam through a column of water under controlled conditions demonstrates the effect of radiation scatter.* (B) *The resulting radiograph shows a dark core of primary radiation exposure surrounded by a gray field of stray radiation exposure. (From Sweeney RJ: Some factors affecting image clarity and detail perception in the radiograph. Radiol Technol 46, No. 6, 1975)*

FIG. 9-3. (A) *If we place lead numbers outside the water column, their images are poorly defined in the resulting radiograph* (B). *(From Sweeney RJ: Some factors affecting image clarity and detail perception in the radiograph. Radiol Technol 46, No. 6, 1975)*

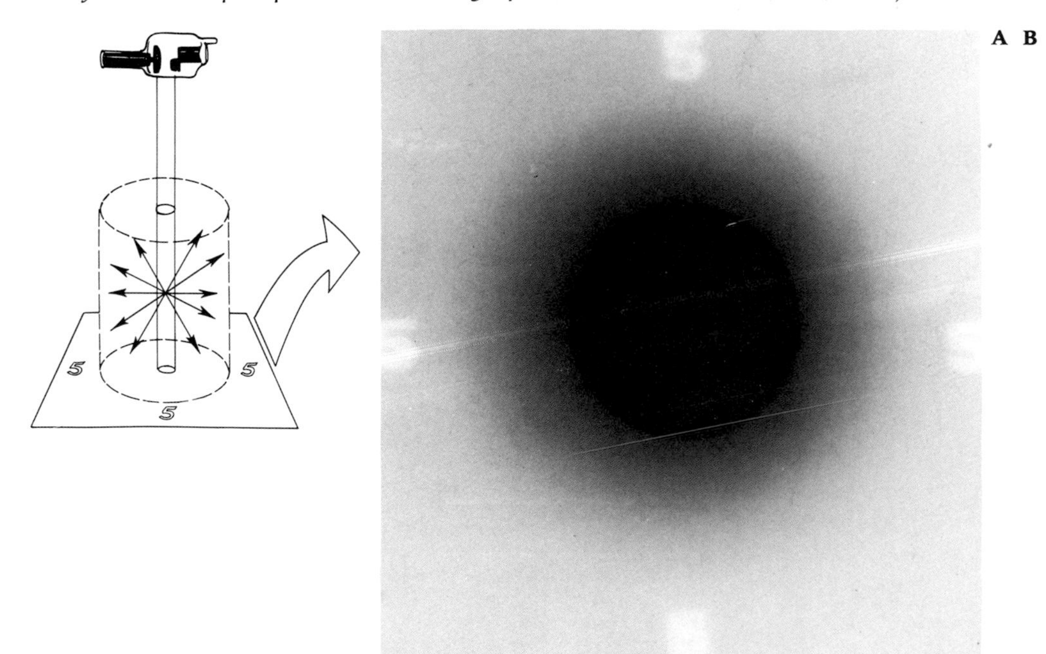

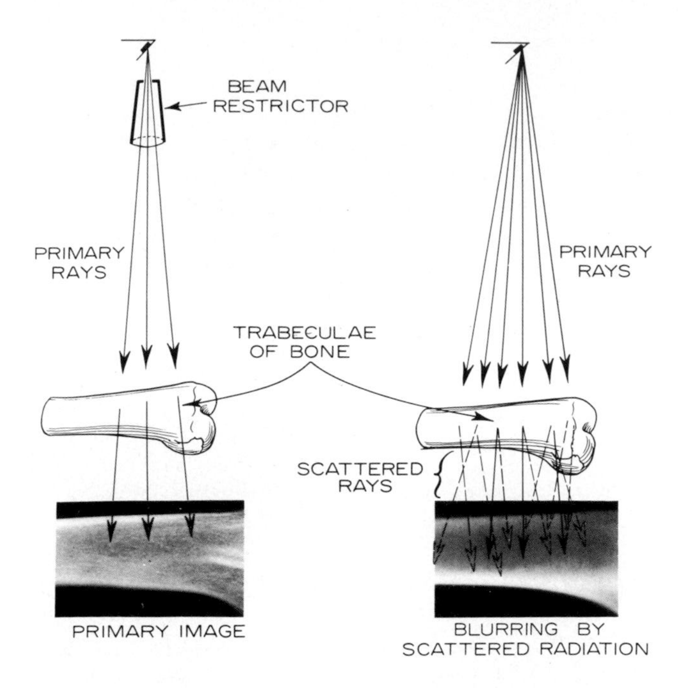

FIG. 9-4. *The influence of scattered radiation on radiographic contrast and image sharpness is analogous to the formation of a penumbra by geometric factors.*

The Importance of X-ray Beam Limitation

One need only compare a radiographic study made with and without x-ray beam limitation to notice that when the primary beam is not restricted, image degradation occurs (Fig. 9-5). Limitation of the primary beam is accomplished by manual or automatic adjustment of a collimating device attached to the x-ray tube housing. However, if the collimator—by design or makeshift installation—cannot be positioned close to the x-ray tube port, off-focus radiation will be responsible for producing an image outside the margin of the collimated field (Fig. 9-6). Notice that the image of the lungs and ribs in Figure 9-6 appears out of focus. This is because the radiation responsible for the appearance of these images did not originate from a point source within the x-ray tube. This condition can be controlled by installing a collimator that can be positioned close to the x-ray tube port (Fig. 9-7). In addition to multiple sets of lead diaphragms, this collimator has a first-stage or external lead shutter that extends inside the port of the x-ray tube to absorb off-focus radiation.

- *X-ray Leakage*

Although x-ray tubes are enclosed in a lead-lined protective shield, a certain amount of radiation leakage may exist, especially with some of the older units. The National Council on Radiation Protection and Measurement requires that "the leakage radiation measured at a distance of 1 meter from the source . . . not exceed 100 mR (milliroentgens) in 1 hour when the tube is operated at its maximum continuous rated current for the maximum rated potential."[2]

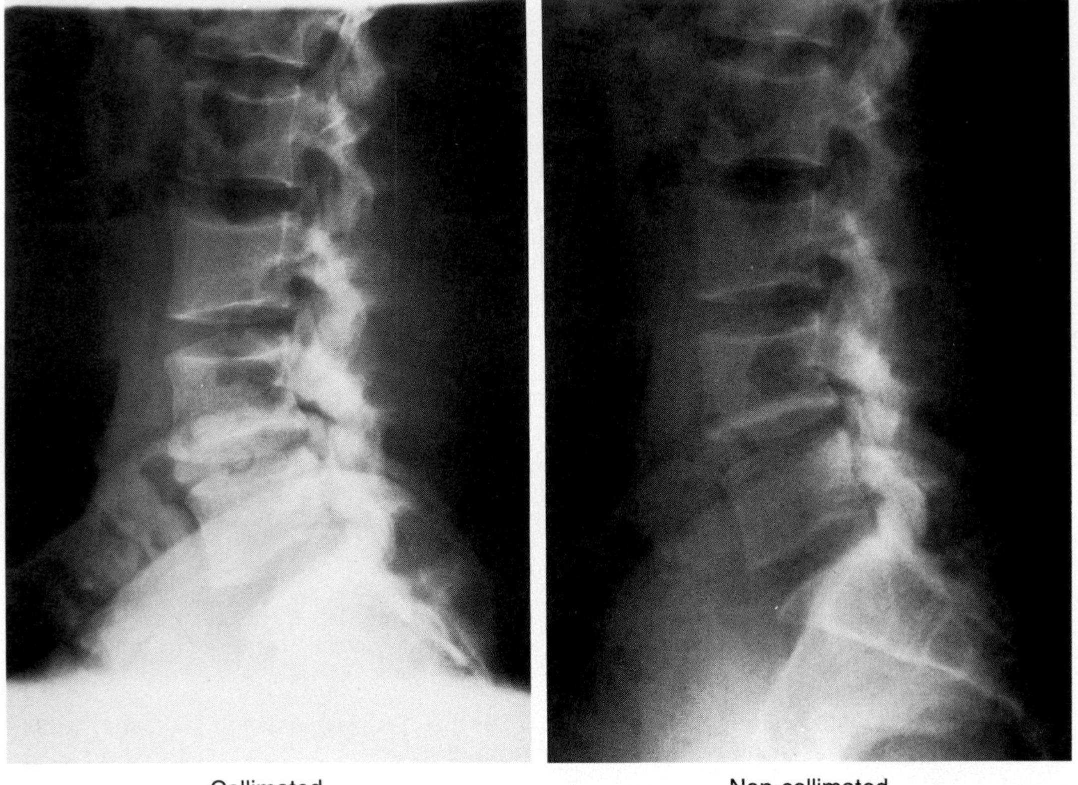

FIG. 9-5. *The importance of limiting the focus of the x-ray beam (collimation) is readily apparent when you compare a radiograph made with a non-collimated beam with a radiograph made with a collimated beam.*

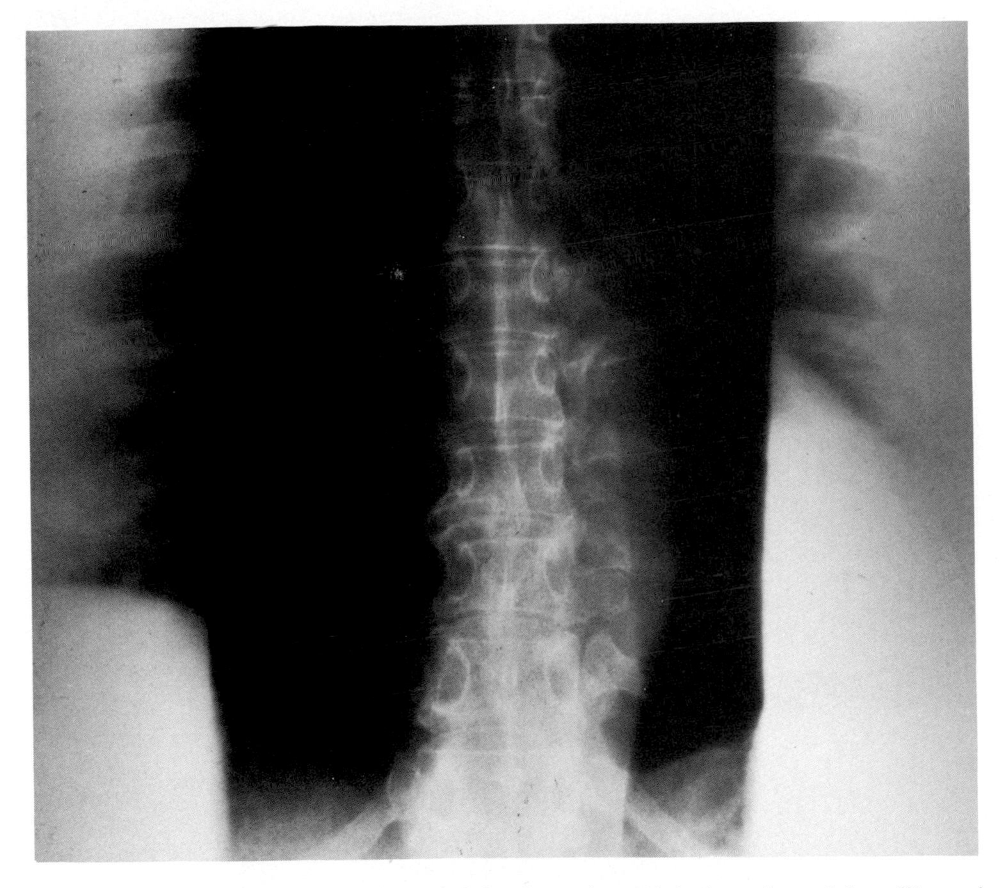

FIG. 9-6. *The blurring of this image of the thorax beyond the boundary of the collimated field resulted from off-focus radiation. (See Cullinan JE, Cullinan AM: Illustrated Guide to X-ray Technics, 2nd ed, pp 28–29. Philadelphia, JB Lippincott, 1980)*

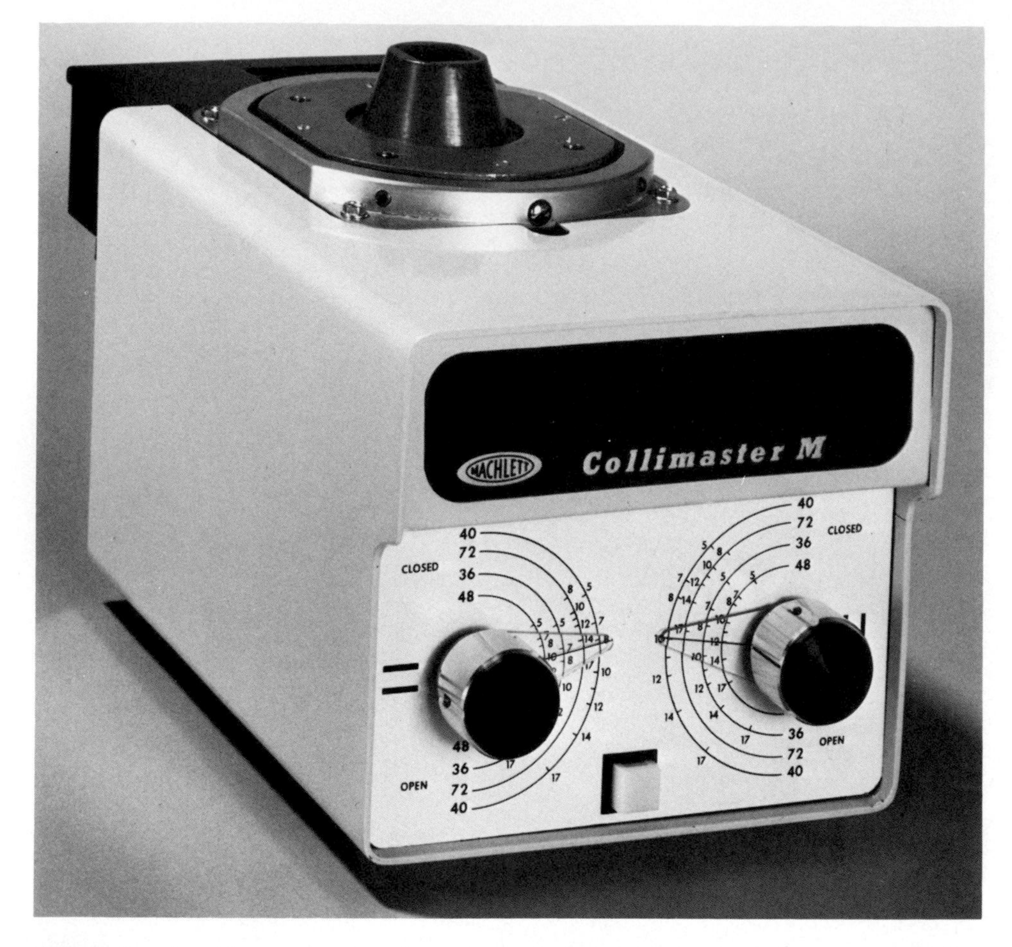

FIG. 9-7. *A collimator, used to focus the x-ray beam. Notice the appearance of the first stage or external shutter mounted at the upper surface. This lead device extends into the tube port in order to eliminate off-focus or stem radiation. (Courtesy of Machlett Laboratories, Stamford, CT)*

FIG. 9-8. *A test exposure obtained in order to demonstrate that off-focus radiation was coming from the port of the x-ray tube. Notice the rather ill-defined border of beam limitation as well as the appearance of the lead number* (arrow) *located on the underside of the intensifying screen for the purpose of identifying the cassette (see page 126). The appearance of the number indicates that this portion of the screen was exposed to radiation outside the field of collimation. The failure to install the collimator properly to the port of the x-ray tube can be responsible for such problems.*

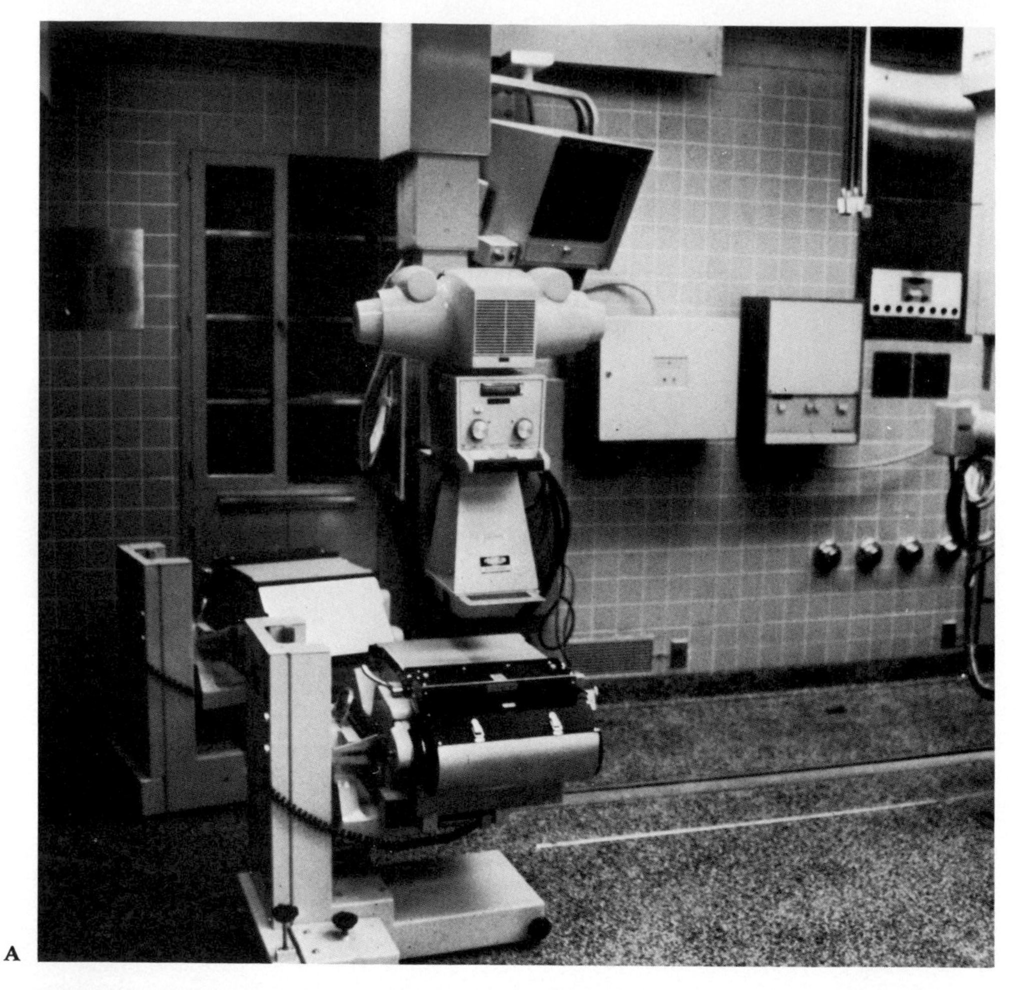

A

FIG. 9-9. *Although refinements have been made over the years in the devices used to limit the field of exposure, we have come full circle by combining many of the original methods with the more modern systems in use today.* (A) *Notice that in this radiographic unit, in addition to the collimator, a diaphragm is attached to the edge of a cone in order to restrict the field of exposure. Although the use of such devices obviously enhances the contrast of the radiograph, the improper use of a cone in conjunction with a collimator can result in radiation exposure outside the boundary of the cone.* (B) *This radiograph of the facial bones was made with an extension cone attached to the undersurface of the collimator. Failure to restrict the lead shutters in the collimator was responsible for the image outside the boundary of the cone. (See Cullinan JE, Cullinan AM: Illustrated Guide to X-ray Technics, 2nd ed, pp 31–32. Philadelphia, JB Lippincott, 1980)*

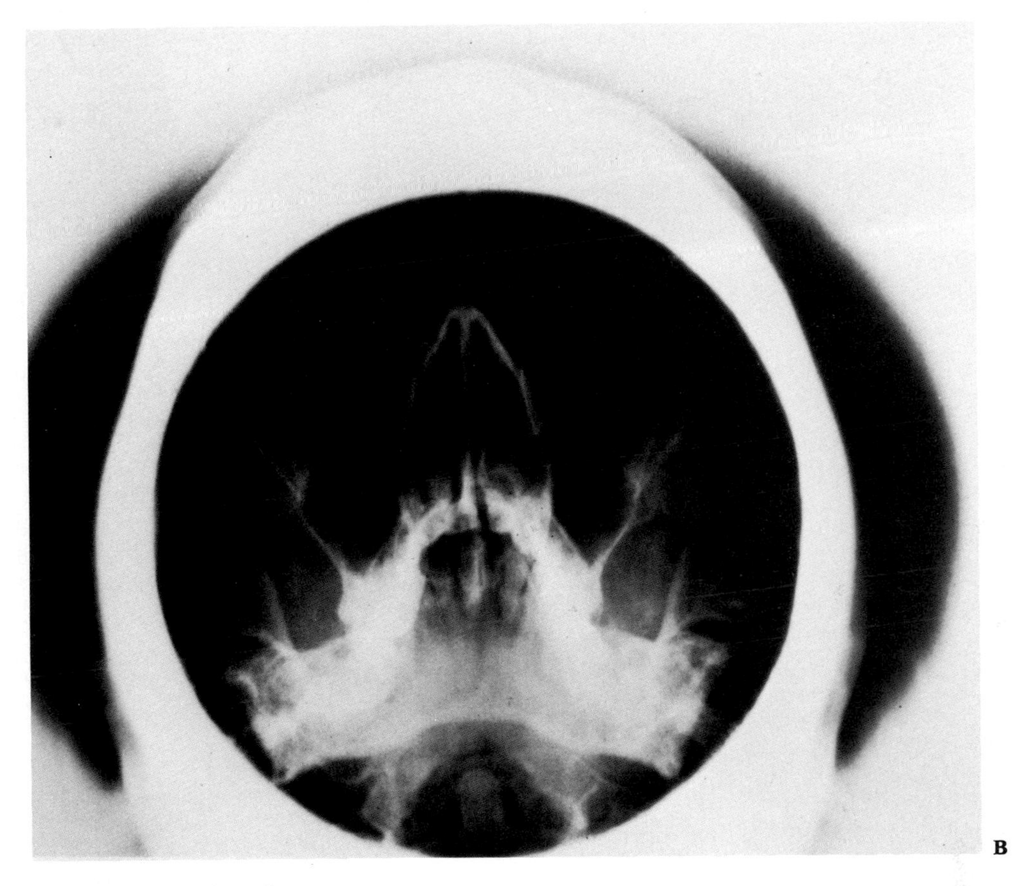

FIG. 9-9. *continued*

FIG. 9-10. (A) *This radiograph of the chest is fogged and demonstrates a rather unsharp circular density with a well-defined center.* (B) *The artifact occurred when the cassette containing the latent image of the chest* (arrow) *was left under the housing of the x-ray tube while an additional radiograph of the chest was being obtained. The unsharp circular density is a magnified image of the metal cap at the end of the housing of the x-ray tube. The image near the center is of the bolt that holds the metal cap in place.*

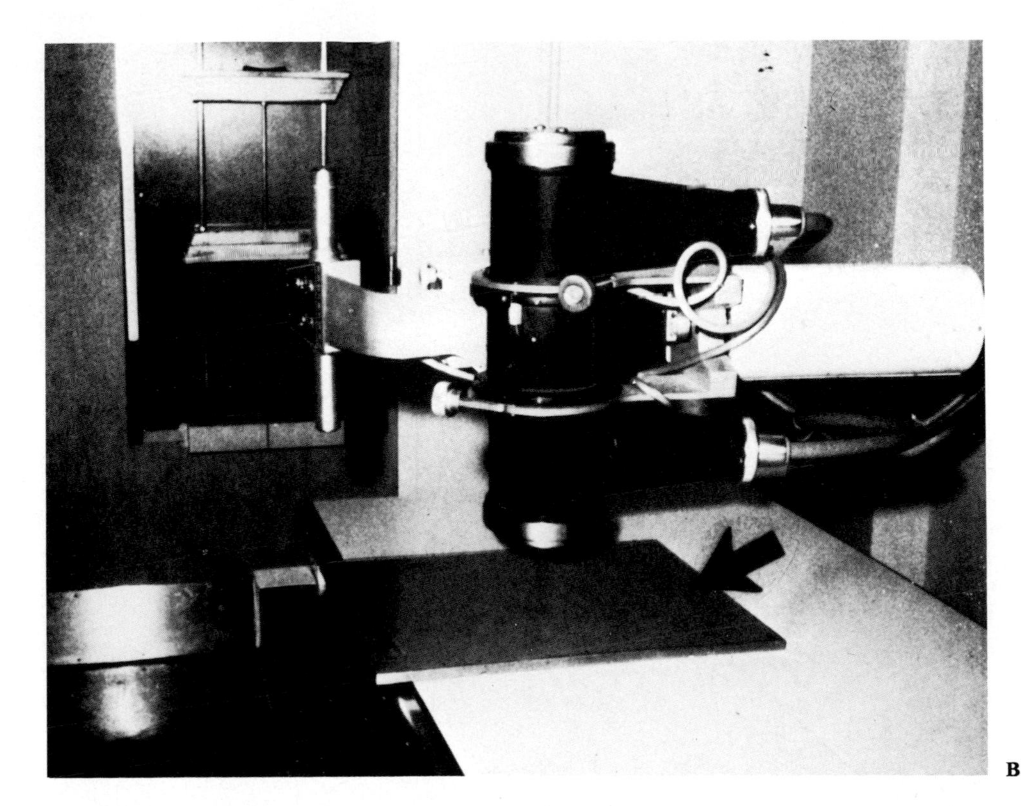

B

FIG. 9-10. *continued*

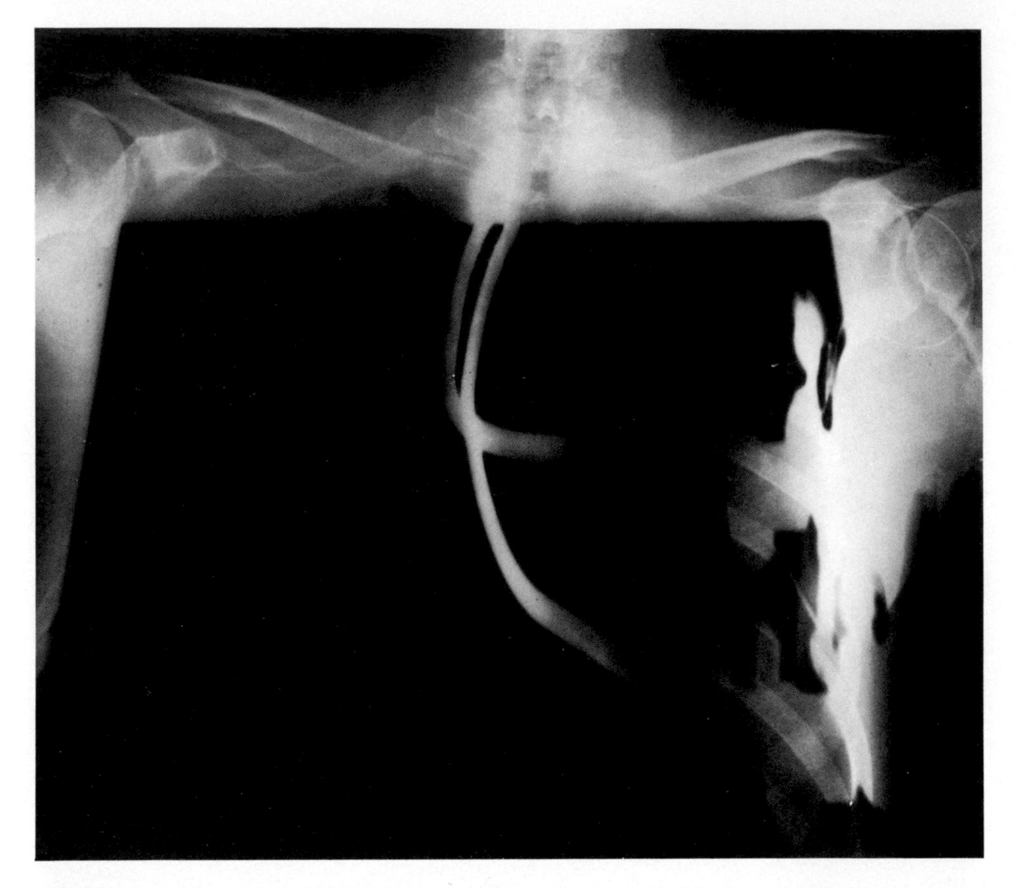

FIG. 9-11. *The absence of the heart and lung shadows in this chest radiograph suggests that the patient is powered by the two large electrodes imaged in the radiograph. This study was obtained at bedside, and following the examination the cassette containing this film was exposed to radiation while located in the storage drawer of the mobile unit. The artifact occurred when the technologist made a test exposure with the x-ray tube directed toward the cassette storage compartment. The electrical components imaged in the radiograph were located in the base of the mobile unit. (Courtesy of Ralph Coates, R.T. [R])*

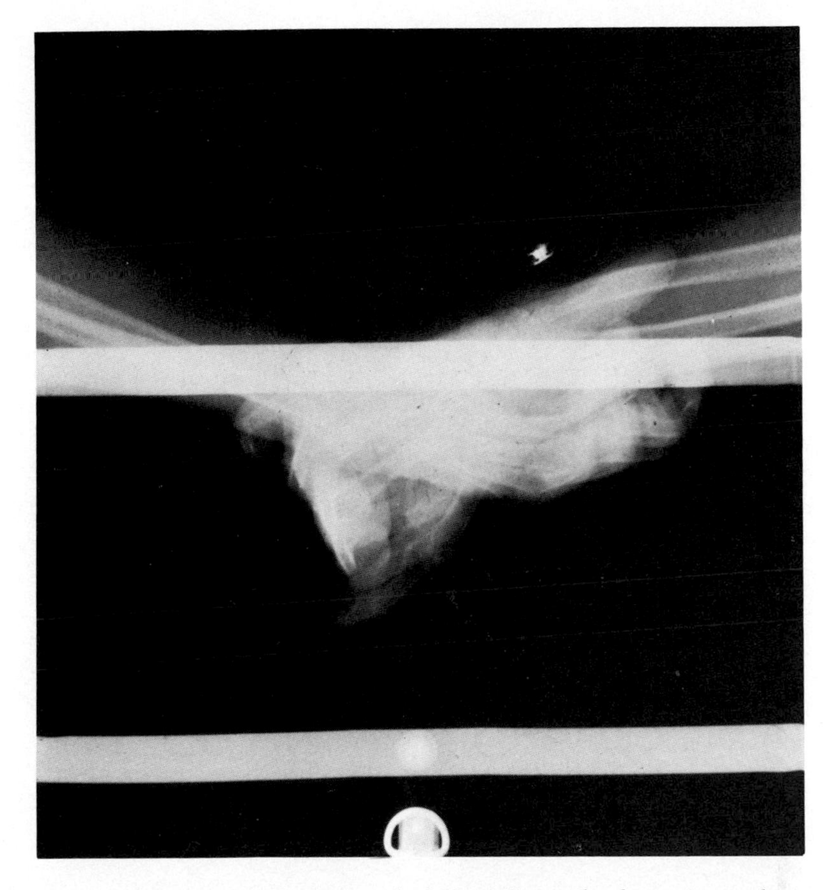

FIG. 9-12. *Here we see an attempt to obtain a bedside radiograph of a patient's chest. The patient is positioned correctly, but the x-ray tube was centered on the wrong side of the cassette. (This radiograph was made by a student technologist who is now a radiologist. I have kept this film over the years and on occasion have reminded him of it whenever I need a favor.)*

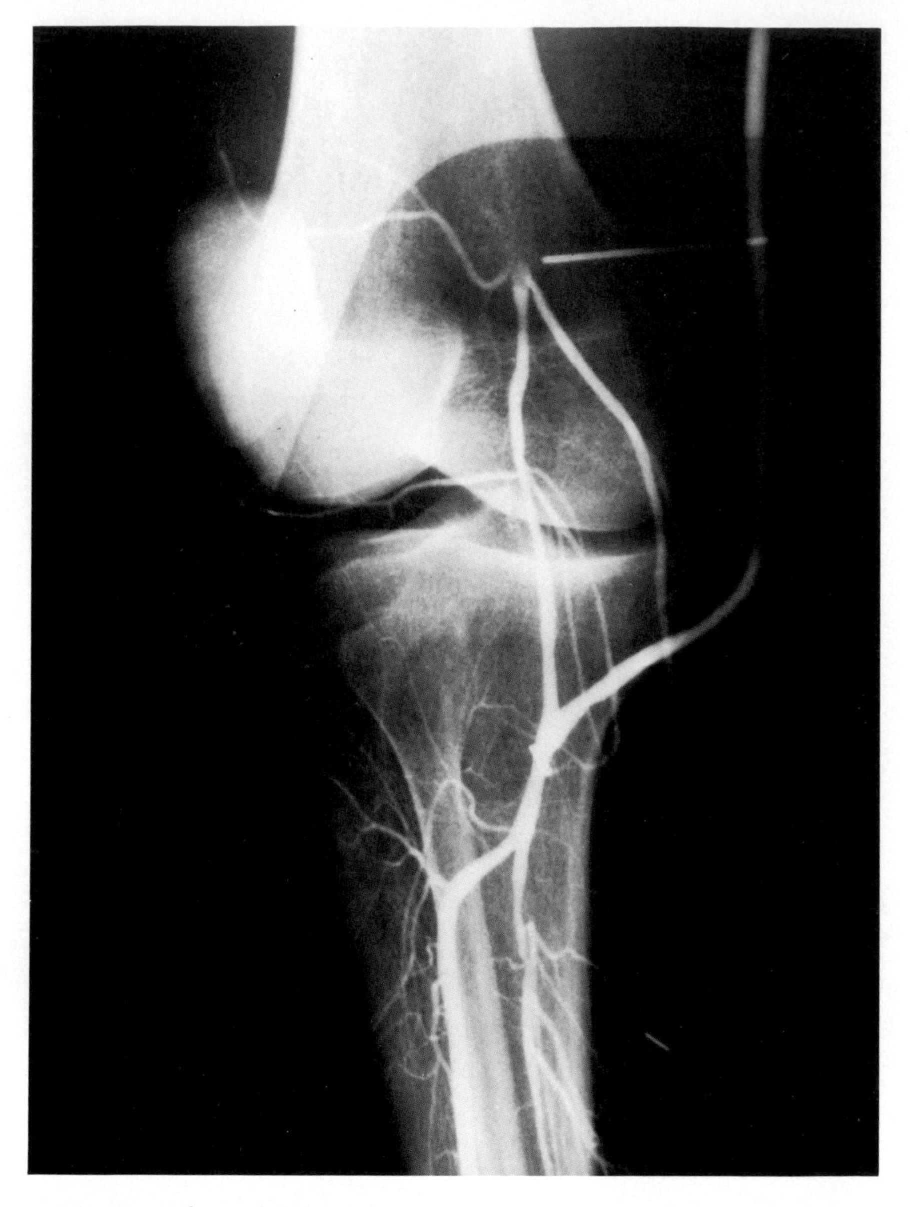

FIG. 9-13. *The well-defined border of the artifact imaged over the distal femur in this radiograph suggests that the film was exposed to light. However, it occurred when a portion of the pad on the operating table was removed in order to position the cassette under the extremity. The lack of radiation absorption in this region was responsible for the appearance of the image.*

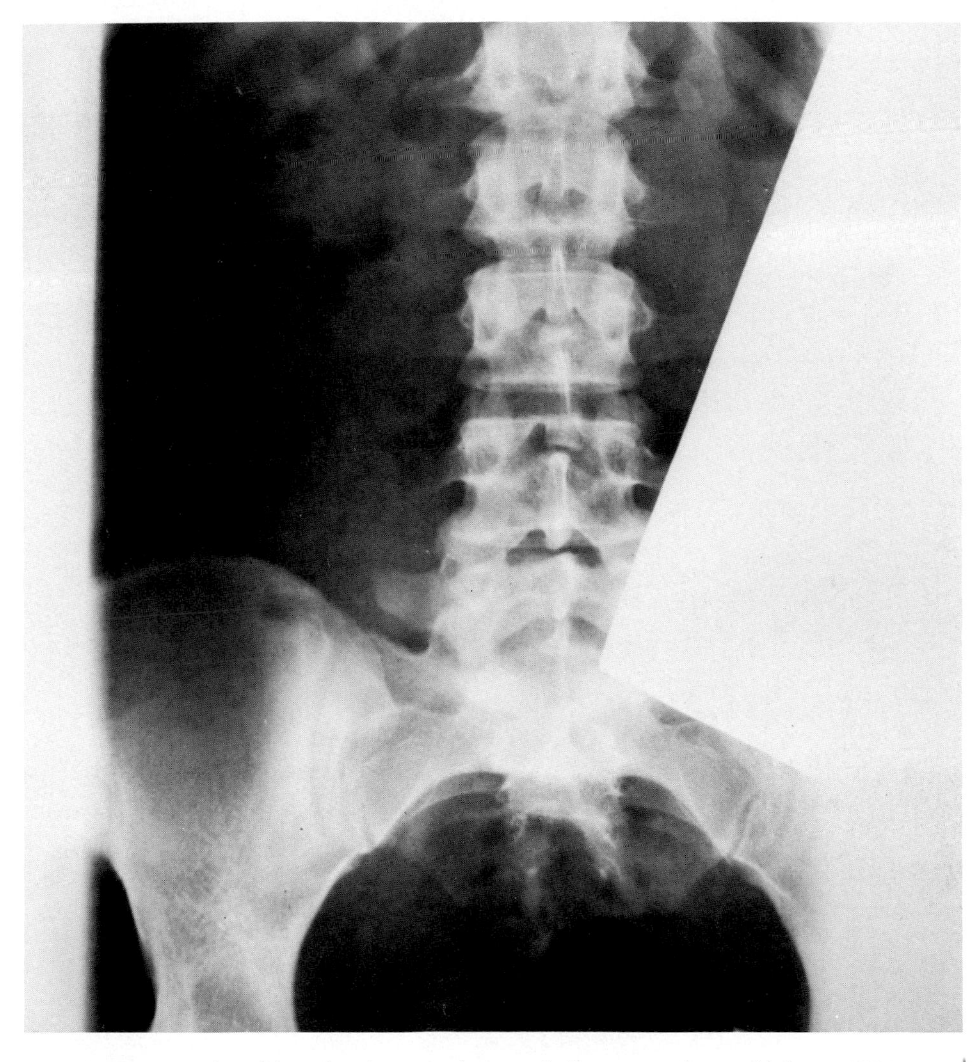

FIG. 9-14. *Lead–rubber sheeting, used to mask the cassette for multiple projections and to shield it from radiation scatter in certain procedures, was responsible for the artifact in this radiograph. The image of minus density is a portion of a lead masking blocker that was accidentally left under a sheet on the radiographic table. Although it is obvious that the opaque material interfered with the visualization of the anatomic region of interest, it could lead to more serious results if such an opaque barrier obscures the photocell in an automated exposure-control system. Such a technical error can destroy or at least seriously damage an x-ray tube if a backup timer is not incorporated in the unit. (See Cullinan JE, Cullinan AM: Illustrated Guide to X-ray Technics, 2nd ed, pp 82–83. Philadelphia, JB Lippincott, 1980)*

Grids and Grid-Related Problems

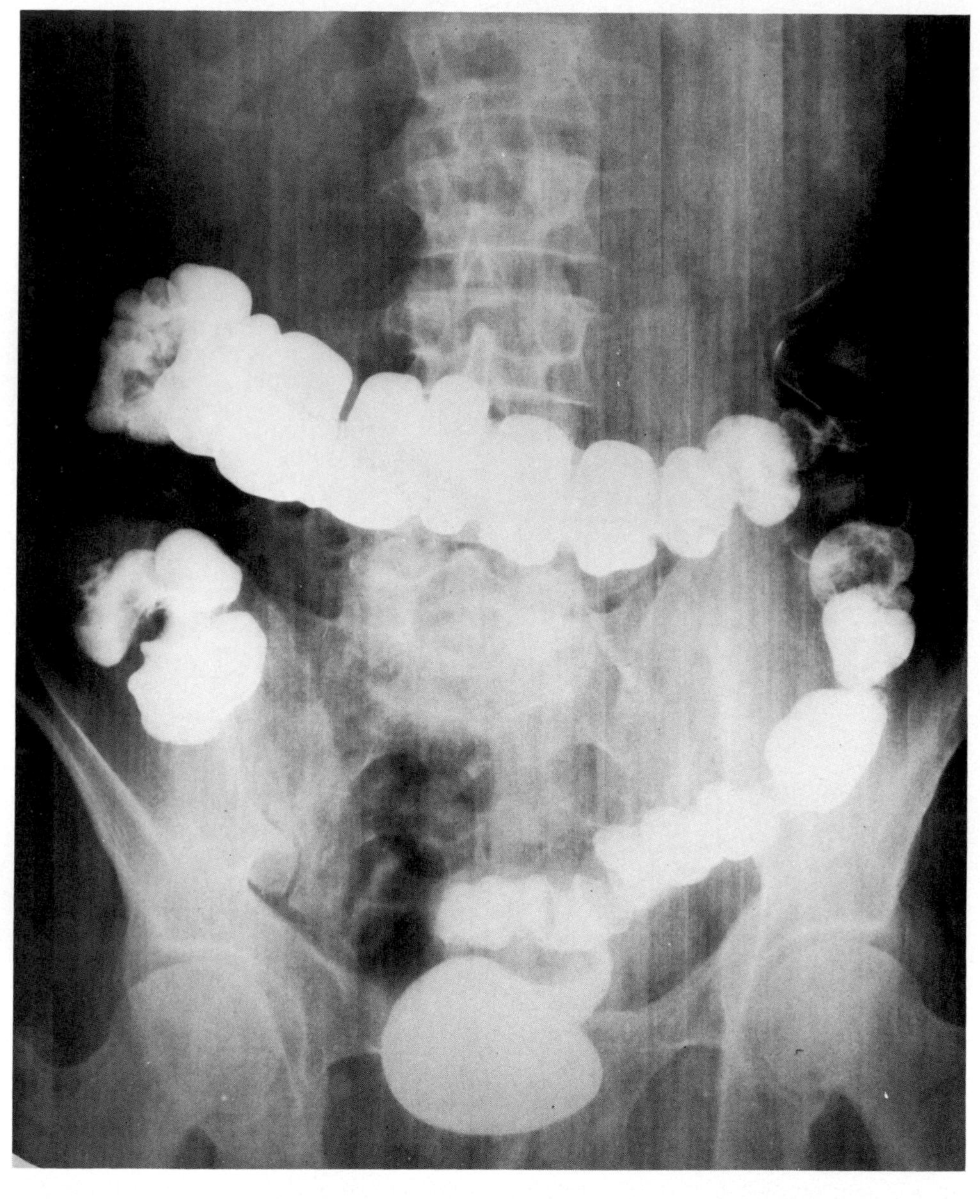

FIG. 9-15. *The irregular vertical pattern in this abdominal radiograph occurred when the lead lines in a moving-type grid were imaged while the grid was stationary.*

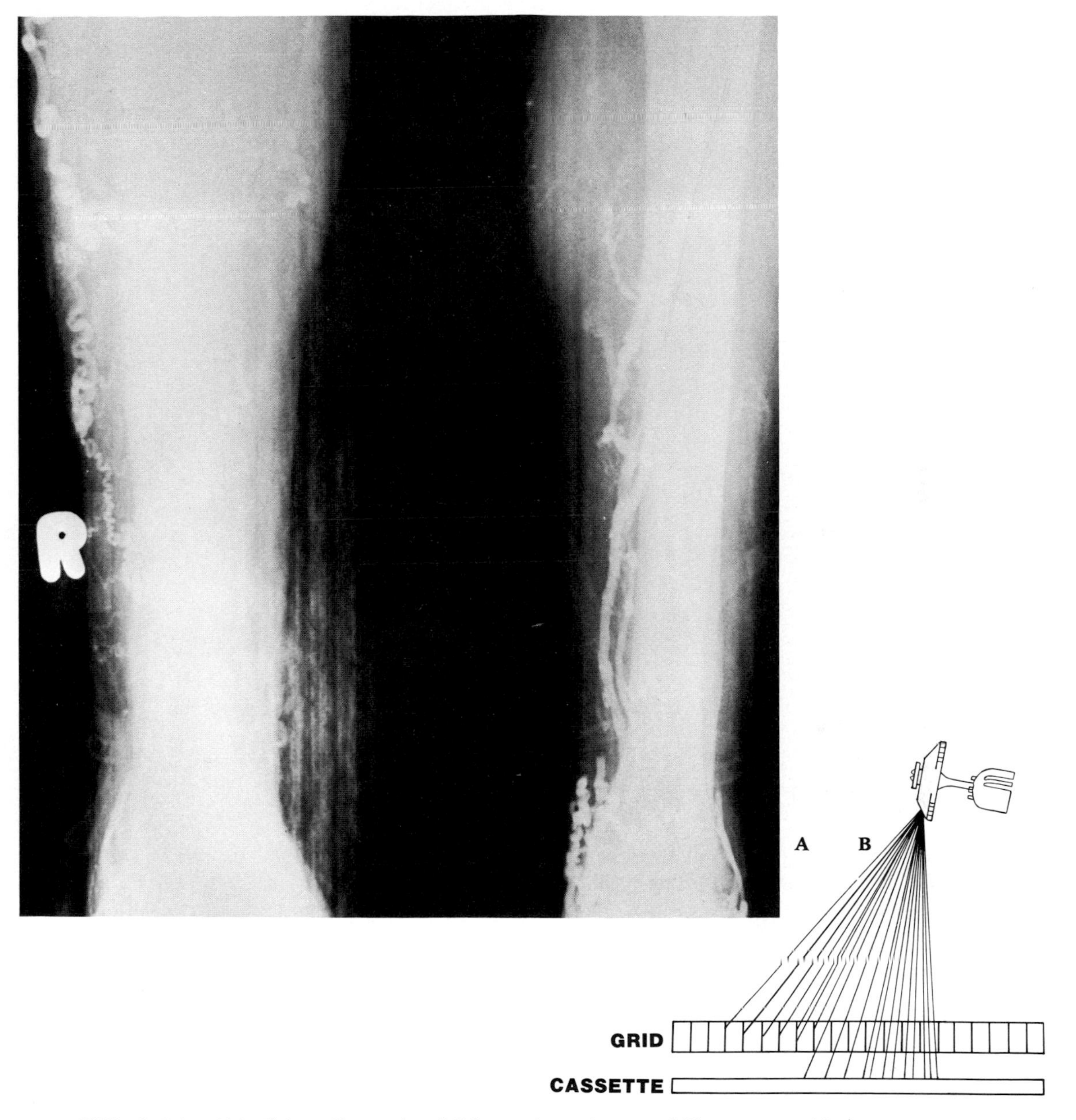

FIG. 9-16. (A) *This radiograph exhibits an irregular streaklike pattern with more pronounced image degradation toward the right extremity. This artifact occurred when the x-ray tube was accidentally angled across the surface of the grid. The lead lines are more visible toward the right extremity because the x-ray tube was angled from the left of the table to the right* (B). *In order to avoid such imaging problems, the technologist should periodically check the alignment of the x-ray tube with the surface of the grid. (See Gray JE, Winkler NT, Stears J, Frank ED: Quality Control in Diagnostic Imaging, pp 84–89. Baltimore, University Park Press, 1983)*

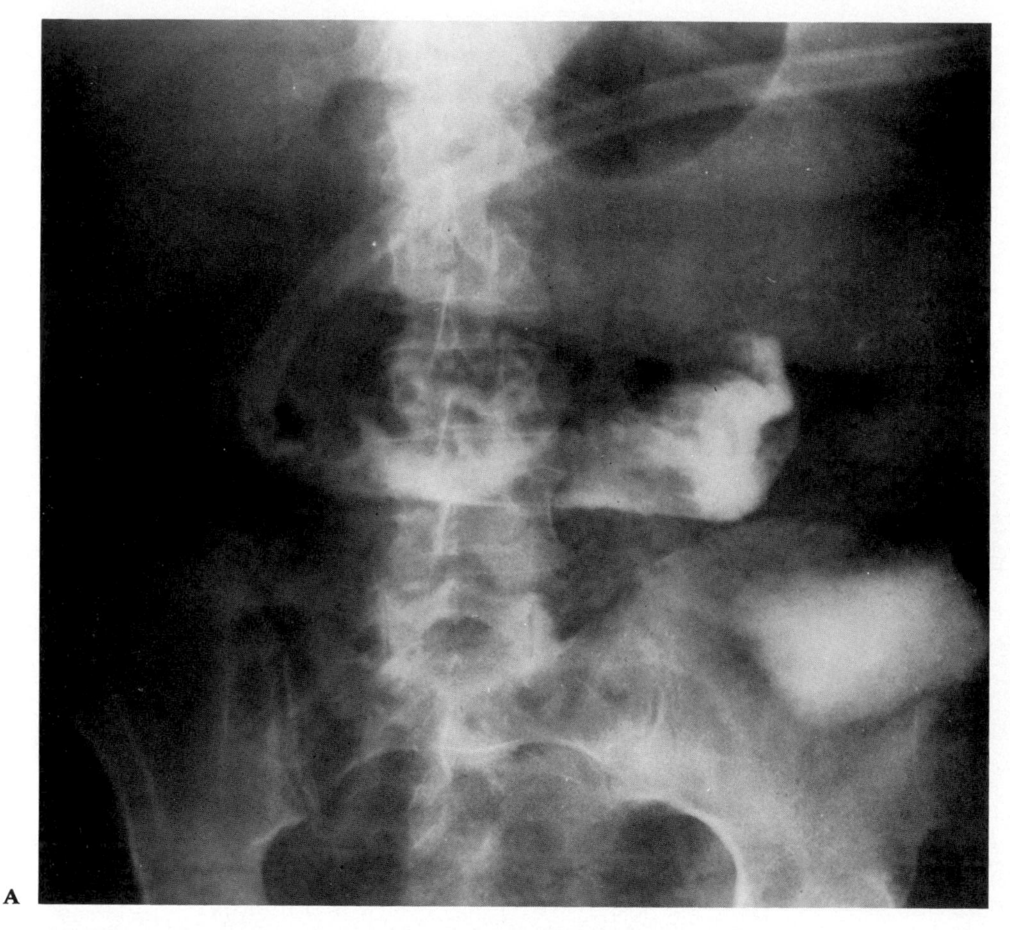

FIG. 9-17. (A) *This radiographic study of a fistula injected with an iodinated substance lacks contrast owing to the fogging caused by radiation scatter.* (B) *A repeat exposure using a crosshatched grid results in a dramatic improvement in image quality. This was accomplished by using an R-8 grid cassette in conjunction with the R-12 grid in the radiographic table. The grid cassette was placed in a transverse position in order to form a crosshatch or right-angled relationship with the lead lines of the grid in the table* (C). *As a result, the effective ratio of the accessory increased to that of an R-20 grid. Although the benefits of this technique are obvious, it is often used incorrectly and results in some unusual imaging patterns (see Figs. 9-23 and 9-24). (See Cullinan JE, Cullinan AM: Illustrated Guide to X-ray Technics, 2nd ed, pp 37–40. Philadelphia, JB Lippincott, 1980)*

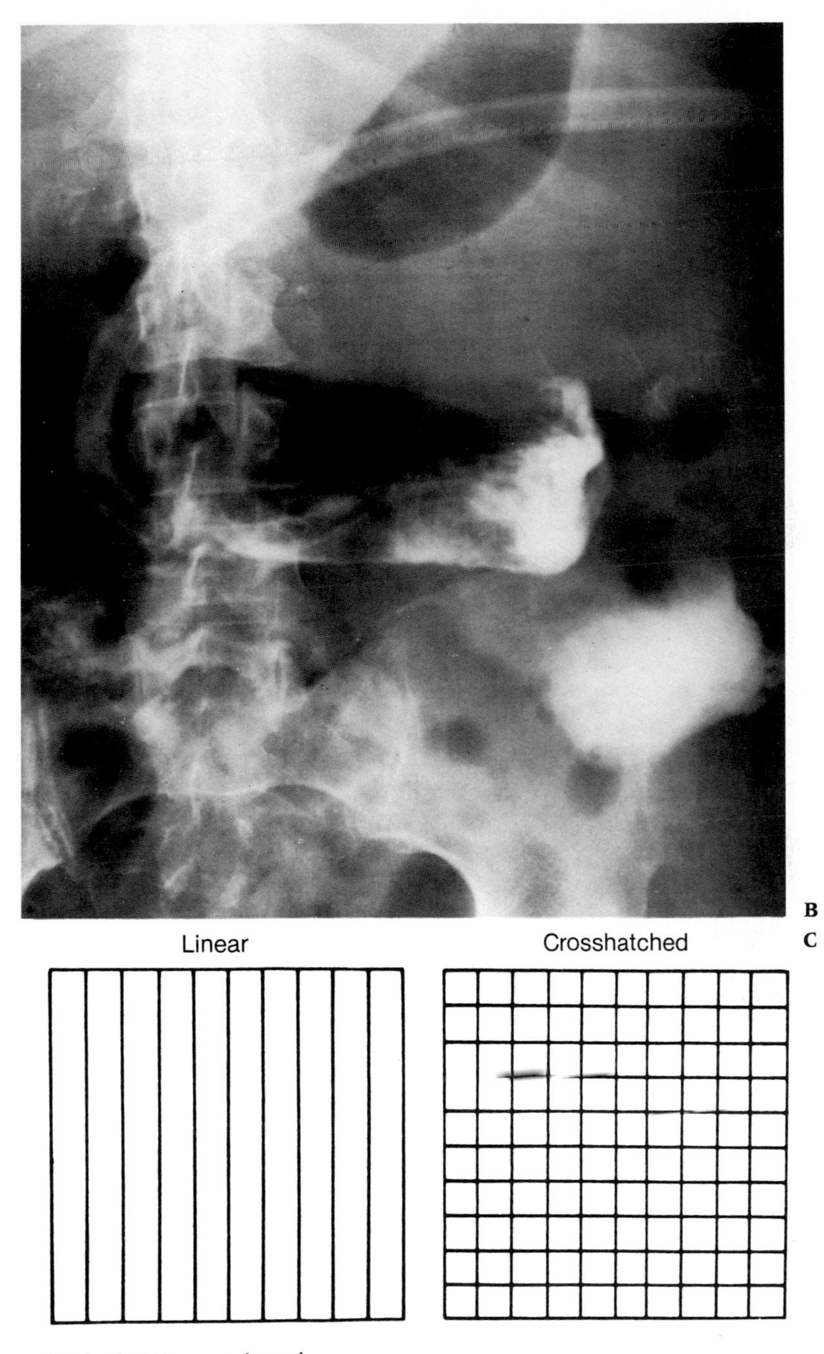

FIG. 9-17. *continued*

A

FIG. 9-18. *Notice in this photograph of the tube side of a wafer grid that the decal that indicates the grid's ratio and focal range is missing. Although it is difficult to imagine how the technologist could accurately compensate for adjustments in technique without knowing the ratio of the grid, technical assessment is probably based on the fact that the grid is of low ratio and that a multiple of three to four times the mAs is routinely employed whenever this grid is used. Such compensation may be within acceptable limits, but not knowing the focal range and which is the tube side of the grid can result in some unusual imaging problems. Notice in* B, *for example, the absorption of radiation in the lateral borders of the chest radiograph, as well as the appearance of lead lines. This condition occurred when a focus-type grid was accidentally used in an inverted position.*

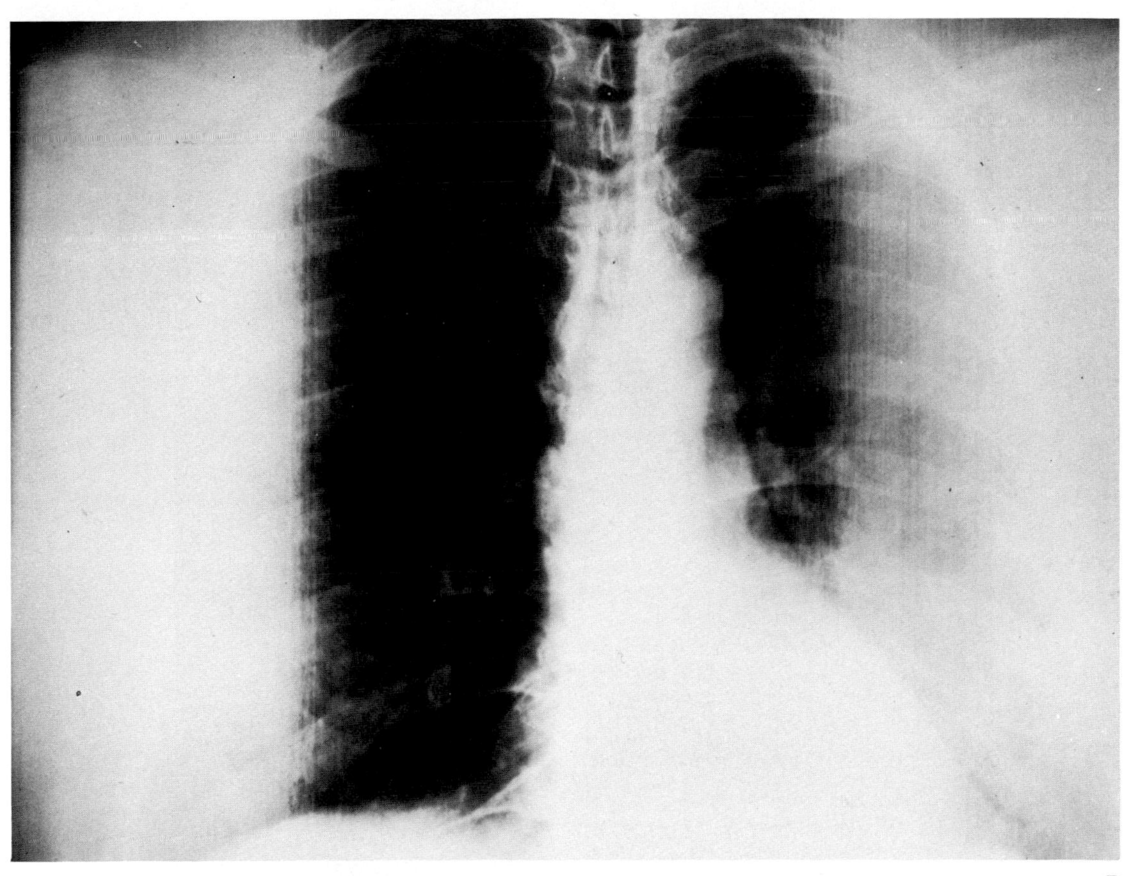

B

FIG. 9-18. *continued*

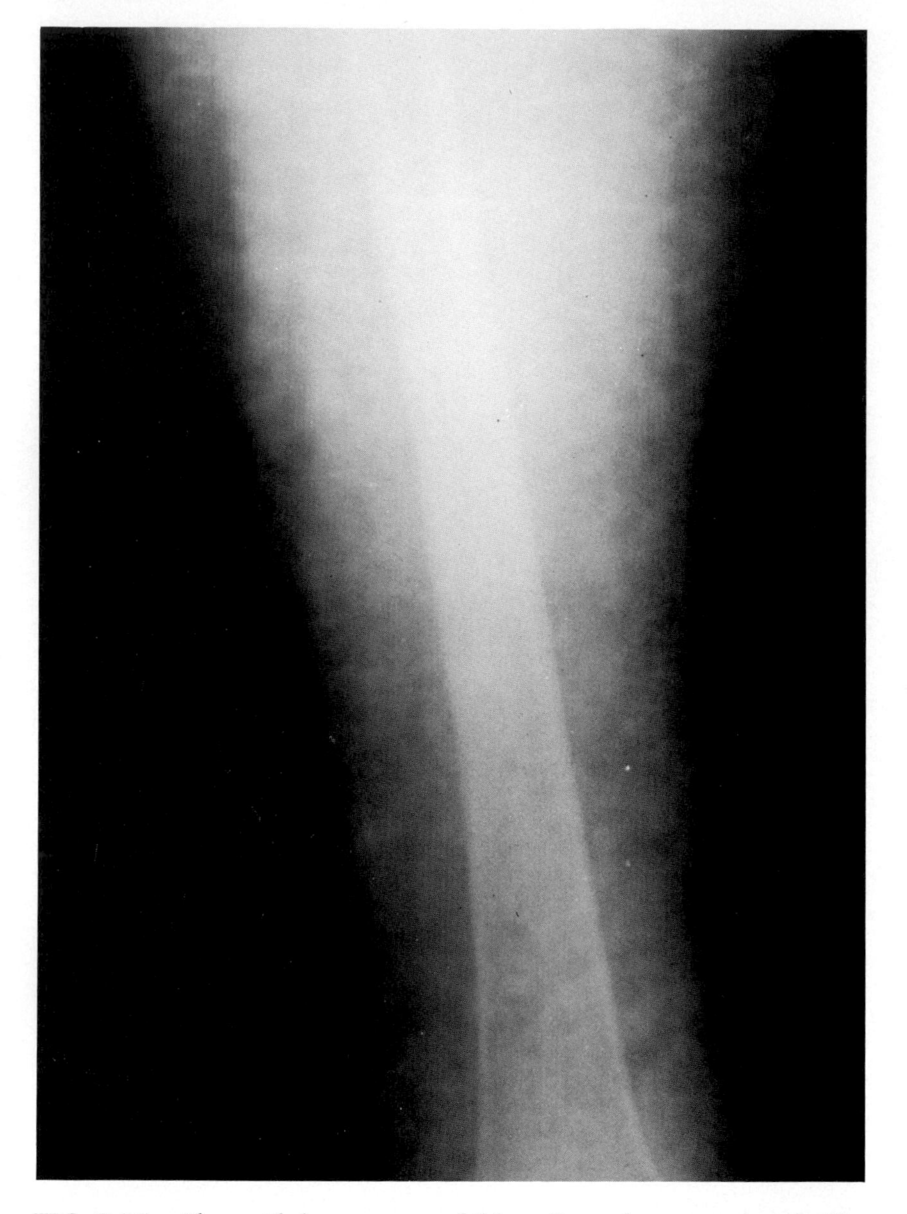

FIG. 9-19. *The mottled appearance of this radiograph suggests that the film could have been damaged or that the patient's trousers were not removed for the radiographic examination. However, the artifact occurred when the x-ray tube was improperly centered and slightly angled in relation to the surface of the radiographic table. The mottled image is that of the grid located on the underside of the table. Cullinan and Cullinan referred to a similar type of pattern as a "corduroy" artifact and indicated that it is associated with grid synchronization problems. (See Cullinan JE, Cullinan AM: Illustrated Guide to X-ray Technics, 2nd ed, pp 44–45. Philadelphia, JB Lippincott, 1980)*

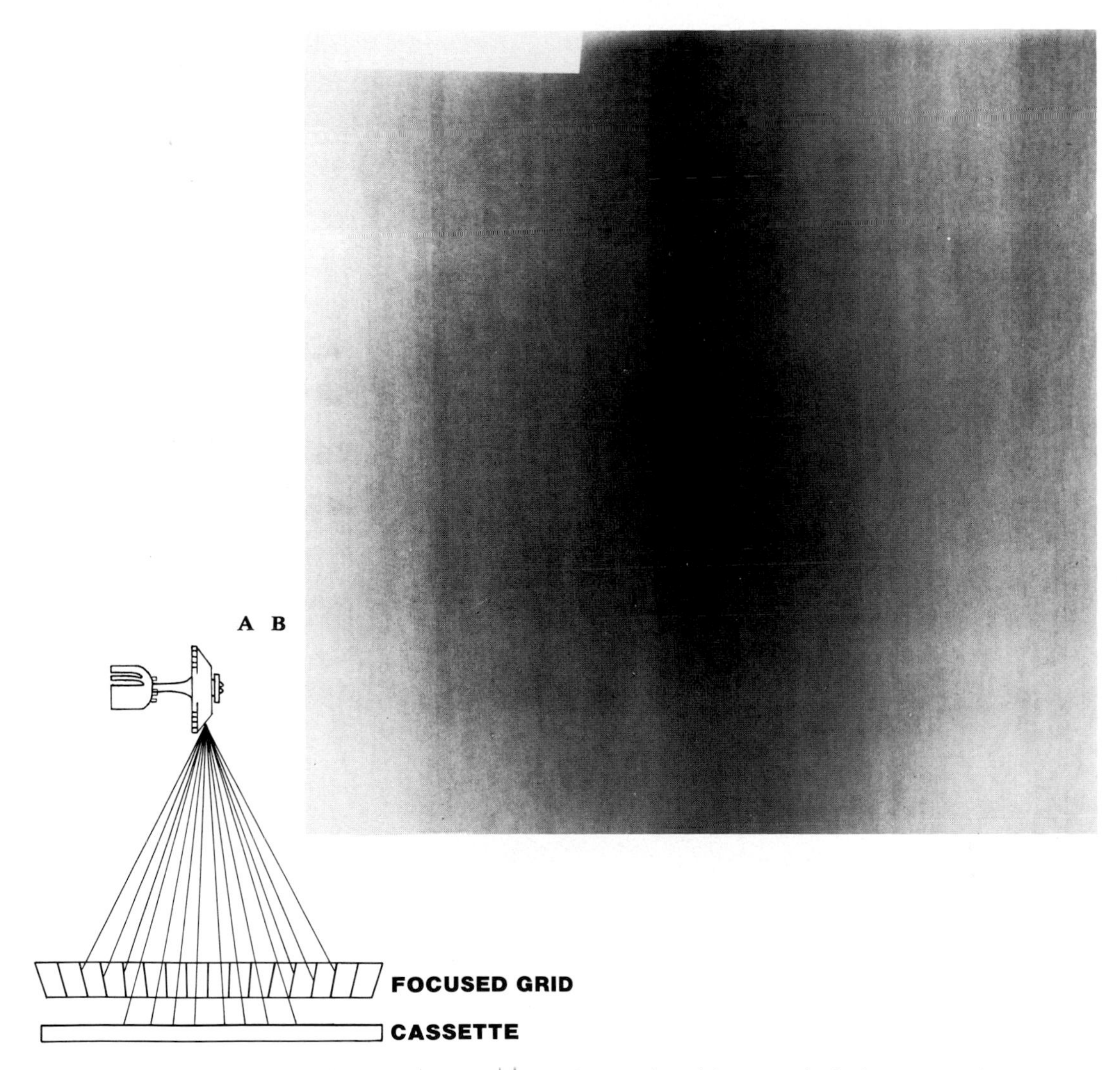

FIG. 9-20. (A) *If a focus-type grid is used in an inverted position, marked absorption of radiation will occur.* (B) *Notice in this radiograph taken through an inverted grid that x-rays were transmitted through the central portion of the grid. The lateral borders exhibit the appearance of lead lines.*

FIG. 9-21. *A test radiograph of a focused grid taken at a source-image distance that exceeded the focal range of the grid. Notice that the radiographic density decreases from the central axis of the image to a point at which the radiation absorption (grid cutoff) becomes more pronounced at the periphery.*

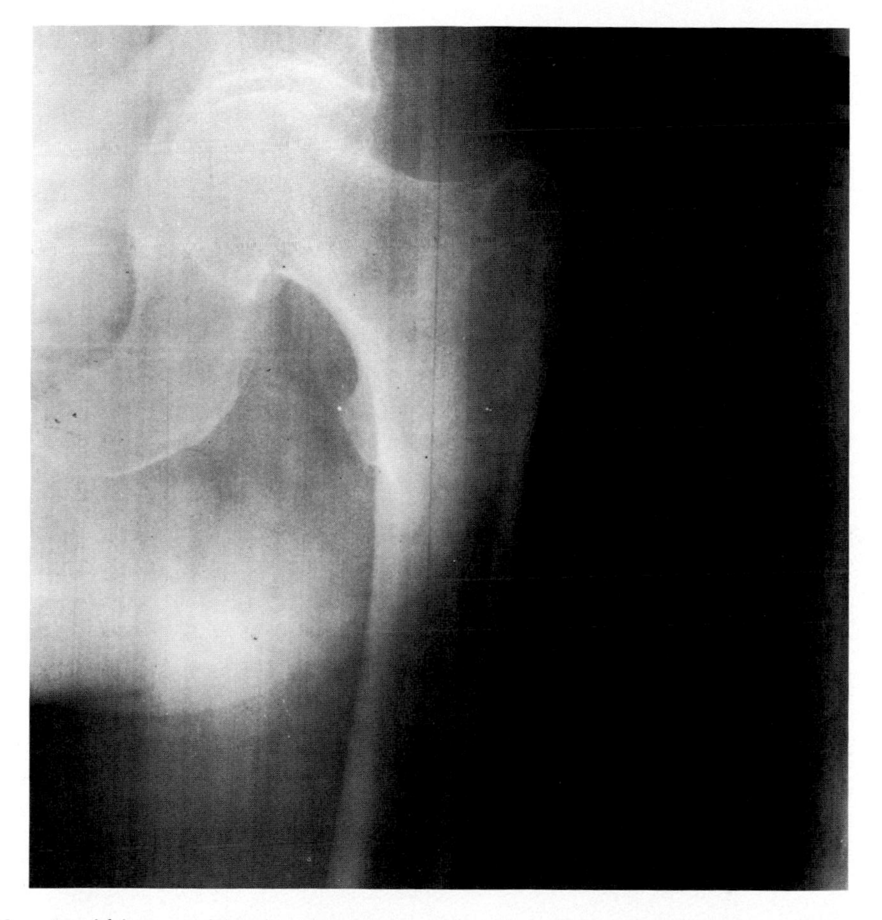

FIG. 9-22. *A grid is a precision instrument, and if not handled carefully, it can easily be damaged. This damage will result in various imaging patterns in the radiograph. In this bedside radiograph of the hip, the black line superimposed over the long axis of the femur, as well as the minus-density streaks, occurred when a damaged grid was used for the exposure. (Courtesy of Ralph Coates, R.T. [R])*

MOIRÉ ARTIFACTS

The wavy patterns seen in Figures 9-23, *B*, and 9-24, *B*, are called *moiré artifacts*. The term is borrowed from the manufacturing of textiles, in which it refers to an irregular, wavy pattern of fabric. In radiography, moiré artifacts occur when the technologist attempts to superimpose two grids with linear patterns running in the same direction.*[,1,3]

If it were possible to superimpose precisely the lead lines of two grids of identical quality, they would function as one. However, because of the variation in the thickness of the materials used in the construction of a grid, the number of lead lines per inch is not constant over the entire area of the grid. As a result, it is impossible to superimpose the linear patterns, and a moiré artifact will occur if you try to do so .

Figure 9-23, *A*, demonstrates that where the lead lines of the grids were not quite parallel and crossed each other, long light shadows (Fig. 9-23, *B*) were produced. In the area in which the lead lines were not crossed, a greater amount of x-ray was transmitted, producing darker areas.*

If you would like to demonstrate various moiré patterns, try the following experiment. Make a tabletop exposure of a wafer grid, and following processing, either cut the radiograph in half or fold it in the manner shown in Figure 9-23, *A*, to produce a wide variety of patterns.

*Personal communication: Otto E. W. Schmidt, Liebel–Flarsheim Company, Cincinnati, OH, May 1970

A

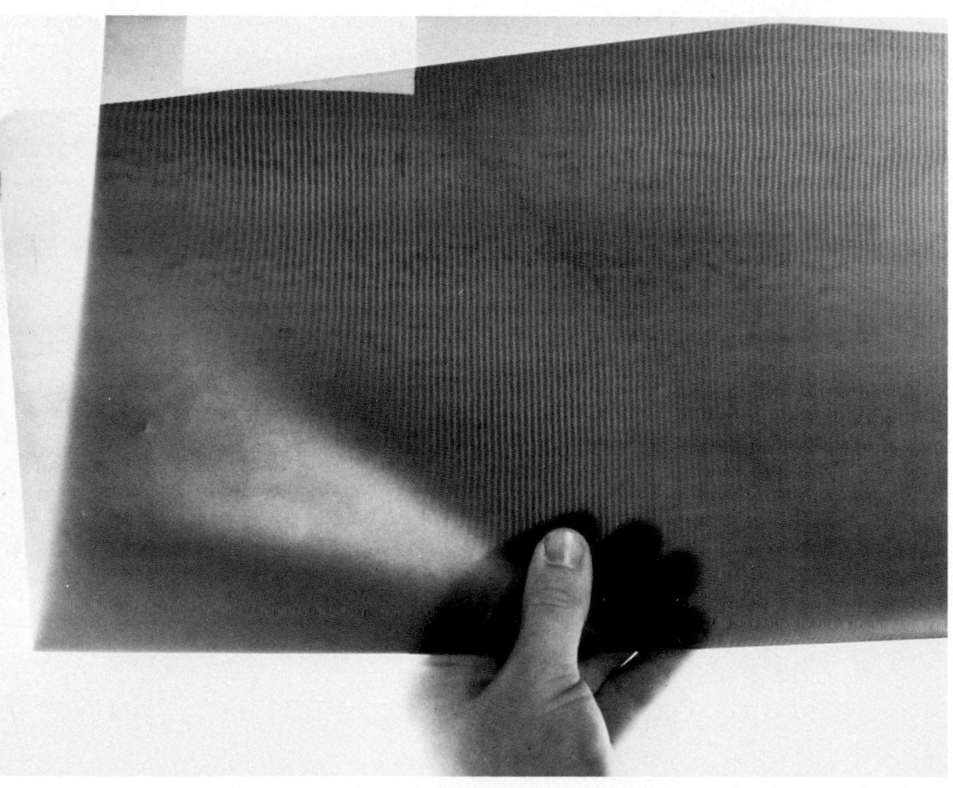

FIG. 9-23. *continued*

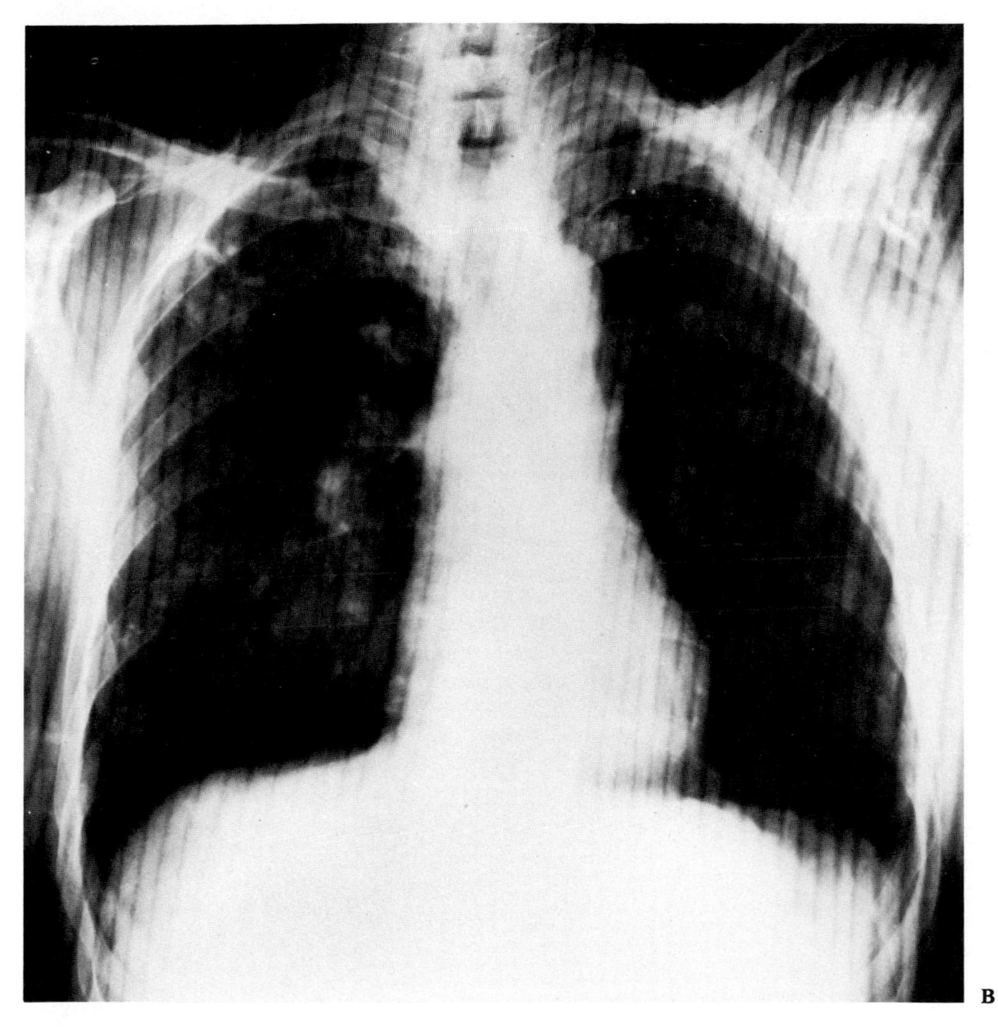

B

FIG. 9-23. *A moiré pattern* (B) *caused by attempting to superimpose two grids with the lead lines running in the same direction* (A). *(See Cullinan JE, Cullinan AM: Illustrated Guide to X-ray Technics, 2nd ed, pp 39–40. Philadelphia, JB Lippincott, 1980)*

A

FIG. 9-24. (A) *The two stationary grids in this radiographic table manufactured by Research Instrumentation, Ltd. (Basel, Switzerland) are supposed to abut on each other, as illustrated in the photograph. However, the technologist accidentally superimposed one of these grids over the other* (arrow) *while attempting a preliminary radiograph prior to a vascular study. The result was a moiré pattern in the radiograph* (B). *(From On the Technical Side. Radiol Technol 48, No. 4, 1977)*

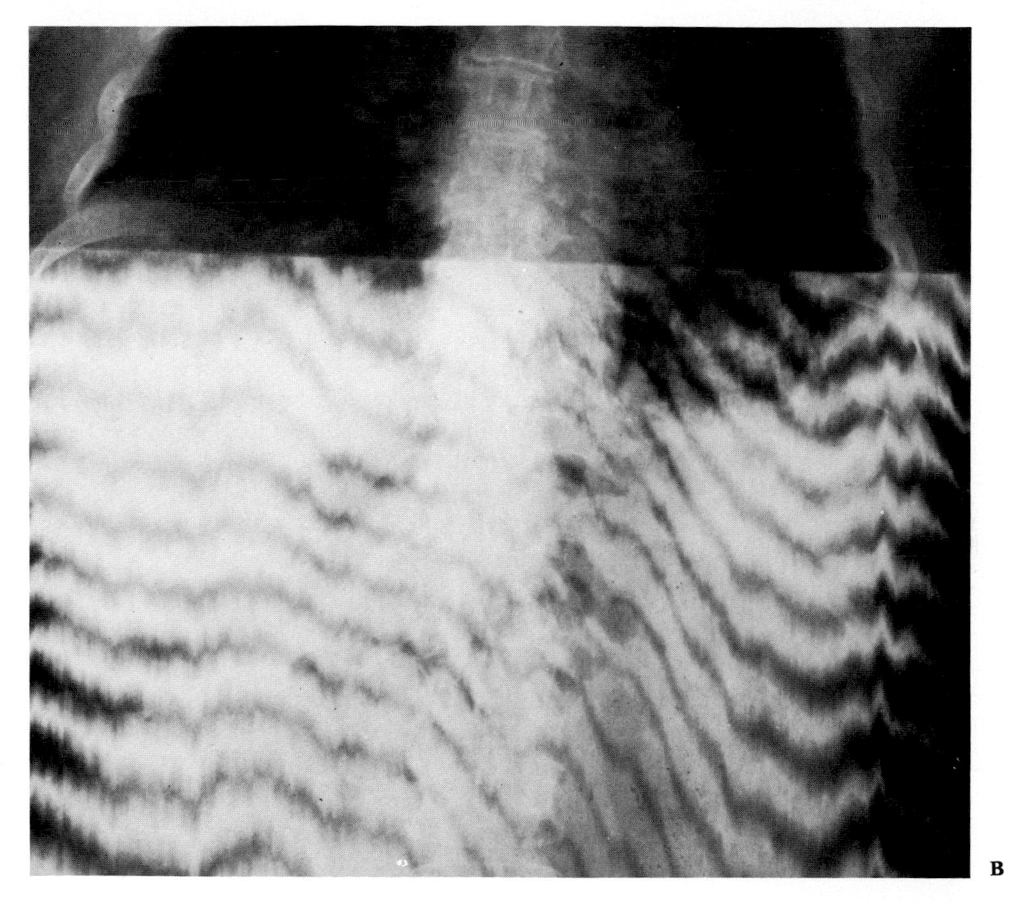

FIG. 9-24. *continued*

PROBLEMS WITH USING AN INVERTED CASSETTE AS AN IMPROVISED GRID

Although using an inverted cassette as an improvised grid is a perfectly acceptable technique, the technologist should be careful not to use a cassette with a damaged lead-foil backing in this manner. To do so can result in some bizarre artifactual images (Fig. 9-25). Because the damage to the lead-foil backing usually occurs when the technologist attempts to rescreen the cassette, the best way to avoid such problems is to return the cassette to the manufacturer for rescreening or repair.

Incidentally, aside from the artifacts caused by damage to the lead foil material in Figure 9-25, the lack of image quality should not be interpreted as the type of result one can expect when deliberately using a Kodak cassette as an improvised grid. Although the manufacturer did not design the cassette to be used in this manner, I have found it to be an extremely versatile method of coping with radiation scatter in select cases. Its application is limited, however, to radiographic examinations that border between grid and non-grid techniques. The absorption of radiation scatter by these improvised grids is less than that of a low-ratio grid. Consequently, it should be thought of as a filter in controlling the effects of radiation scatter (Figs. 9-26 through 9-28).[5]

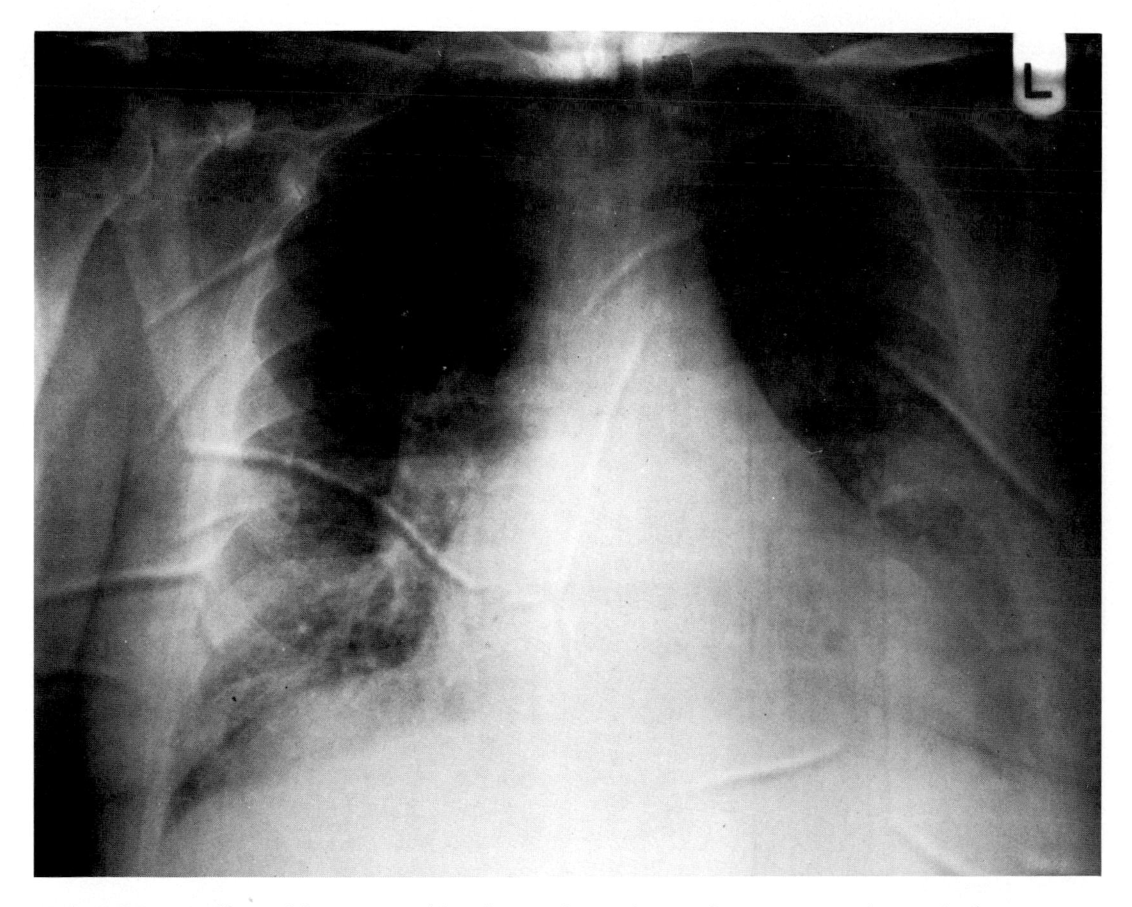

FIG. 9-25. *Artifactual images resulting from using a damaged cassette as an improvised grid. The radiograph was made with an X-Omatic cassette in an inverted position. The image is not only grossly overexposed, but in addition, it exhibits creaselike defects. These artifacts occurred because the lead foil back in the cassette was damaged during the process of replacing the intensifying screens. In an attempt to correct the condition, the lead foil was folded and overlapped behind the screen. When the cassette was deliberately inverted for use as an improvised grid, the damaged lead backing resulted in the unusual imaging pattern. (Courtesy of Marie Rochette, R.T. [R])*

A B

FIG. 9-26. *In these radiographs of an X-Omatic cassette in an open position, compare the radiolucent image of the front surface* (A) *with the filtration effect of the back surface* (B).

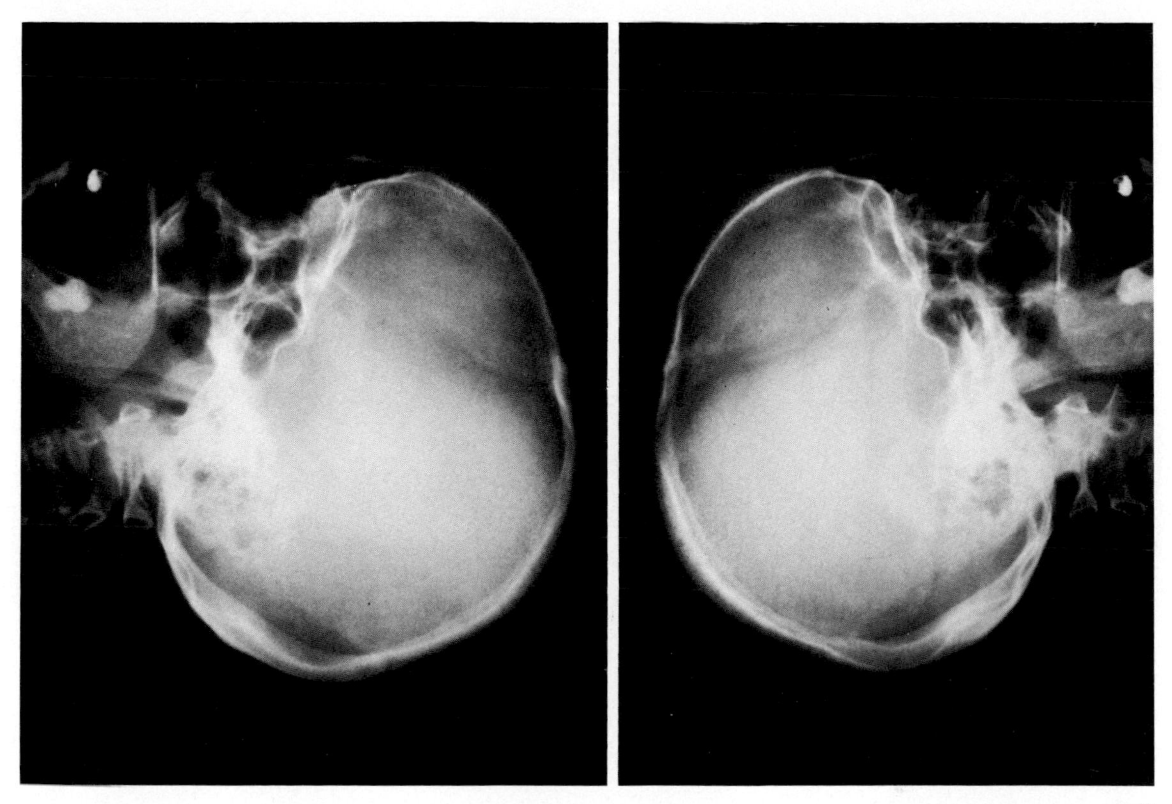

FIG. 9-27. *In these radiographs of the skull, compare the grid exposure* (A) *with the non-grid exposure obtained with inverted X-Omatic cassette* (B). *(From Sweeney RJ: The use of an inverted Kodak X-Omatic cassette as an improvised grid. Radiol Technol 49, No. 3, 1977)*

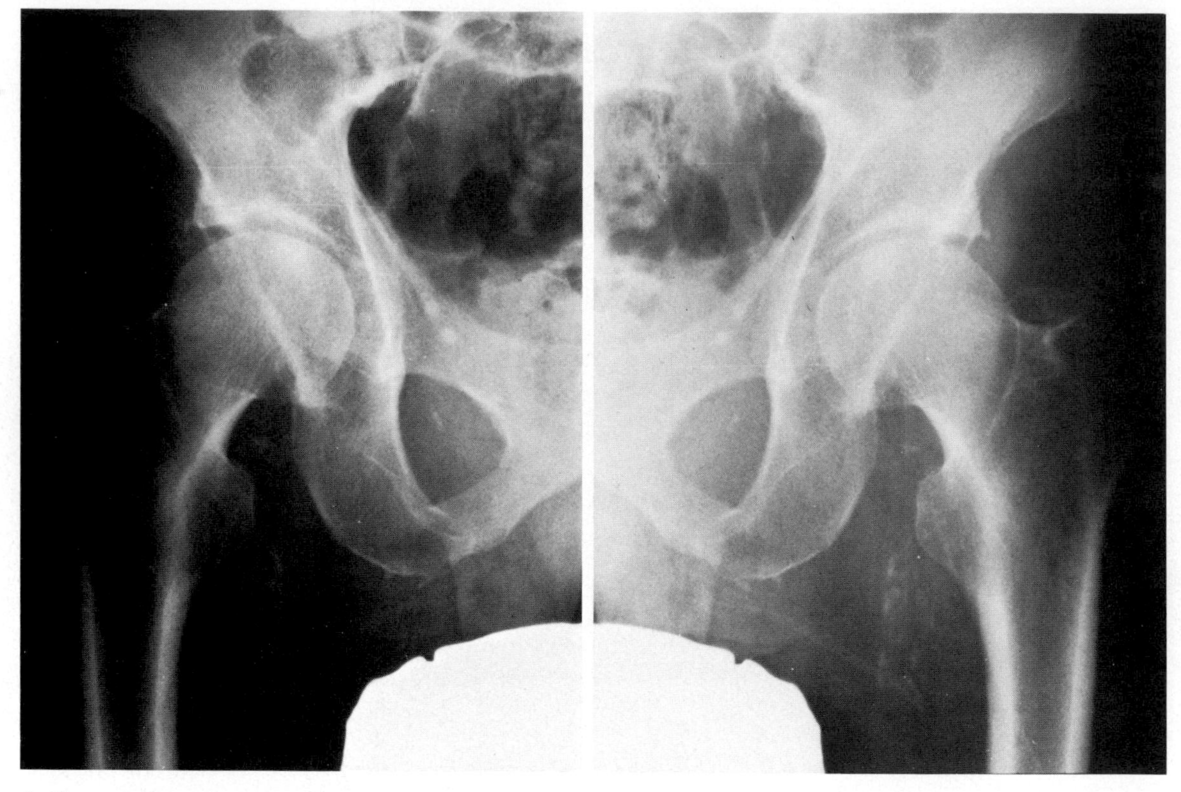

FIG. 9-28. *Both radiographs* (A *and* B) *were made with a grid. However,* B *was made with the X-Omatic cassette inverted. Notice the influence of this cassette as a compensating filter. (From Sweeney RJ: The use of an inverted Kodak X-Omatic cassette as an improvised grid. Radiol Technol 49, No. 3, 1977)*

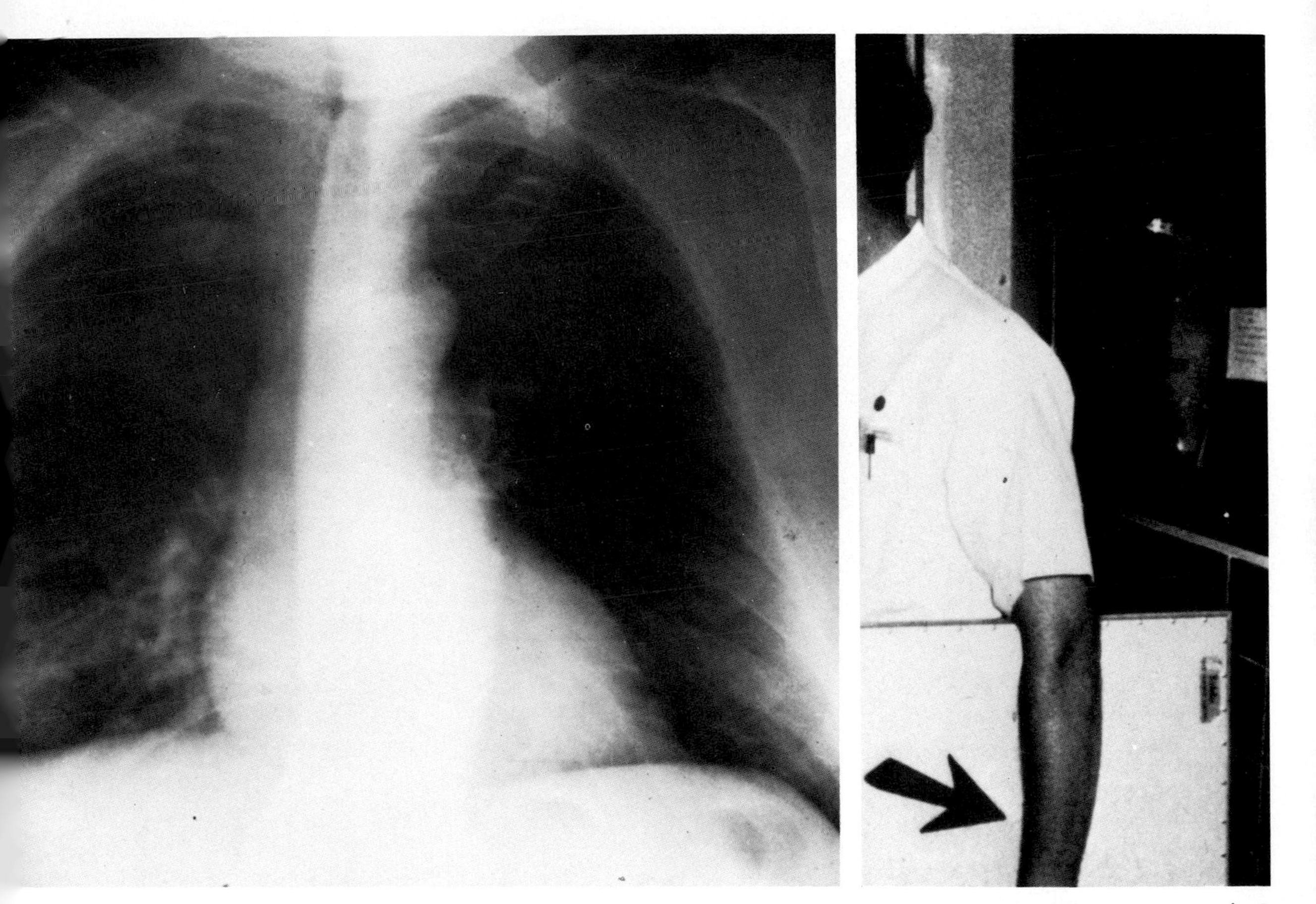

A B

FIG. 9-29. (A) *The silhouette imaged in the central portion of this radiograph is the soft-tissue outline of an arm imaged on the film when the technologist stood in the opening of the control booth while an additional radiograph of the patient's chest was being obtained* (B, arrow). *The cassette that he is holding contained the exposed film. Incidentally, this cassette had been positioned in a transverse direction for the anterior view of the chest. If it had not, the image of the technologist's arm would be across the transverse axis of the thorax. Obviously, a couple of improper conditions are associated with the appearance of this image. First, the technologist should not have stood in the opening of the control booth when the exposure was being made. (The physical size of the control booth, as well as overcrowding, could lead to such problems.) Second, the placement of the wall-mounted cassette-holding device is incorrect: radiation scatter of this intensity should never reach the control area. Although the initial design of the room may have conformed to radiation safety standards, the size of a new cassette-holding device required that it be mounted against a different wall. As a result, its new location allowed a greater intensity of radiation scatter to reach the control booth. This condition was corrected by installing a lead-lined interlock door at the opening of the control booth.*

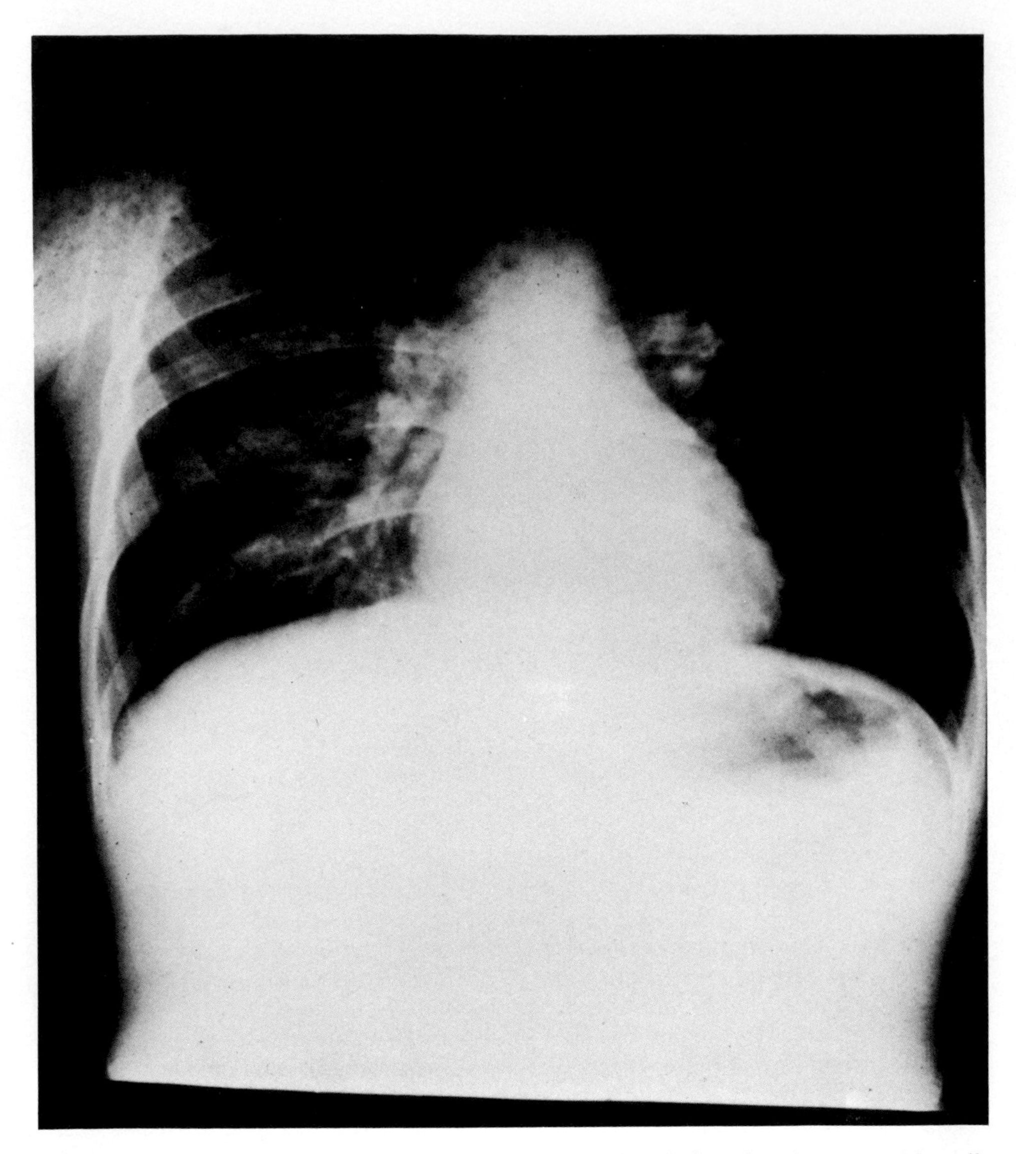

FIG. 9-30. *A portion of this film was exposed to white light when it was accidentally dropped onto the floor. The light leak was discovered to be coming from the bottom of the darkroom door.*

FIG. 9-31. *A cassette used to record computerized tomographic images was loaded with a sheet of single-emulsion film and subjected to light exposure in order to prove that the cassette was responsible for a light leak. Although such damage could not be detected upon visual inspection of the cassette, it is obvious in the radiograph that the cassette is in need of repair or replacement.*

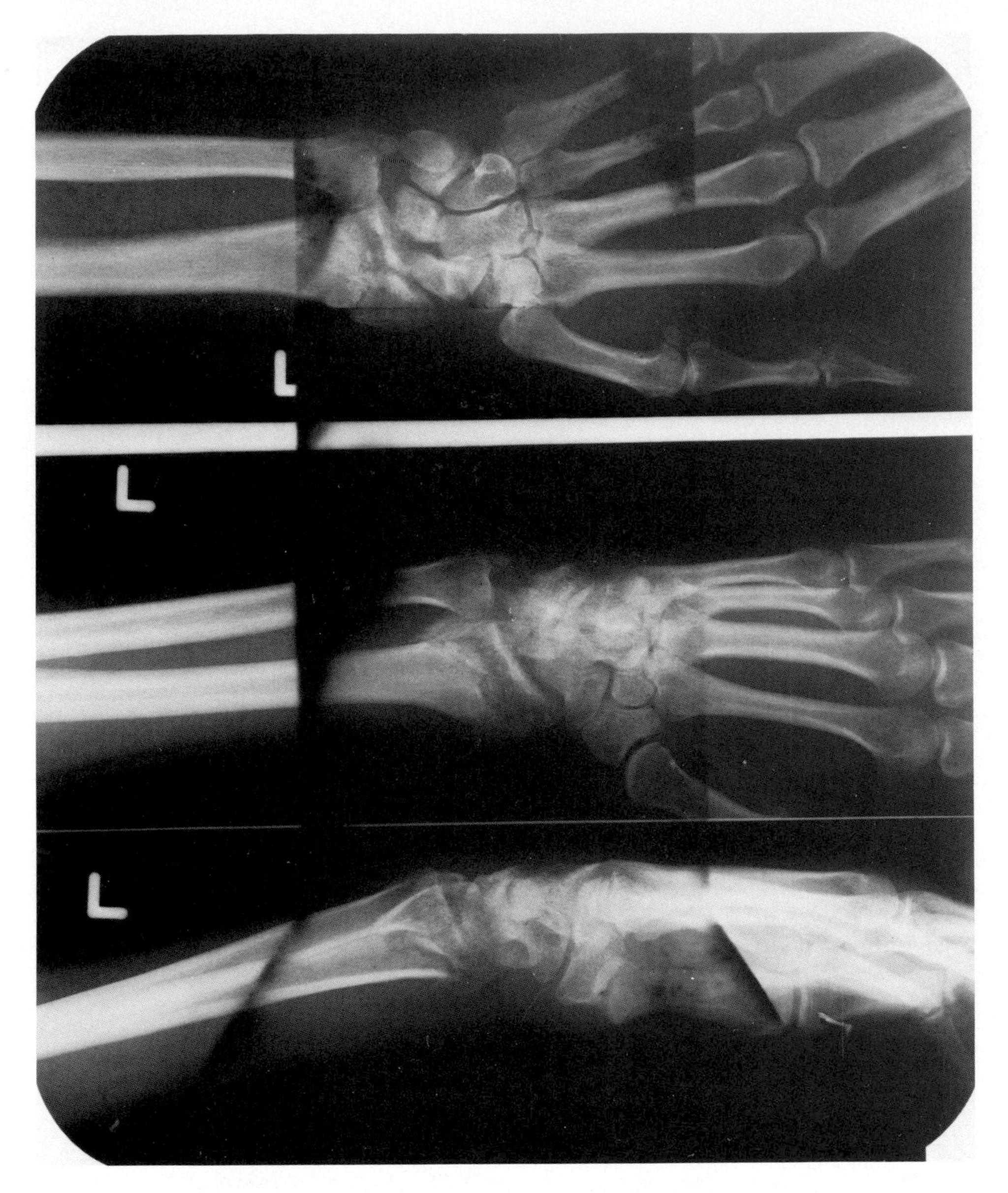

FIG. 9-32. *Failure to load this film properly in the light-tight pouch within a cardboard exposure holder was responsible for this columnlike image. Extraneous light from each side of the holder exposed the midportion of the film, but the flaps on the light-tight pouch protected the lateral borders of the film.*

FIG. 9-33. (A) *The black streaklike pattern outside the edge of the cone field was caused by a light leak due to the lack of a light-opaque substance within the plastic covering of a vacuum-type cassette holder* (B). *(A, courtesy of William A. Conklin, R.T., F.A.S.R.T.)* ▷

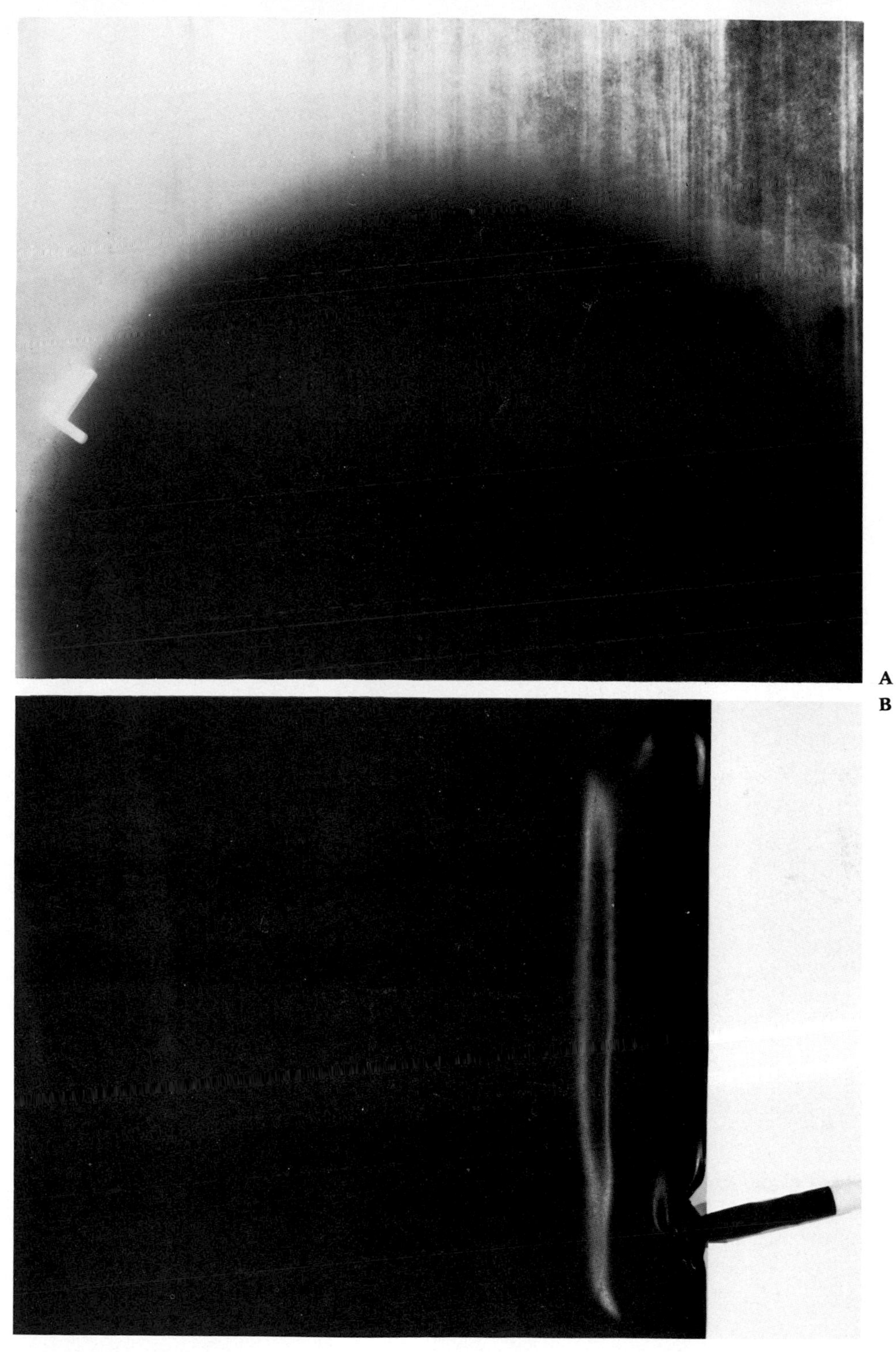

A
B

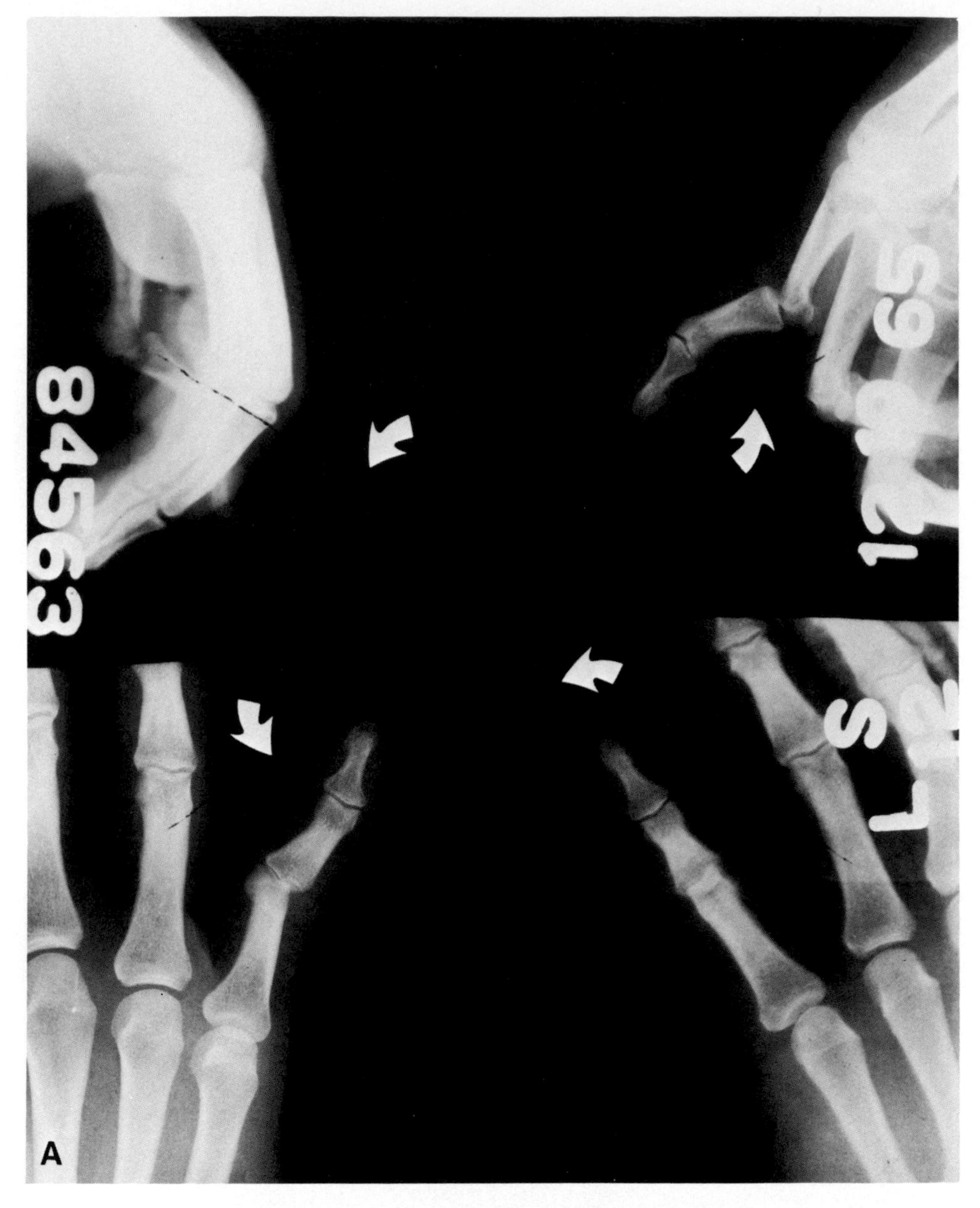

FIG. 9-34. *The lines* (arrows) *in each quadrant of this radiograph resulted from a tear in the lead masking material used for the examination.* (B) *Notice the tear in a radiograph of the lead masking substance. (Courtesy of Ralph Coates, R.T. [R])*

FIG. 9-34. *continued*

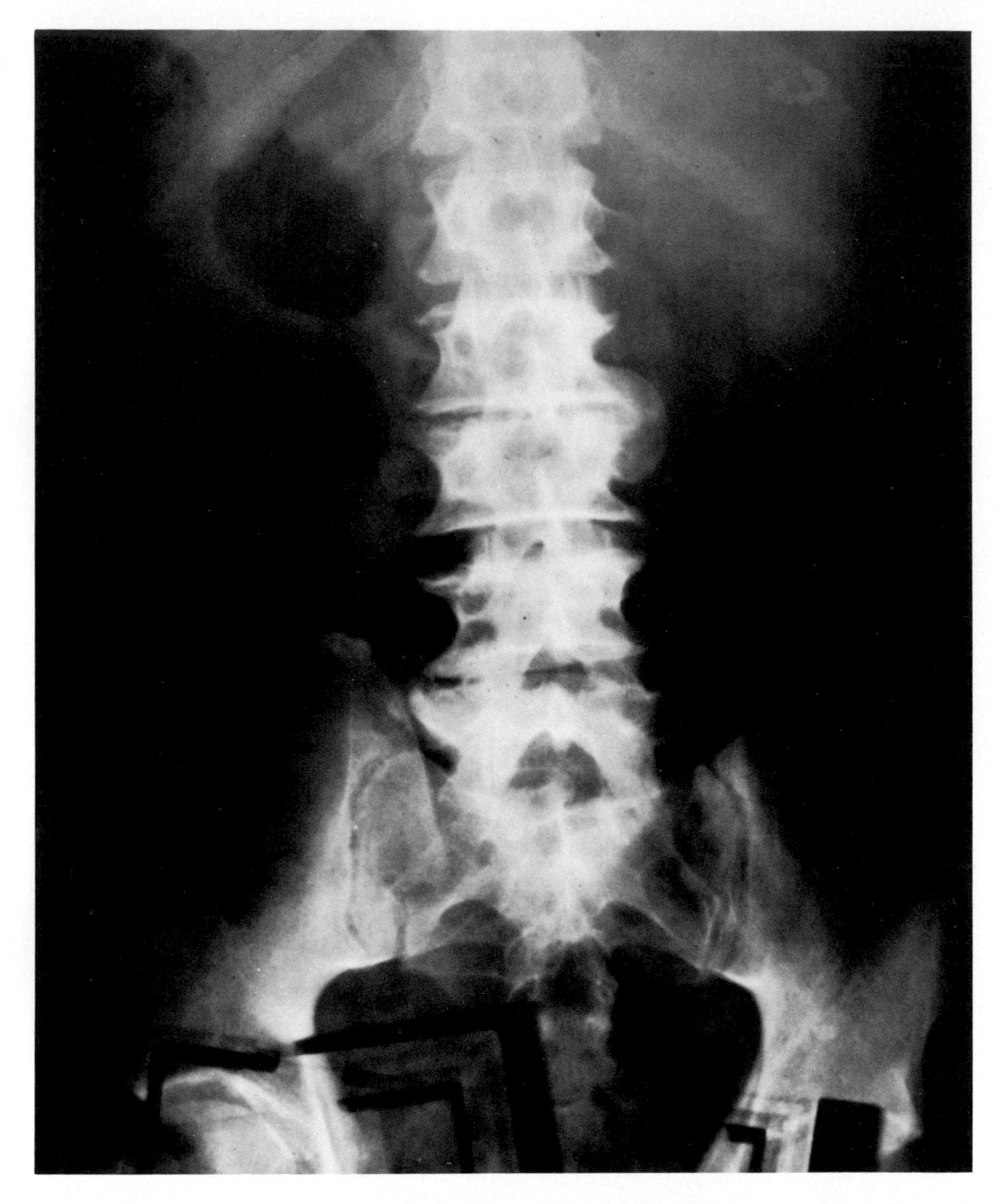

FIG. 9-35. *The multiple image in the lower region of the abdomen in this radiograph occurred during a tabletop examination of the extremities. The technologist performing the examination was unaware that a 14"- × -17" cassette was already in the Bucky tray for radiography of another patient. Consequently, the border image of the cassettes used for the radiographic study of the extremities appeared in the radiograph of the abdomen.*

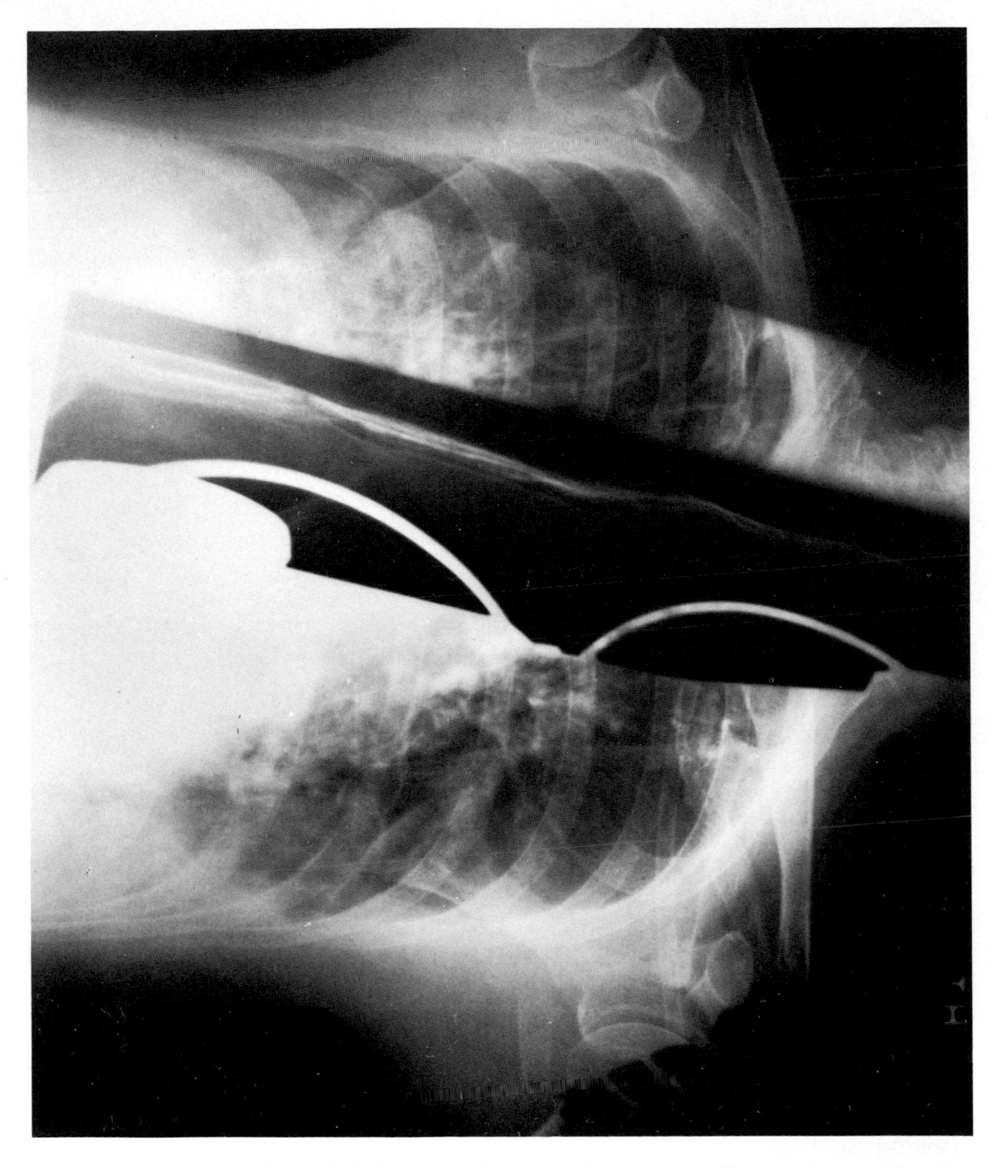

FIG. 9-36. *This radiograph is horizontally oriented in order to illustrate an artifact that occurred when a patient in a four-bed ward had a radiograph of his chest and a lower extremity taken. After obtaining the chest x-ray, the technologist placed the exposed cassette between the mattress and the bed-rail assembly of an adjacent bed. (The adjacent bed was unoccupied, and the cassette was propped between the bed rail and the mattress on the side of the bed.) While in the process of obtaining a lateral projection of the lower extremity with the x-ray tube directed horizontally, the technologist accidentally exposed the cassette containing this film to x-ray. The image is that of the box spring and fabric under the mattress. The irregular boundary of the image and the lack of exposure of portions of the chest were due to the absorption of radiation by the bed-rail assembly.*

A

B

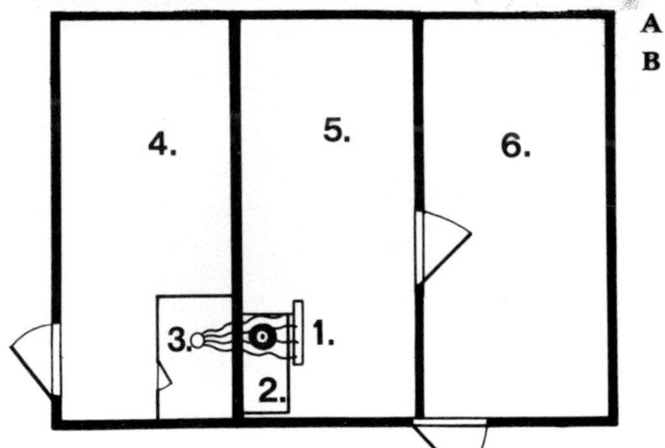

1. Box of film
2. Vacuum film changer
3. Cesium source
4. Hot lab
5. Storage area
6. Special procedure room

FIG. 9-37. (A) *The artifactual image in this radiograph is the outline of a motor in a vacuum-type cassette device. The radiation responsible for the image is from an unshielded cesium source in the closet of a hot lab next door.* B *illustrates the relationship of the radiation source to the object responsible for the artifact in* A. (A, *courtesy of Leo M. Fox)*

References

1. Cullinan JE, Cullinan AM: Illustrated Guide to X-ray Technics, 2nd ed, pp 39–40. Philadelphia, JB Lippincott, 1980
2. National Council on Radiation Protection and Measurements (NCRP). Medical X-ray and Gamma Ray Protection for Energies up to 10 MeV: Equipment Use and Design, NCRP Report No. 33, p 37. Washington, February 1, 1968
3. Riebel FA: The moiré effect in radiography. Am J Roentengol 115, No. 3:641–643, 1972
4. Sweeney RJ: Some factors affecting image clarity and detail perception in the radiograph. Radiol Technol 46:443–451, 1975
5. Sweeney RJ: The use of an inverted Kodak X-Omatic cassette as an improvised grid. Radiol Technol 49, No. 3:257–261, 1977

Suggested Reading

GRIDS AND GRID-RELATED PROBLEMS

Bushong SC: Radiologic Science for Technologists, 2nd ed. St Louis, CV Mosby, 1980

Characteristics and Applications of X-Ray Grids, Revised ed. Cincinnati, Liebel–Flarsheim, 1968

Cullinan JE, Cullinan AM: Illustrated Guide to X-Ray Technics, 2nd ed. Philadelphia, JB Lippincott, 1980

Eastman TR: Radiographic Fundamentals and Technique Guide. St Louis, CV Mosby, 1979

Lamel DA, Arcarese JS, Brown R et al: Correlated Lecture–Laboratory Series in Diagnostic Radiological Physics, Bureau of Radiological Health, Public Health Service, Food and Drug Administration, U.S. Department of Health, Education and Welfare, Washington

Selman J: The Fundamentals of X-ray and Radium Physics, 6th ed. Springfield, IL, Charles C Thomas, 1979

Thompson TT: Cahoon's Formulating X-ray Techniques, 9th ed. Durham, Duke University Press, 1979

10 SAFELIGHT PROBLEMS

Maintaining proper safelight illumination in the darkroom involves the wattage of the bulb, the distance between the lamp and film, and the color of the filter. The technologist should periodically evaluate the safelight as part of a quality-control program, as well as whenever considering using a different type of film. The test involves subjecting a presensitized sheet of film to various times of exposure under safelight illumination (Figs. 10-1 and 10-2). Following processing, the test film should be visually inspected for evidence of exposure. If the safelight can pass a 4-minute test, it is in excellent condition. If the safelight is judged to be unsafe, check that the following are true:

No light is leaking from the safelight.

The wattage of the bulb is not greater than specified.

The safelight filters are suited to the film being used.

The safelight filters are not faded.

In addition, while checking the condition of the safelight, the technologist should spend a few extra minutes in the darkroom in order to detect the location of light leaks. During such an evaluation of our darkrooms, we found that even though the safelights were within acceptable limits, light leaks were discovered around the film processor and doorjambs, and in one instance from an improperly installed heating vent. For a more detailed explanation of some of the problems and methods of testing the level of safelight illumination in the darkroom, refer to the excellent articles listed at the end of this chapter. The figures that follow illustrate a variety of artifacts that can result from safelight problems.

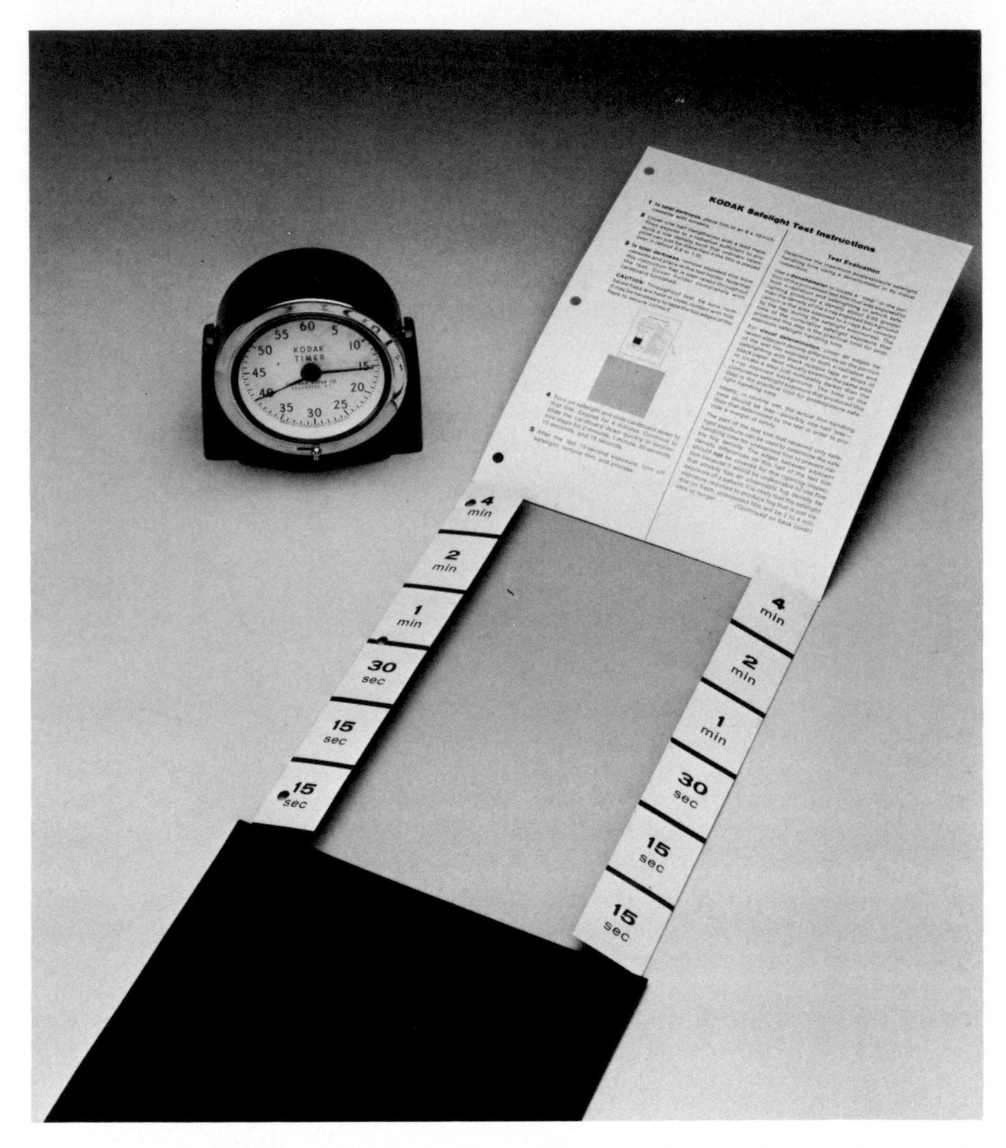

FIG. 10-1. *A safelight test film holder. Notice that a set of instructions is printed on the flap of the holder in order to ensure that the test is conducted accurately. (Courtesy of Eastman Kodak. Copyright © by Eastman Kodak Company, Rochester, NY)*

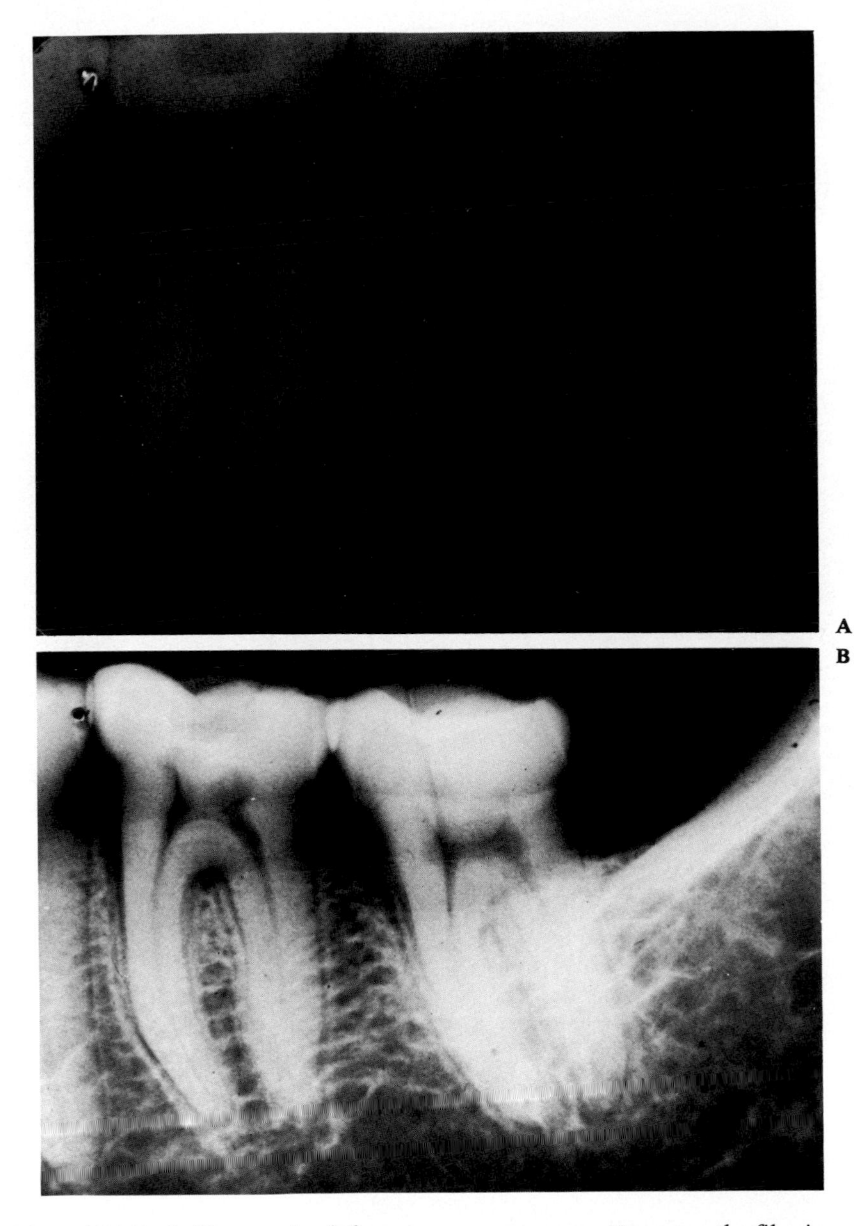

Fig. 10-2. (A *and* B) *Both films received the same x-ray exposure. However, the film in* A *was subjected to prolonged safelight exposure. Even though this film was left under the safelight for an excessive period of time, the high density and lack of contrast of the image demonstrate the importance of determining the actual amount of time a film may be safely handled under normal conditions.*

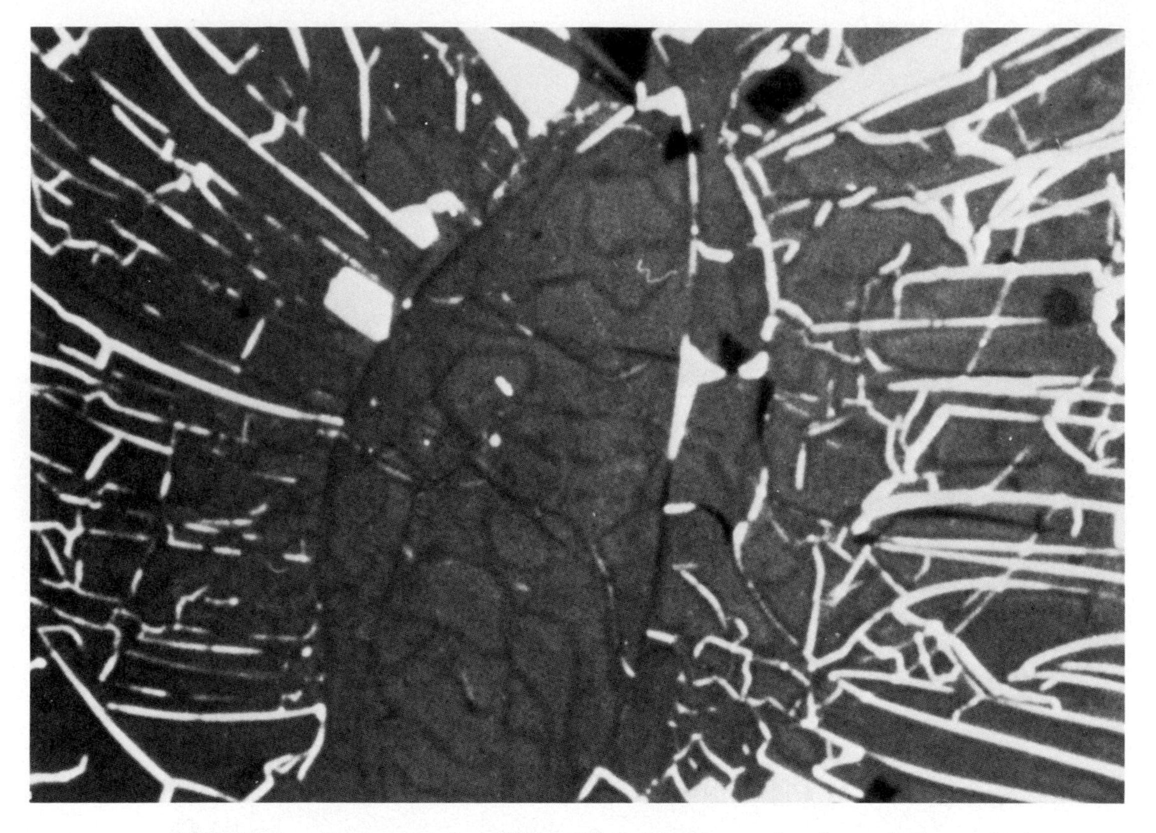

FIG. 10-3. *A segment of a safelight filter that became deteriorated when a high-wattage bulb was used in the fixture. Difficult as it may be to believe, this filter was only recently removed from a safelight in the darkroom of a large metropolitan hospital.*

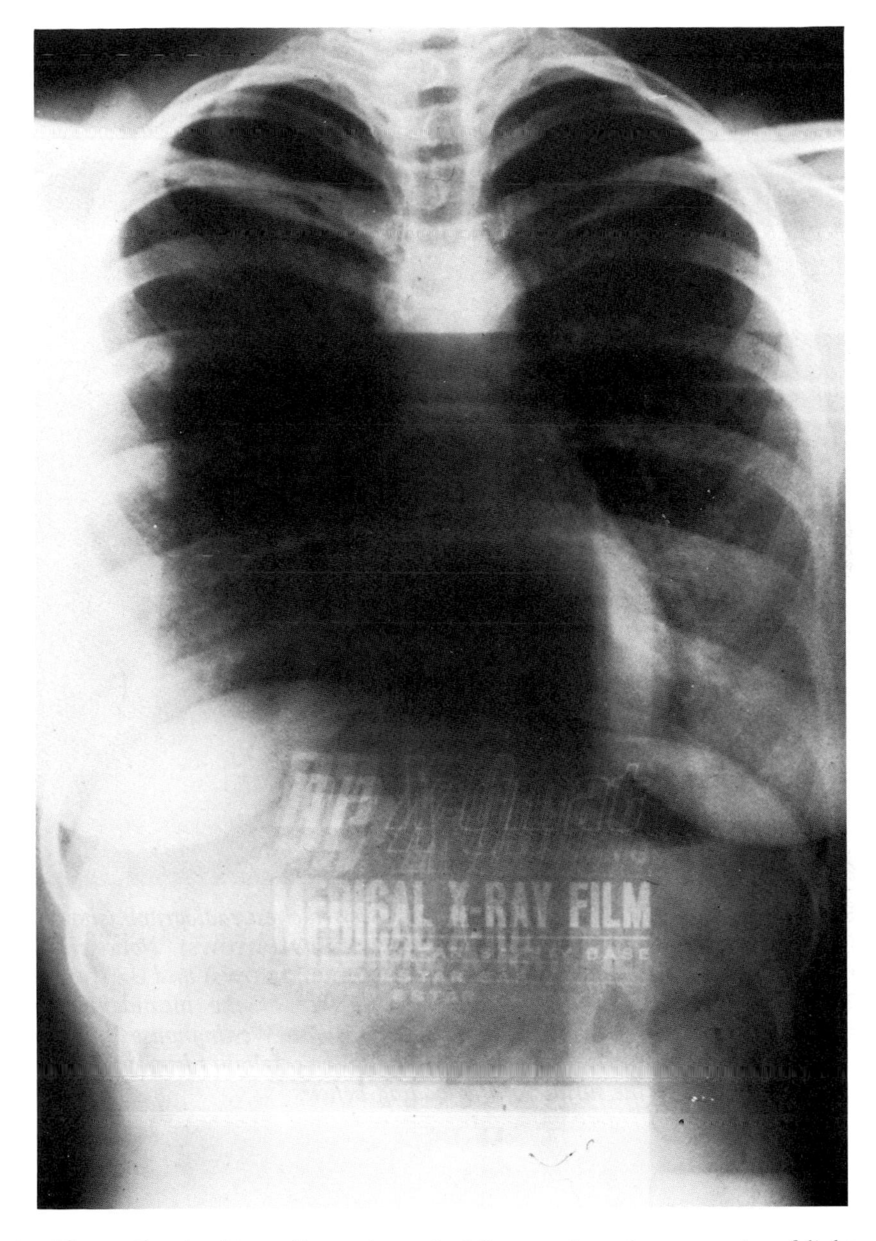

FIG. 10-4. *The artifact in this radiograph resulted from prolonged exposure to safelight illumination. The technologist left this film on top of an empty film carton while he proceeded to load the film bin. Notice that the upper portion of the image was not affected by safelight exposure because it was protected from illumination by an overhanging shelf located above the loading bench. The reason for the ghostlike appearance of the artifact is that the film was moved about several times in relation to the front of the carton.*

Fig. 10-6. (A *and* B) *Sensitometric test films. Notice in* A *the pressurelike marks and increased density along the border* (arrow). *Following evaluation of the films, it was discovered that an improper safelight was responsible for the artifacts in* A. *The retaining cap on the lens had not been secured following replacement of the bulb. This resulted in light fogging during the test exposure. The white areas resembling pressure marks are the shadow images of the fingers of the technologist performing the sensitometric test. The test film shown in* B *indicates that the fogging problem caused by the safelight has been resolved.* (C) *The sensitometer used to perform these tests. Notice that by design the film can be inserted only part way into the unit for exposure. This results in a normally unexposed border where the base-plus-fog reading is made. (C, courtesy of EI du Pont de Nemours & Co, Inc., Wilmington, DE)*

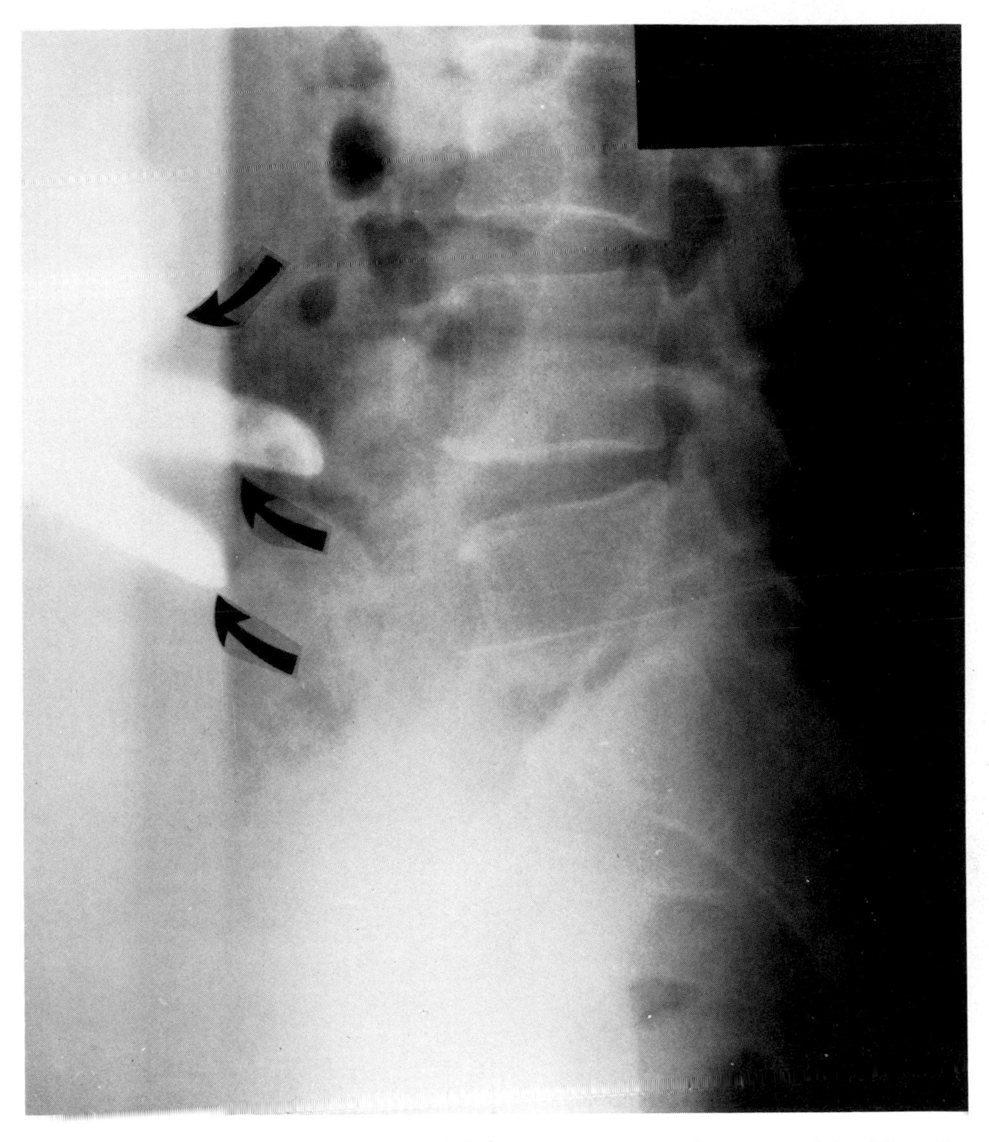

FIG. 10-7. *The effect of improper safelight illumination is demonstrated in this radiograph of the spine. Notice the shadow image of the fingers along the border of the film. This imaging problem occurred when the technologist handled this sheet of orthochromatic film under a safelight not recommended for this type of film.*

Suggested Reading

Hurtgen TP: Safelighting in the automated radiographic darkroom. Med Radiogr Photogr 54, No. 2:32–38, 1978

SAFELIGHT FOG

Gray JE: Light fog on radiographic films: How to measure it properly. Radiology 115:225–227, 1975

Gray JE, Winkler NT, Stears J, Frank ED: Quality Control in Diagnostic Imaging, pp 34–35. Baltimore, University Park Press, 1983

SAFELIGHT TEST

Gray JE, Winkler NT, Stears J, Frank ED: Quality Control in Diagnostic Imaging, pp 42–43. Baltimore, University Park Press, 1983

11 AUTOMATIC PROCESSING ARTIFACTS

The introduction of automatic processing of radiographic film in 1955 was one of the most significant improvements in radiology during that period. Not only did automatic processing provide a method whereby a high volume of films could be processed in a relatively short time, but in addition, automating the system eliminated many of the variables that existed in manual processing that often were responsible for variations in the quality of radiographic studies. Over the years the control of these variables in automatic processing has resulted in greater refinements in film, processing solutions, replenishment techniques, and temperature control. Despite these improvements, however, an automated system can perform at peak efficiency only when the unit is properly maintained according to the recommendations of the manufacturer. Since the basic principles of processing radiographic film are universal, all of the manufacturers of automatic processors design into their products the same basic features. There may also be some special features incorporated into the design, construction, and parts specifications. For example, a radiograph demonstrating the components of a Kodak RP X-Omat Processor, model M8, is illustrated in Figure 11-1.

Artifact Analysis

Aside from those resulting from failing to clean and adjust properly the components responsible for transporting the film through the processor, the most common processing artifacts arise from an improper relationship between the temperature and chemical activity of the processing solutions. In actual field conditions the following procedures may be helpful in identifying certain problems with these conditions.

TRANSPORT PROBLEMS

Unless the technologist is familiar with the appearance of a particular type of artifact associated with automatic processing, the problem is to isolate the cause. This may require a rather detailed analysis of the various components of the processor. In order to simplify this procedure, a test film should be run through each assembly and inspected for scratches, pressure marks, and the like. In this way you can isolate the particular section responsible for the problem.

CHEMICAL PROBLEMS

In order to maintain the specific activity of the processing solutions, it is important that the solutions be properly mixed and that the temperature of the solutions and the rates of replenishment be according to the manufacturer's recommendations. Whenever a question arises that the processing solutions could be oxidized, contaminated, or old, they should be drained, and after the processor is cleaned and adjusted, fresh chemicals should be installed. The technologist should recognize that in order to detect such problems before they become major ones, the most reliable test is to measure the overall performance of the processor by determining the contrast, speed, and base-plus-fog of a sensitometric test film on a daily basis.

As the various processor components are mentioned throughout this section, it may be helpful to refer back to Figs. 11-1 and 11-6, *B*. This will enable you to become familiar with the appearance and location of some of the major components in the automatic processor.

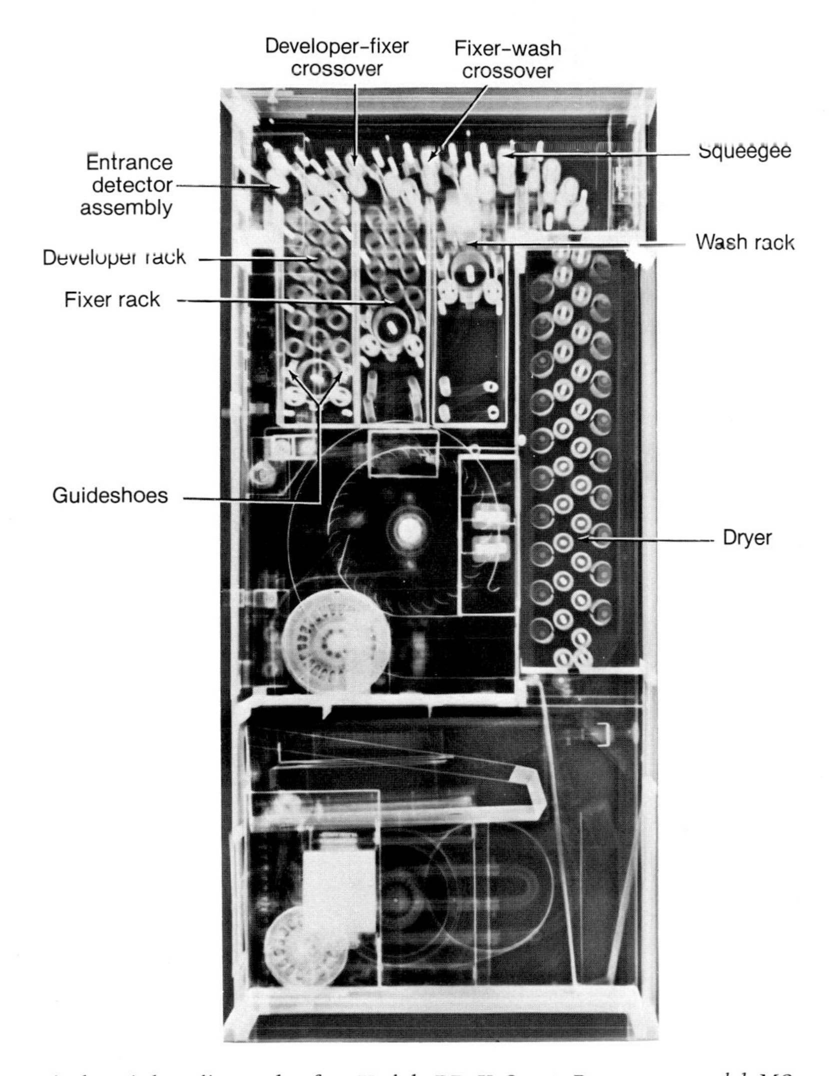

FIG. 11-1 *An industrial radiograph of a Kodak RP X-Omat Processor, model M8. (Courtesy of Eastman Kodak. Copyright © by Eastman Kodak Company, Rochester, NY)*

Classification

An important point to remember in classifying artifacts associated with automatic processing is that if the marks do not run horizontally or vertically in a representative pattern, they probably did not result from a faulty transport mechanism.

The artifact shown in Figure 11-2, however, did occur in the processor when the film jammed in the unit and the technologist ran it through the wash and the dryer section. The reason for the right-angle pattern is that the technologist inserted the film in a different direction on its second trip through the processor. The condition responsible for the artifact was found to be a small paper clip located on the squeegee assembly of the dryer section. The paper clip traveled back and forth like a stylus, scratching the soft emulsion. On many through-the-wall processors, the cover on the processor is used as a writing surface and general catchall. This practice should be avoided, since foreign objects can easily find their way into the processor when the cover is removed for service or cleaning.

Figure 11-3 illustrates an automatic processing artifact that could have been attributed to any of a number of conditions.

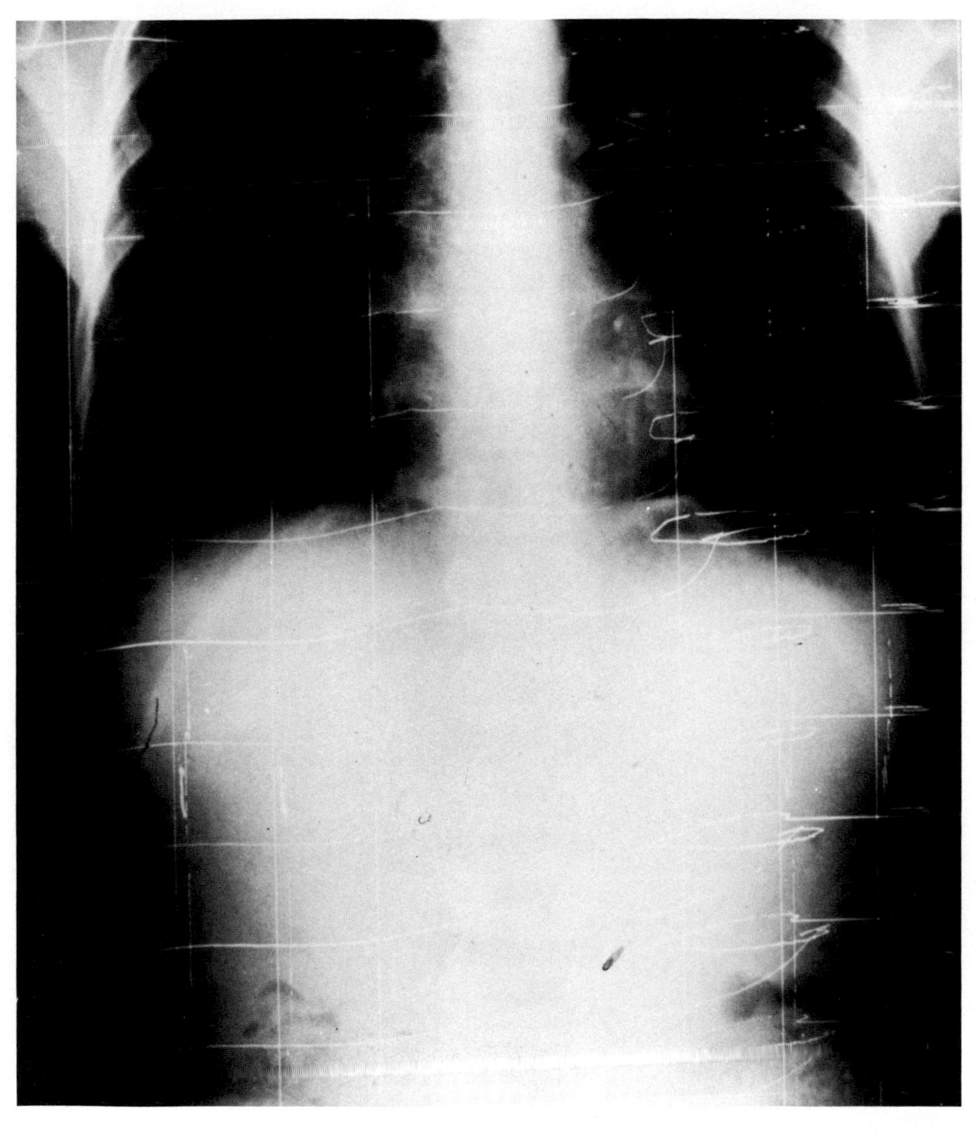

FIG. 11-2. *A grid pattern etched into the emulsion of a radiograph during two runs through the automatic processor.*

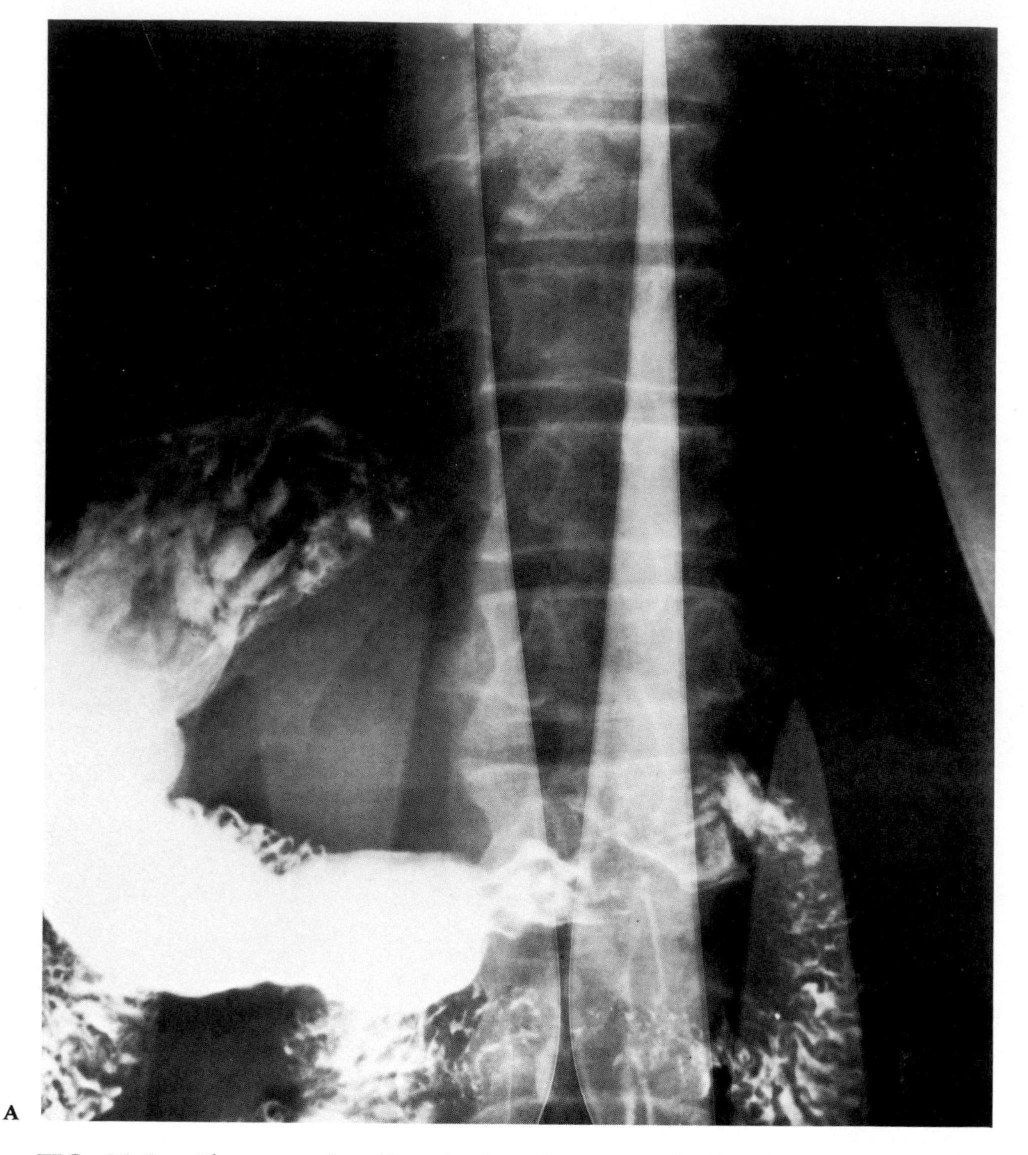

FIG. 11-3. *The unusual artifacts in* A *and* B *occurred when a portion of the film emulsion peeled away from its base during automatic processing. The causes of this condition have been eliminated in the manufacturing of radiographic film today, but the same artifact may also be attributed to other conditions: excessive developer temperature, lack of hardener in the developer and fixer solutions, and misalignment of a roller in the processor.* (A *and* B *courtesy of William A. Conklin, R.T., F.A.S.R.T.)*

FIG. 11-3. *continued*

The Importance of Proper Film Feeding

As a film passes through the detector rollers of the processor, it activates a microswitch assembly that in turn starts the developer and fixer replenishment pumps (Fig. 11-4, *A*). Because these pumps are running while the film is between the detector rollers, it is important to feed the films into the unit in the manner illustrated in Figure 11-4, *B*, in order to avoid overreplenishment. In addition, by establishing a standardized method of running the films, you can determine more easily whether a particular artifact occurred in the processor, based on whether the artifact runs parallel to the direction of film travel. The manner in which the film was inserted in the processor can usually be determined after processing by the identification of short parallel lines that appear near the leading and trailing edges of the film (Fig. 11-4, *C*).

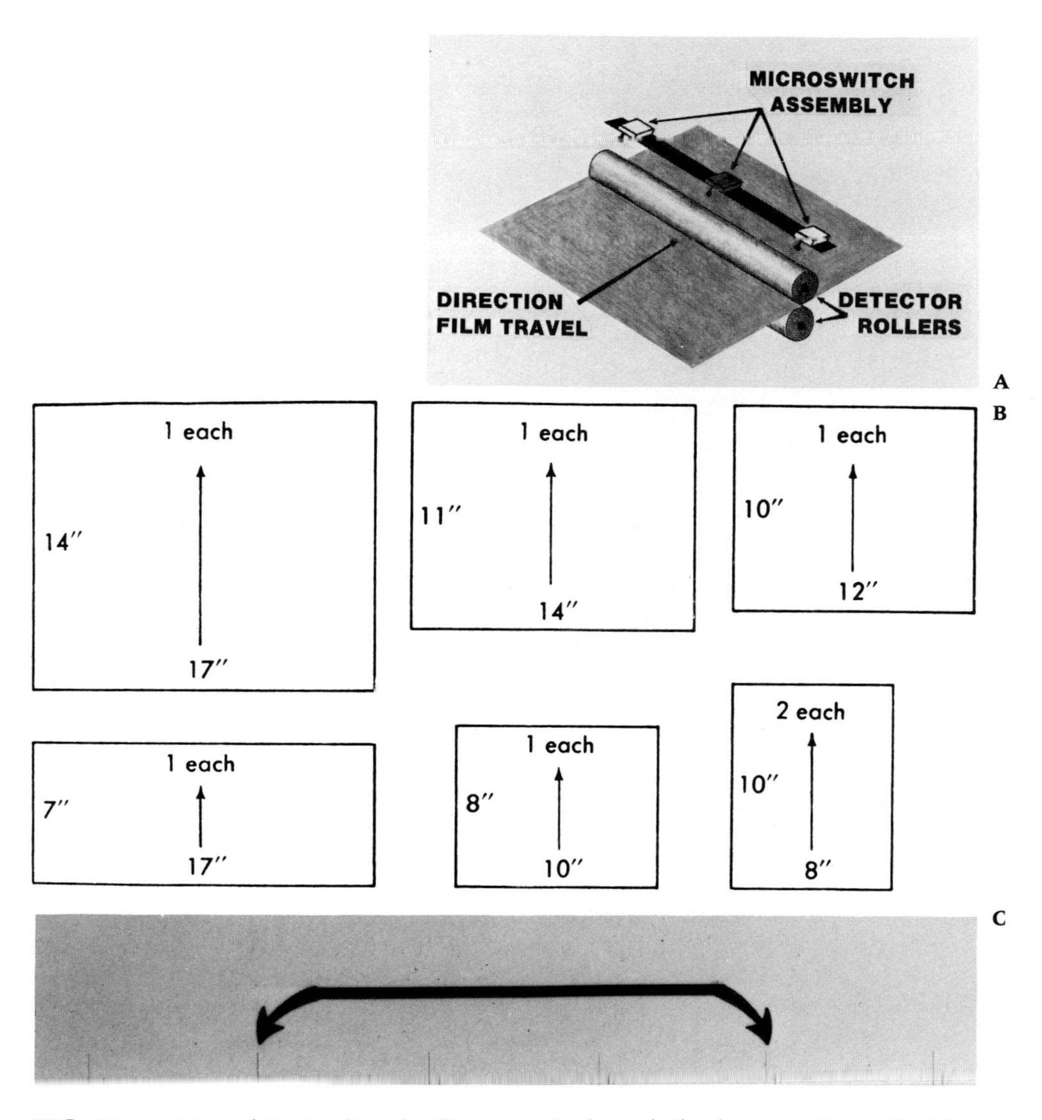

FIG. 11-4. (A *and* B) *Feeding the film properly through the detector rollers will aid in determining whether various artifacts are processor related.* (C) *Parallel lines near the edge of the film indicate its direction of travel through the processor. (*B, *modified from Thompson TT: Cahoon's Formulating X-ray Techniques, 9th ed, p 257. Durham, Duke University Press, 1979.*

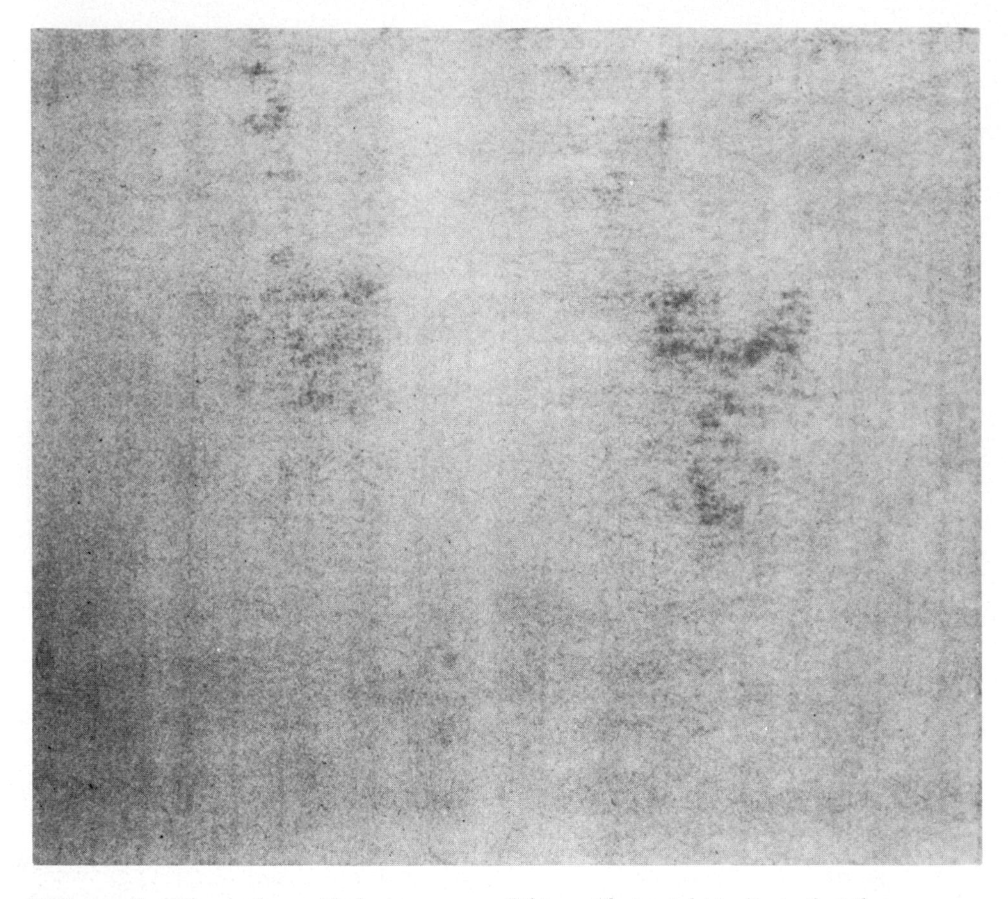

FIG. 11-5. *The dark, mottled appearance of this artifact might indicate that the processor is in need of cleaning. Although such a tracking pattern could be caused by a build-up of chemical by-products on the processor rollers, a lack of hardener in the developer solution was responsible for allowing excessive pressure to be exerted on the soft film emulsion during transit.*

FIG. 11-6. (A) *The uniform pattern in this radiograph is the image of a roller located in the detector crossover assembly of the processor* (B). *The artifact occurred when the trailing edge of the film was exposed to light as the technologist opened the darkroom door before the film entered the processor. (B, From Pub. no. 639166. Rochester, Eastman Kodak. Copyright © by Eastman Kodak Company, Rochester, NY)*

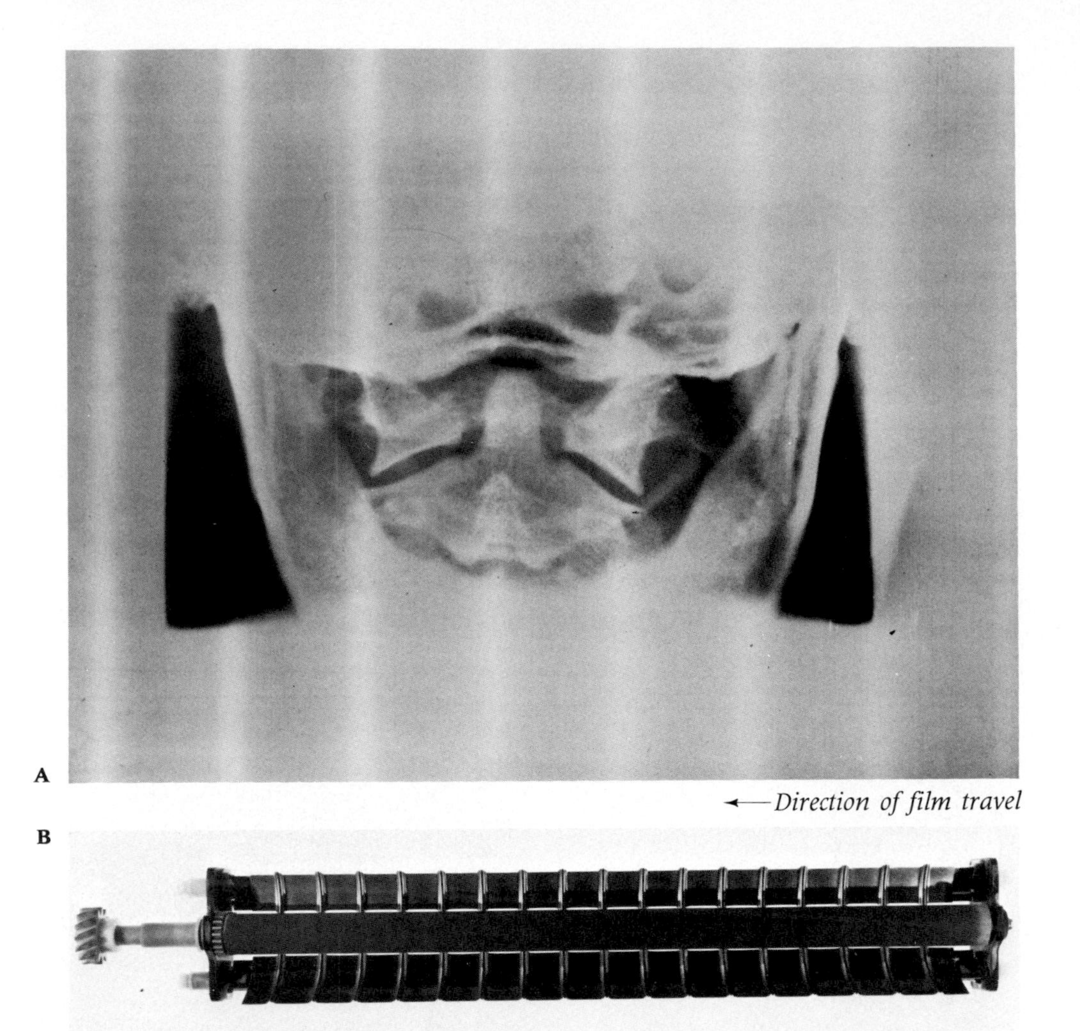

FIG. 11-7. (A) *The artifact in this radiograph occurred when the cover of the processor was accidentally removed while the film was in transit. As a result, white light imaged the pattern of the guideshoe attached to the developer–fixer crossover assembly* (B) *onto the film.*

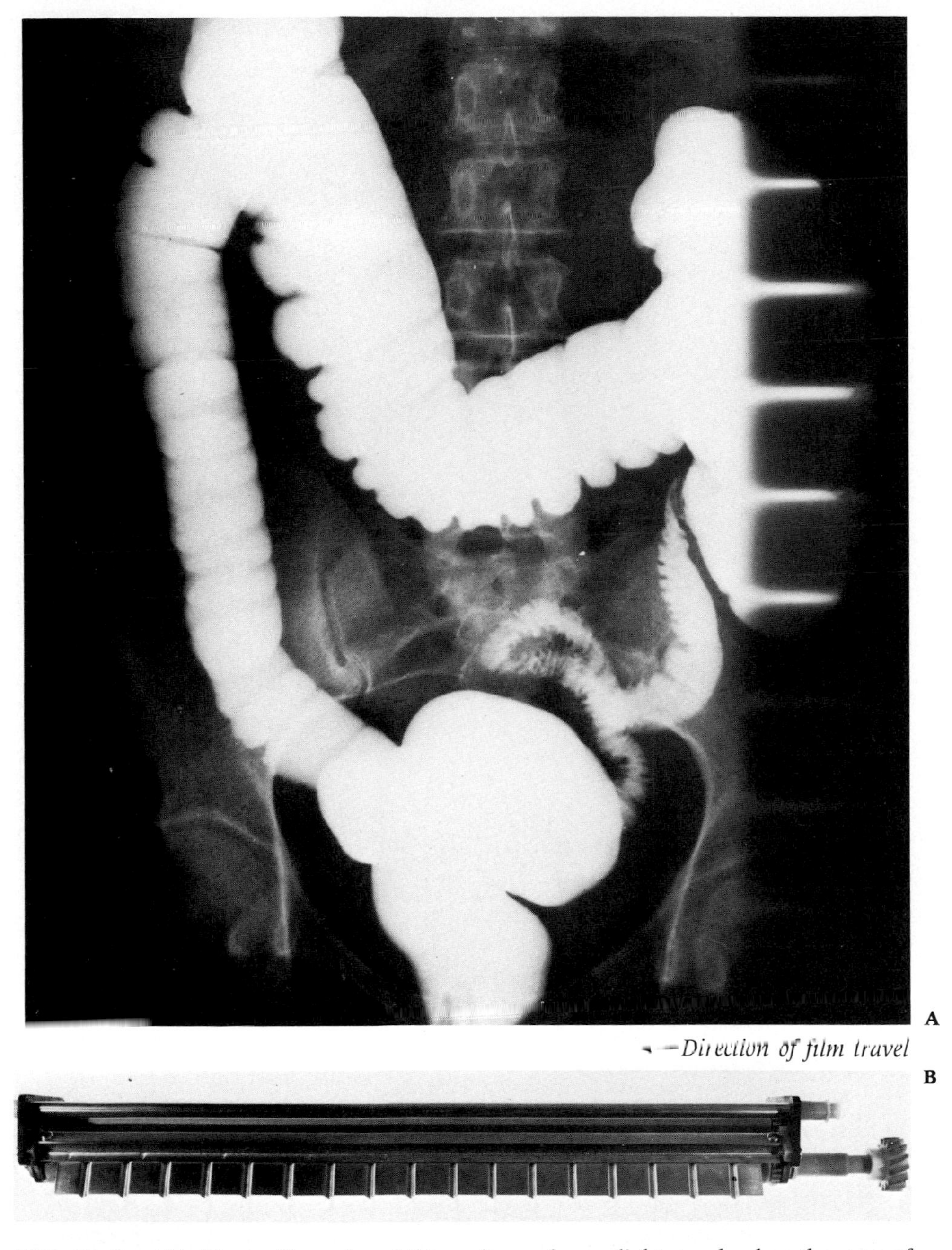

FIG. 11-8. (A) *The trailing edge of this radiograph was light-struck when the cover of the processor was removed while the film was in transit. The image is of a guideshoe located on the detector crossover assembly* (B) *of the processor.*

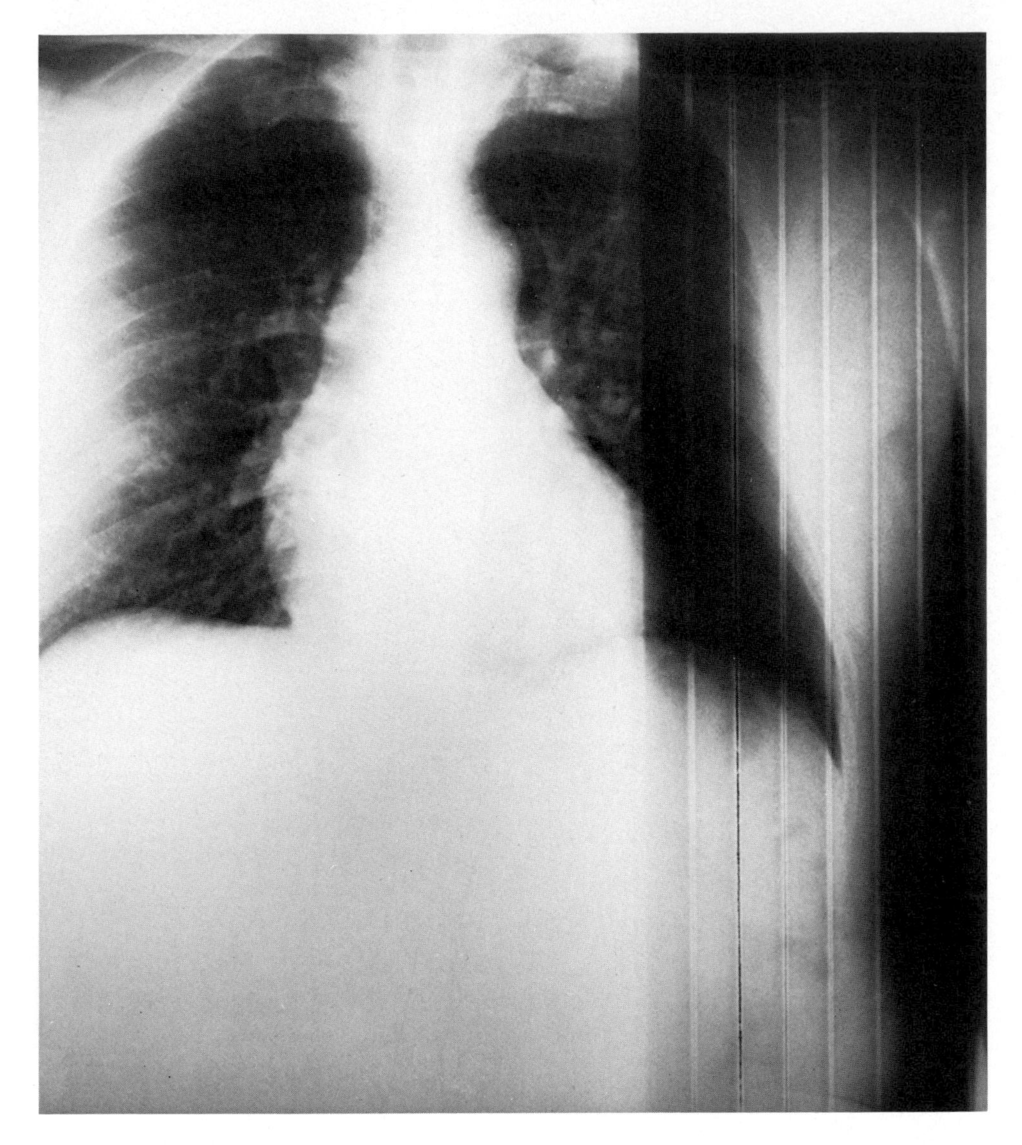

Direction of film travel⟶

FIG. 11-9. *A faulty electrical relay in the automatic processor was responsible for the hesitation of this film while in transit through the entrance-roller assembly. As a result, the rollers exerted pressure on the surface of the film, producing these artifacts.*

FIG. 11-10. *The multiple black lines in* A *and* B *occurred when the films came into contact with solution splashed on the entrance rollers of the processor. Failure to use the splash guard while filling the insert tank can result in such a condition.* (B, *courtesy of Ralph Coates, R.T. [R]*) ▷

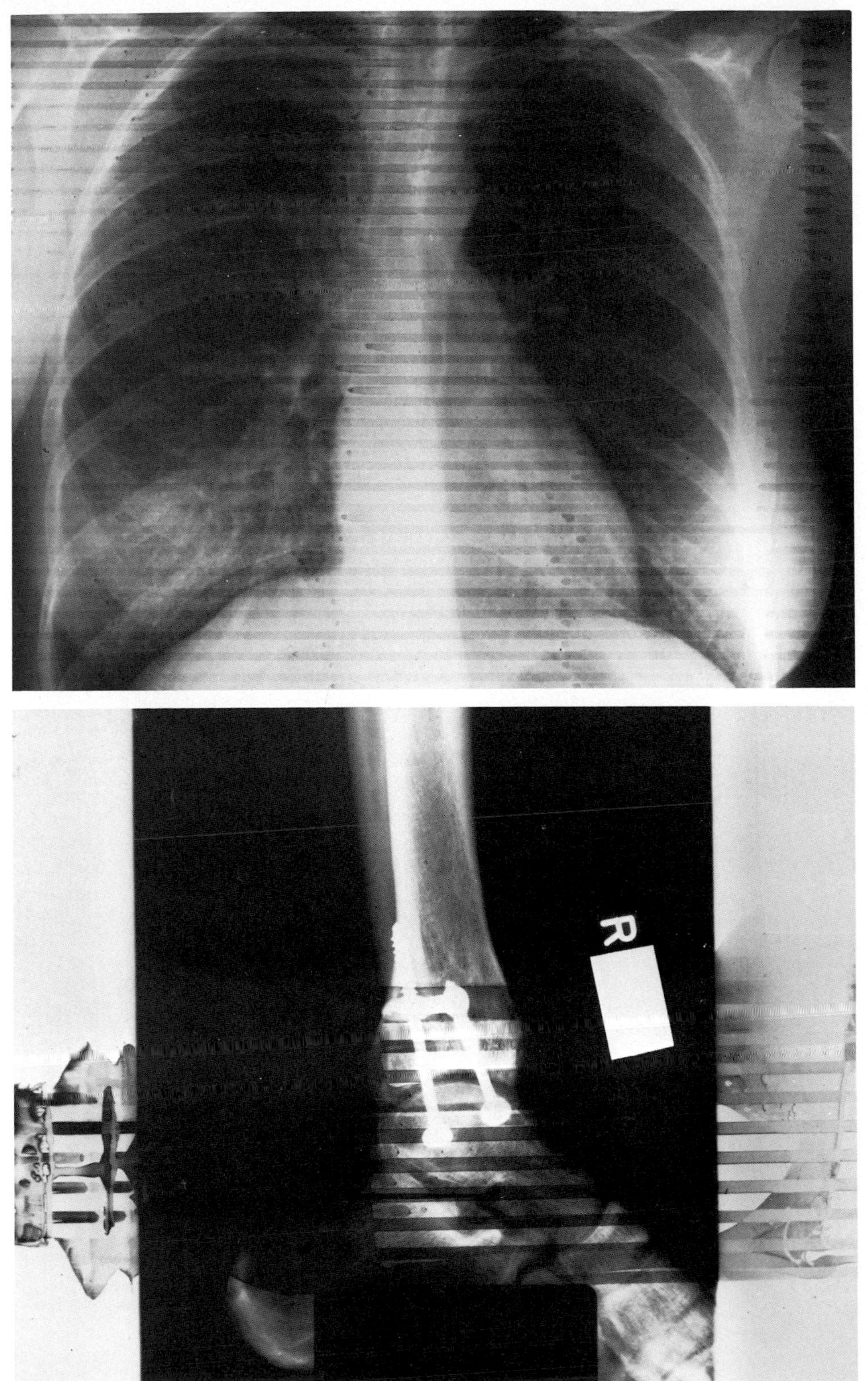

A
B

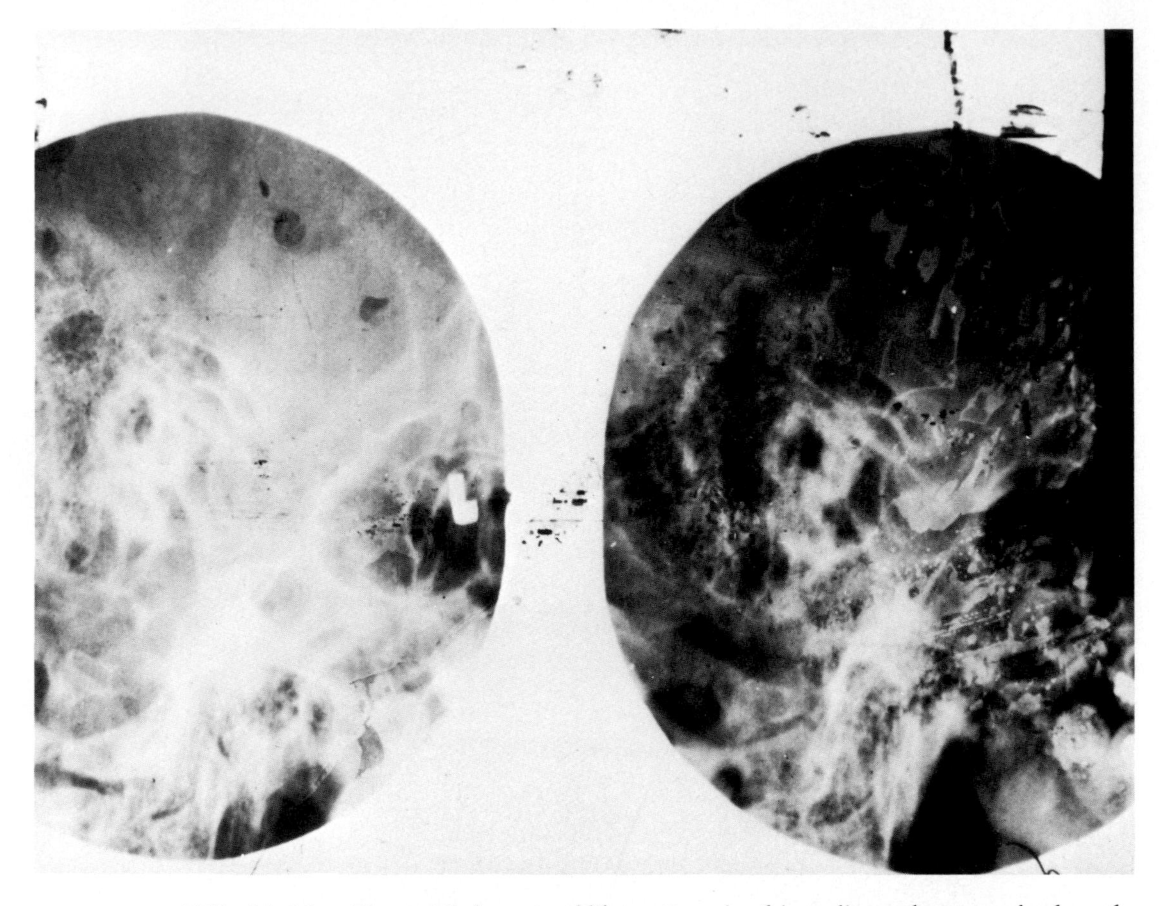

FIG. 11-11. *The mottled or streaklike pattern in this radiograph occurred when the technologist dropped the film while unloading it from the cassette. As a result, the film came into contact with the wet floor. After locating the film, the technologist ran it through the processor.*

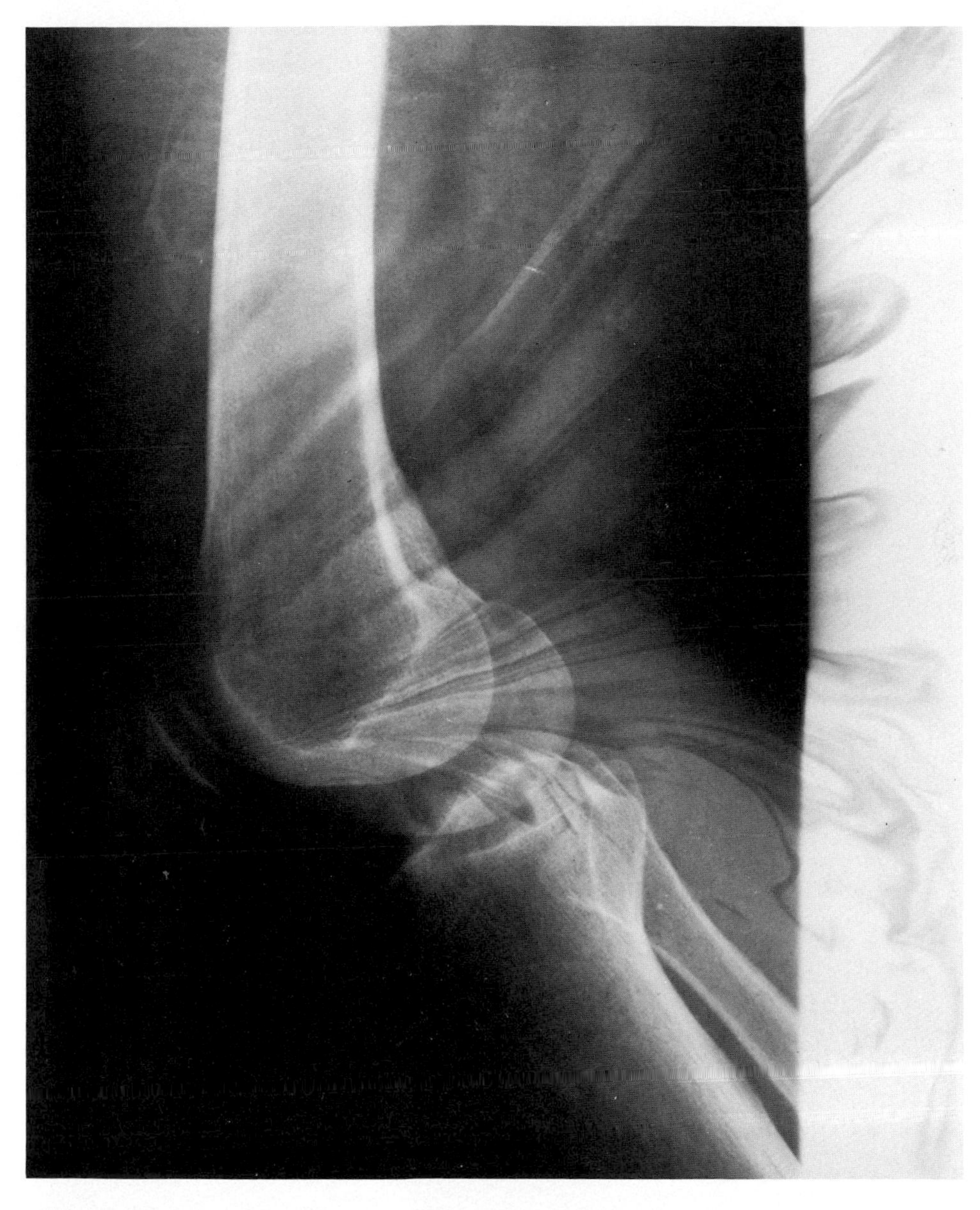

FIG. 11-12. *The rather unusual artifact in this radiograph appears to originate from the edge of the film, and it spreads out in a rather odd pattern. The presence of contaminants floating on the surface of the developer solution and the lack of agitation of the solution were responsible for the appearance of the artifact. (Courtesy of Ralph Coates, R.T. [R]. Analysis based on personal communication: Jon S. Cole and William McKinney, EI du Pont de Nemours & Co., Inc., Clifton, NJ, March 23, 1982)*

FIG. 11-13. *The function of the recirculation pump is to keep the solution in turbulence during processing. Failure of the pump will result in the uneven reaction of the developer chemicals and the appearance of the flamelike pattern seen in this radiograph.*

FIG. 11-14. *The multiple black streaks in this radiograph occurred when the technologist withdrew the film after it had started through the entrance assembly of the processor. The black lines are static discharges caused by friction between the detector rollers and the surface of the film.*

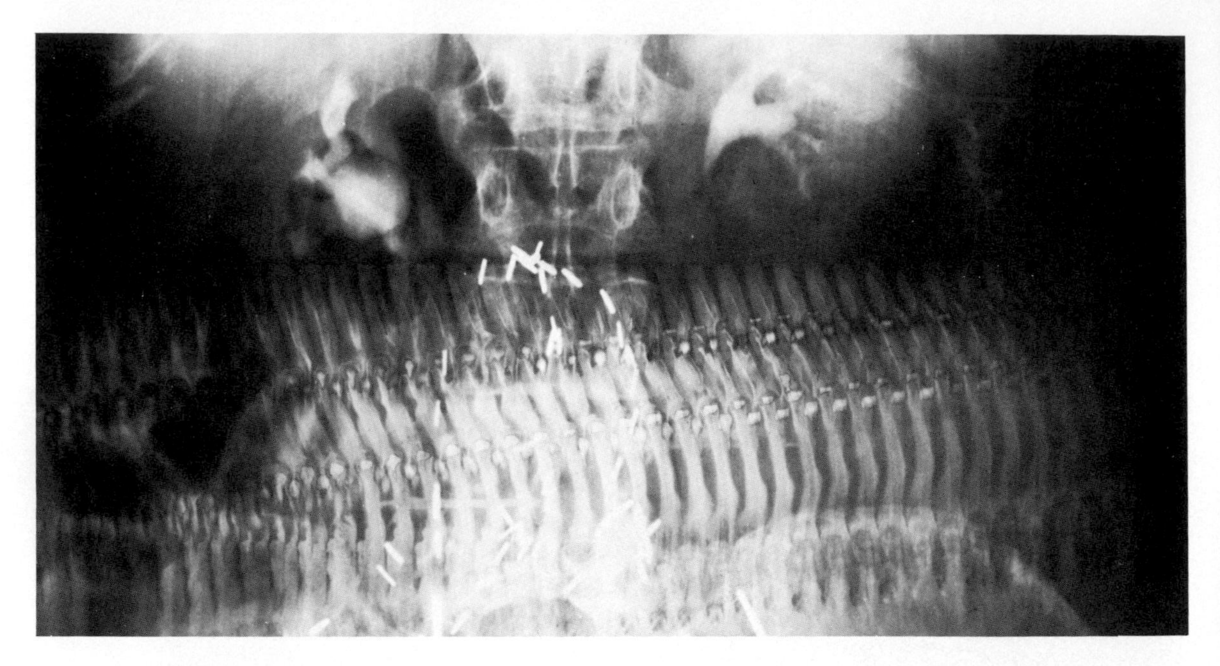

A

FIG. 11-15. (A) *When a particular artifact has not been previously described, the originator probably has the right to name the image as he chooses. Consequently, I have decided to refer to this as a ''boa'' artifact. (Notice the similarity of the artifact to the skeletal configuration of the large boa constrictor in* B.*) This image occurred when the technologist forcibly withdrew the film after it was partway into the processor. As a result, developer solution coated the ribbed roller in the detector crossover assembly (see Fig. 11-6). The rather unusual and repeated pattern of the artifact suggests that the technologist must have had a difficult time in retrieving the film from the entrance rollers.* (A, *courtesy of William A. Conklin, R.T., F.A.S.R.T.)*

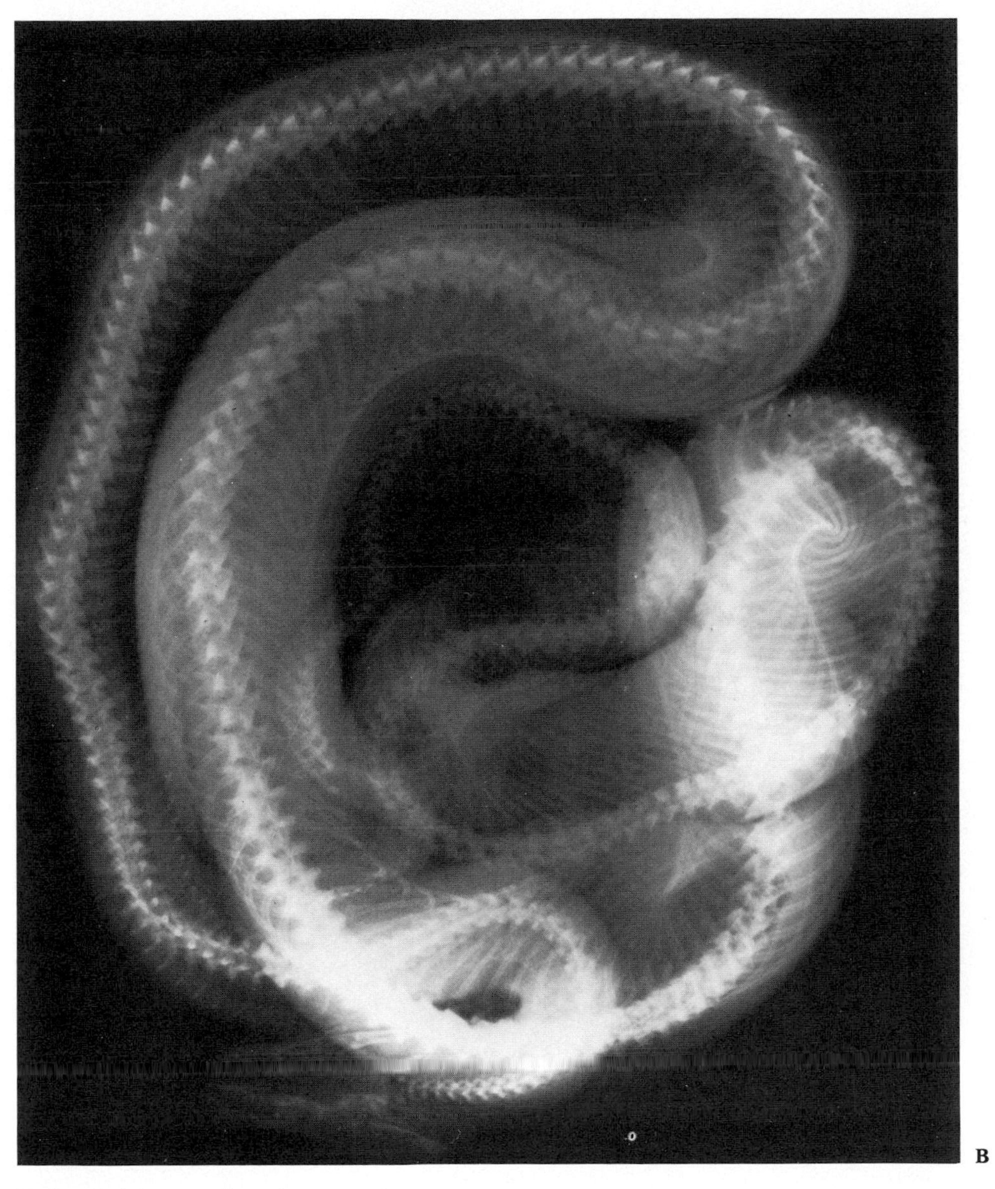

FIG. 11-15. *continued*

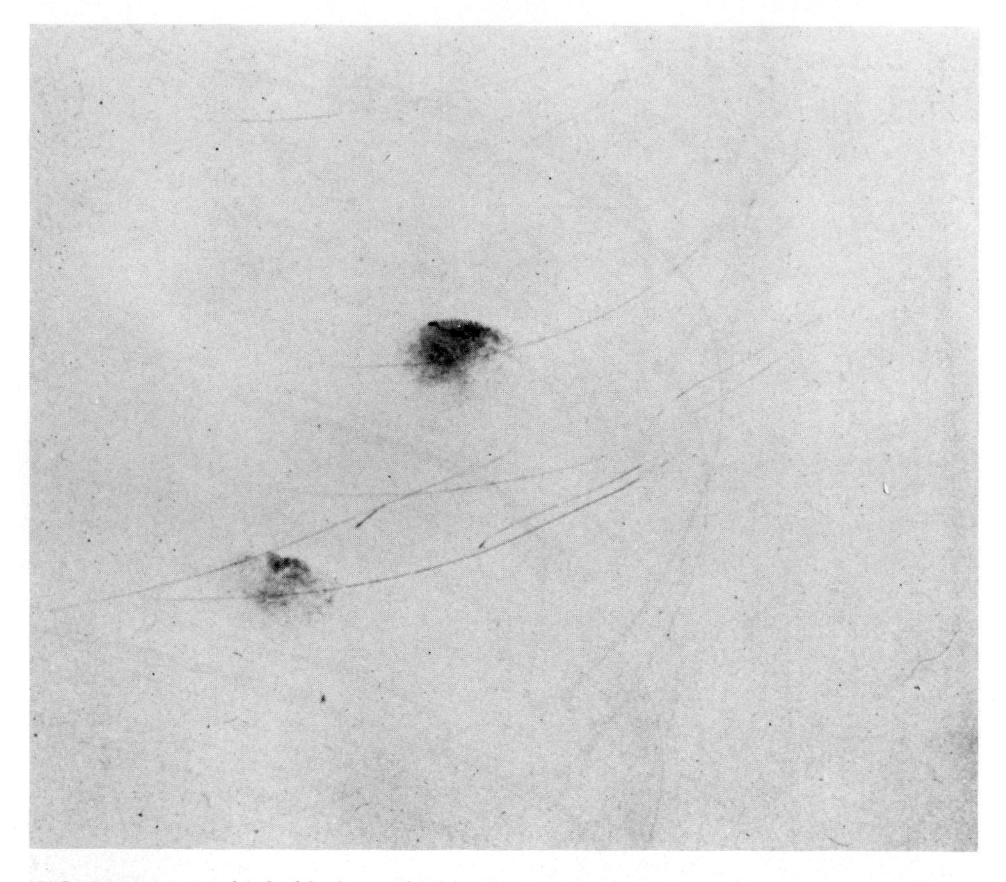

FIG. 11-16. *Multiple black marks like those seen in this radiograph are occasionally blamed on a faulty roller-transport mechanism. However, it is important to remember that if the image of the artifact does not run in a horizontal or vertical pattern, it probably did not occur in the processor. The surface scratches seen in this example occurred when the film was mishandled prior to processing. The film had been slid across the rough surface of the loading bench.*

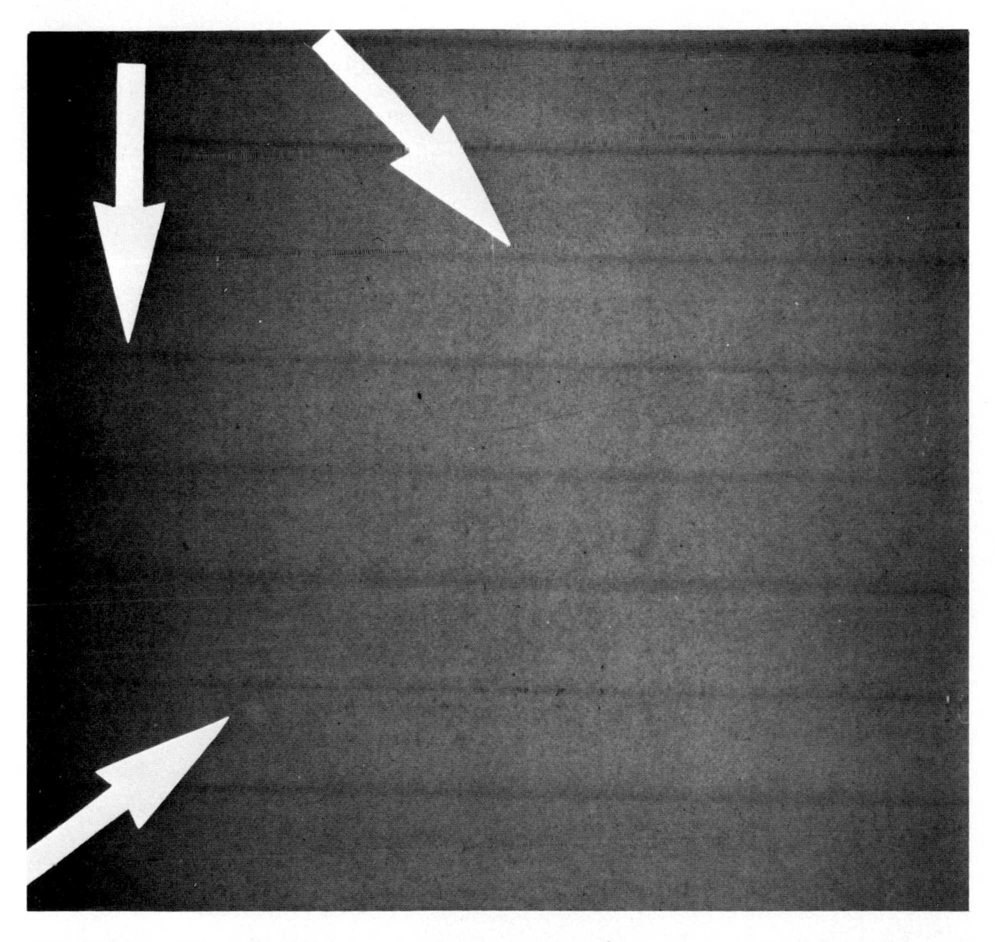

FIG. 11-17. *The dense lines* (arrows) *seen in this radiograph are called hesitation marks and occur in automatic processing when the film pauses in transit. Although the marks are evenly spaced in this illustration, they may also appear in a non-uniform pattern. Such artifacts are usually attributed to the improper adjustment or wear of the components in the transport rack.*

FIG. 11-18. *The thin black line* (arrows) *seen in this radiograph is called a ''pi line.'' It normally appears 3.14 inches (hence the ''pi'') from the leading edge of the film. It is a plus-density line that runs at a right angle to the direction of film travel. The artifact is usually seen following the installation of a new processor, or after the rack assembly has been thoroughly cleaned. The problem usually disappears with repeated use of the processor. (See Artifact Identification Program, Pub. No. 668066, p 73. Rochester, Eastman Kodak, 1977; Thompson TT: Cahoon's Formulating X-ray Techniques, 9th ed, p 273. Durham, Duke University Press, 1979)*

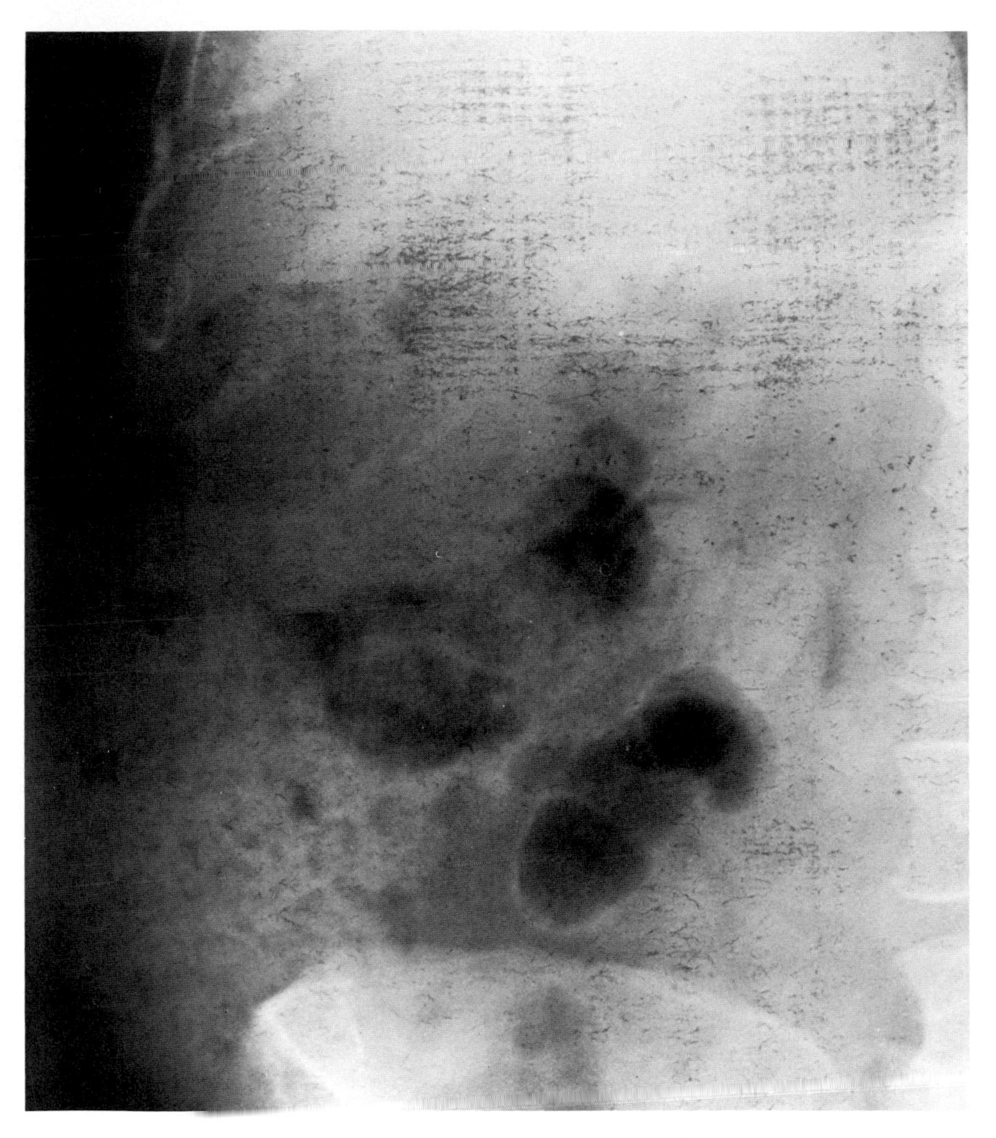

FIG. 11-19. *The tire-tread configuration in this radiograph is in the direction of film travel, and the varied intensity of the pattern suggests the presence of fixer on a warped roller in the developer–fixer crossover assembly. Thorough cleaning or replacement of this component should eliminate the problem. (Courtesy of Ralph Coates, R.T. [R])*

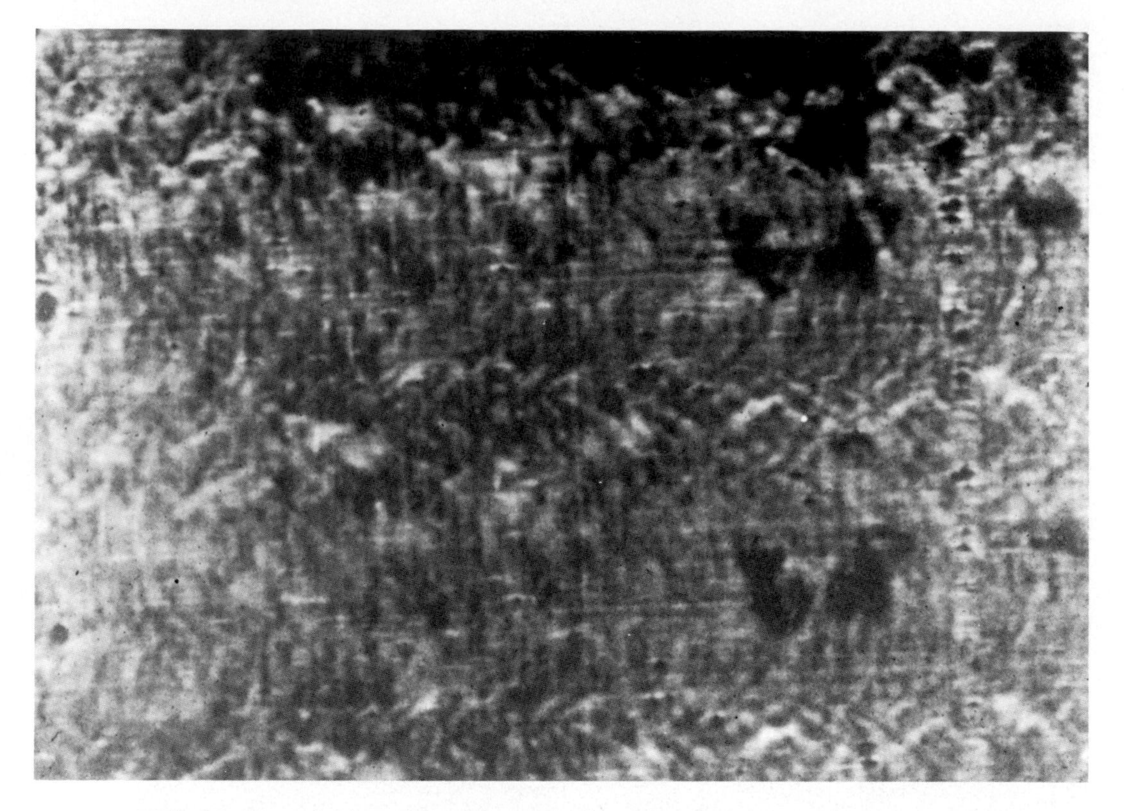

FIG. 11-20. *If this radiograph were printed in color, it would appear as if gold paint coated the film. This image is actually silver and is often seen when maintenance of the rack assembly and the automatic processor as a whole is totally ignored.*

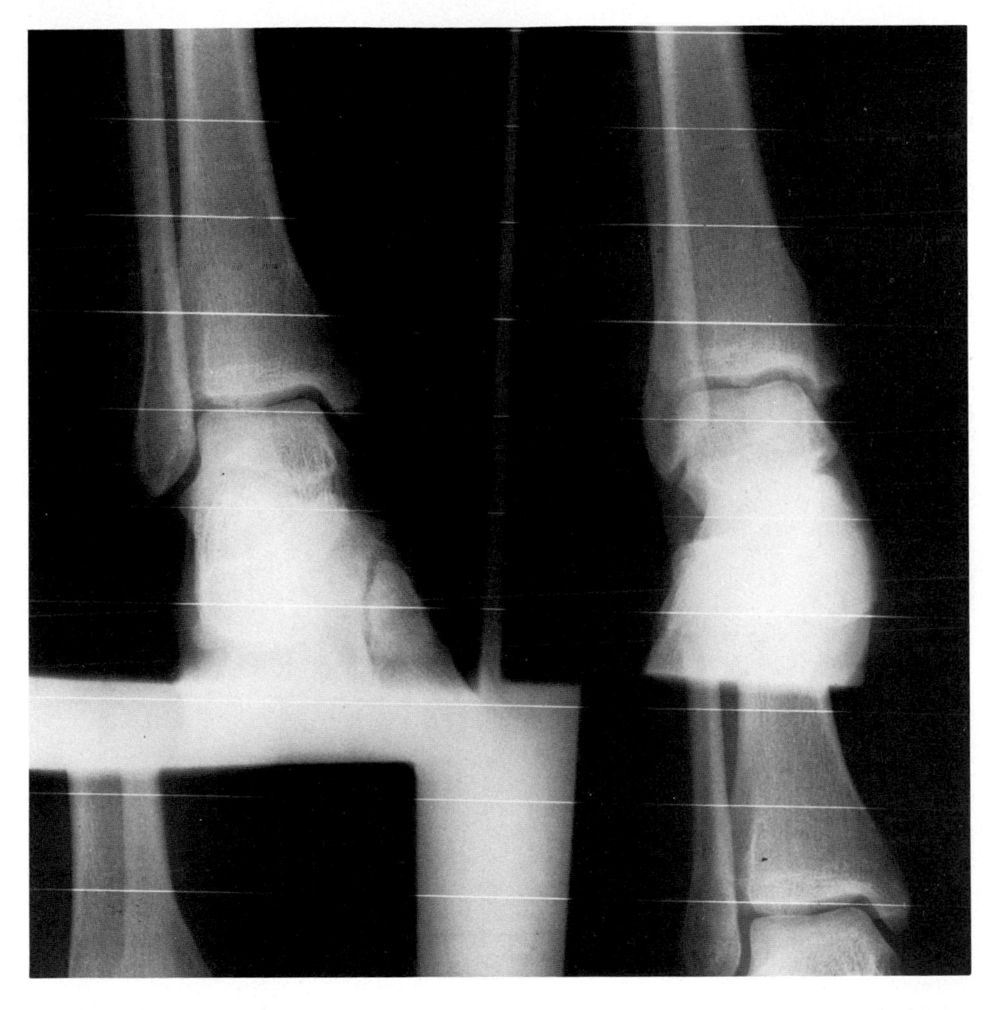

FIG. 11-21. *This artifact was caused by the misalignment of a guideshoe in the turnaround assembly of the developer rack. The evenly spaced scratches are called shoe marks and can be eliminated by adjusting the relationship of the guideshoe to the rollers in the developer rack.*

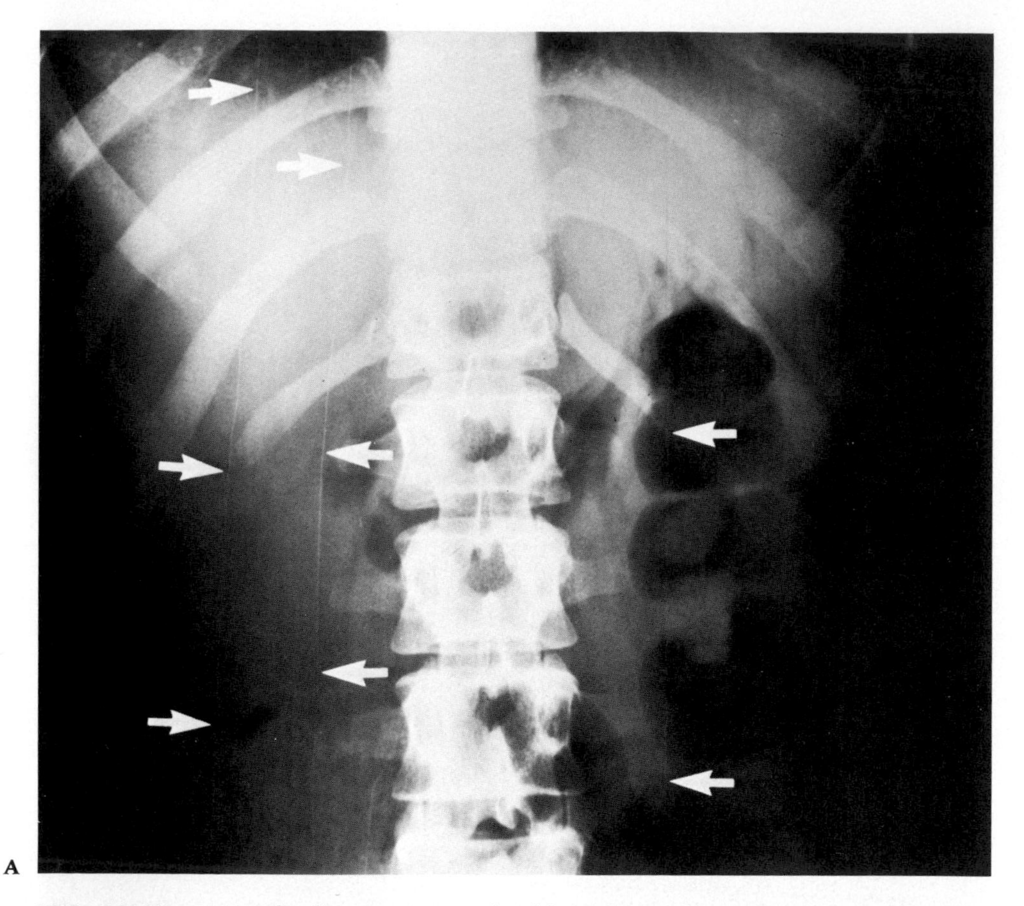

FIG. 11-22. (A) *The lines* (arrows) *in this radiograph suggest that the film was scratched during processing. However, notice that they are not running in the normal direction of film travel and that they are very irregular. This artifact occurred when barium became absorbed within the laminated surface of a wood-grained tabletop, a portion of which is seen in an enlarged radiograph in* B. *Notice in* B *the appearance of the field detectors, as well as the thin line of residual barium within the surface of the table. This streak of barium is the same image as seen just to the right of the spine in* A.

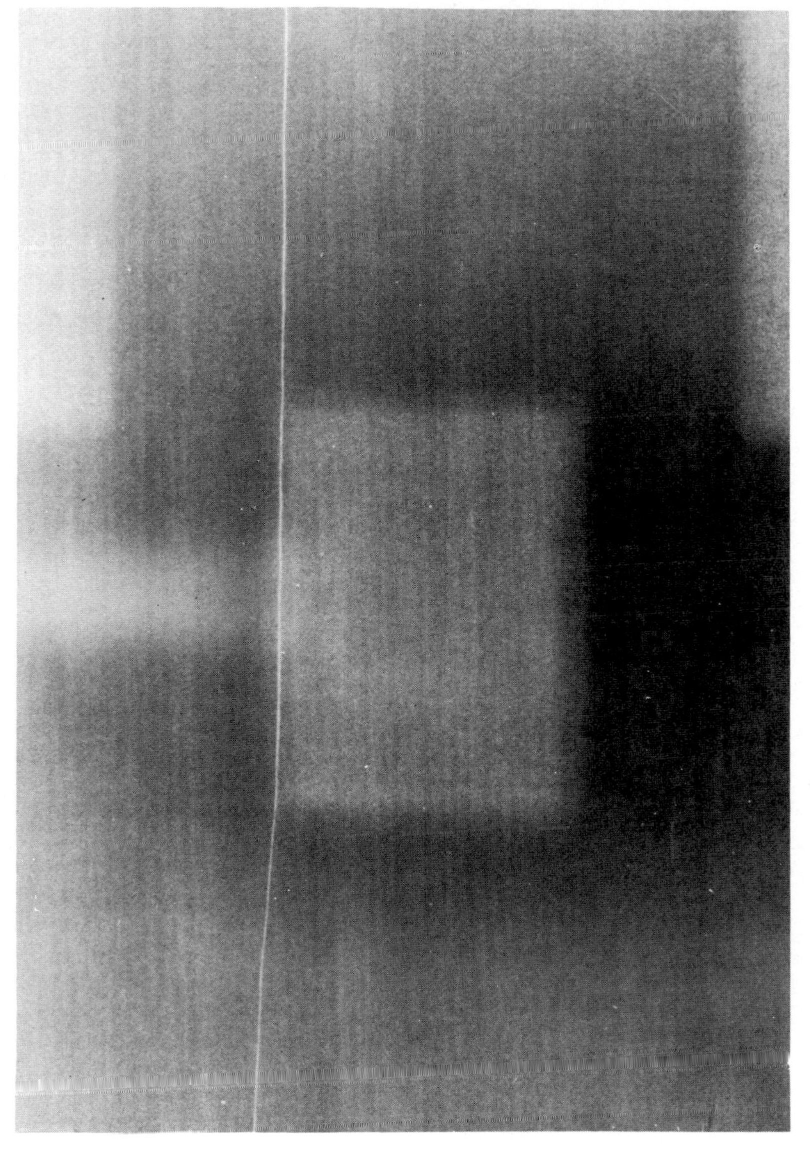

B

FIG. 11-22. *continued*

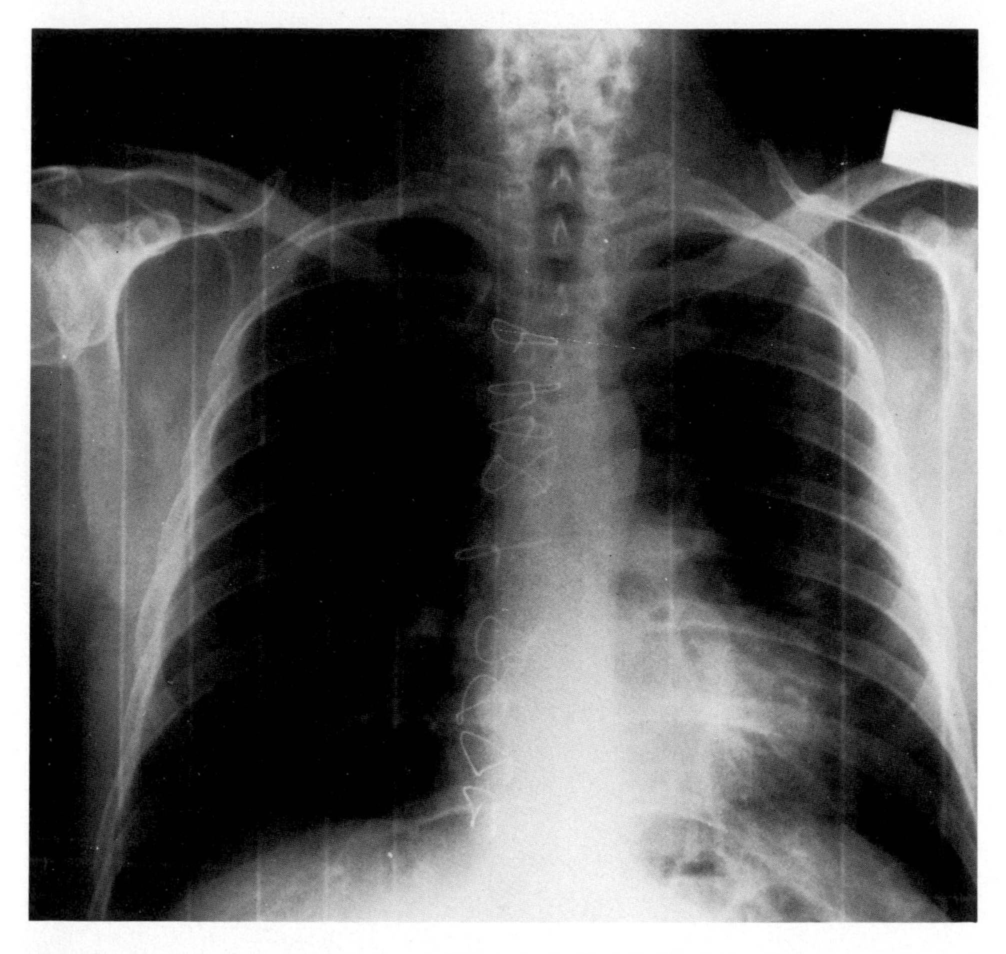

FIG. 11-23. *At first glance, one might conclude that the lines in this chest radiograph were caused by the misalignment of a guideshoe. However, notice that the pattern is not in the normal direction of film travel and that it is unevenly spaced. This image resulted from positioning the cassette on the underside of a thermal mattress used to control the patient's body temperature.*

FIG. 11-24. (A) *A roller located near the turnaround assembly of the developer rack. In some of the earlier processors a heating element was located at the bottom of the insert tank in order to regulate the temperature of the developer solution. This roller was blistered by such an element when the processor was activated for use without solution in the insert tank. Intermittent transport problems, as well as the appearance of an image of plus density* (B), *led to the eventual discovery of the damaged roller.*

FIG. 11-25. *Not only do accumulations of sediments of any kind on the rollers affect the transport of the film through the processor, but if these particles come into contact with the film, characteristic artifacts will result. The artifacts seen in this radiograph resulted from foreign particles that were deposited on the film as it passed through the entrance assembly of the automatic processor. These artifacts are referred to as ''black comets'' and can be eliminated by cleaning the entrance-detector assembly. (See Artifact Identification Program, Pub. No. 668066, p 109. Rochester, Eastman Kodak, 1977)*

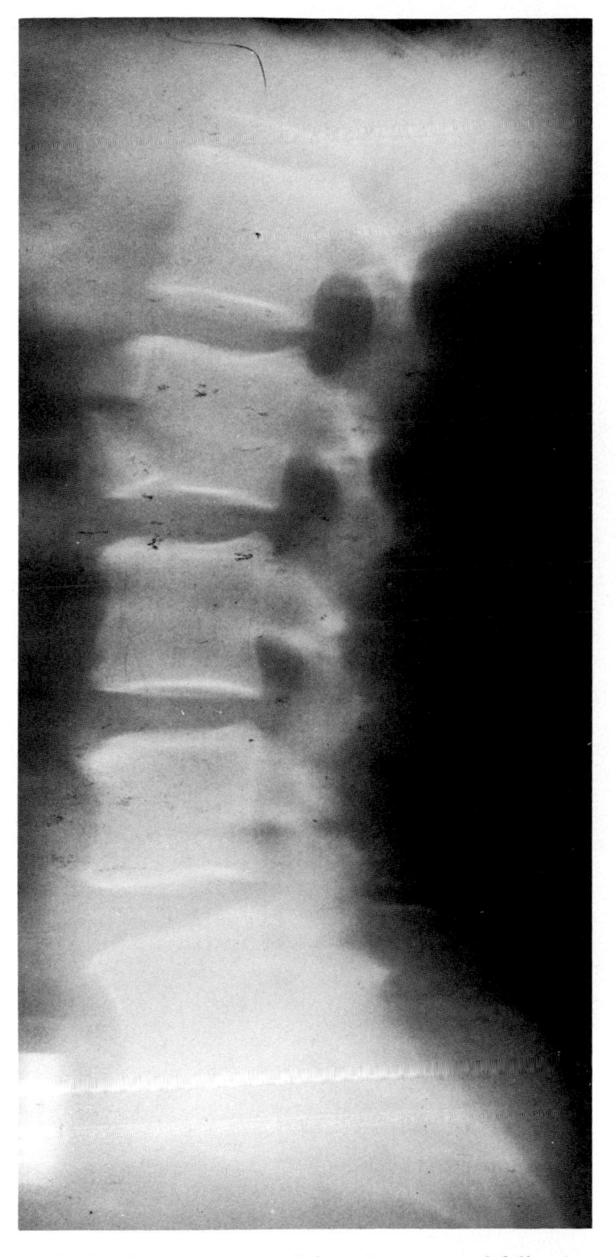

FIG. 11-26. *The multiple black specks in this tomogram of the spine occurred following the shutdown and repair of the water supply line to the processor. As a result, dirt and other foreign particles entered the wash tank and came into contact with the film while in transit.*

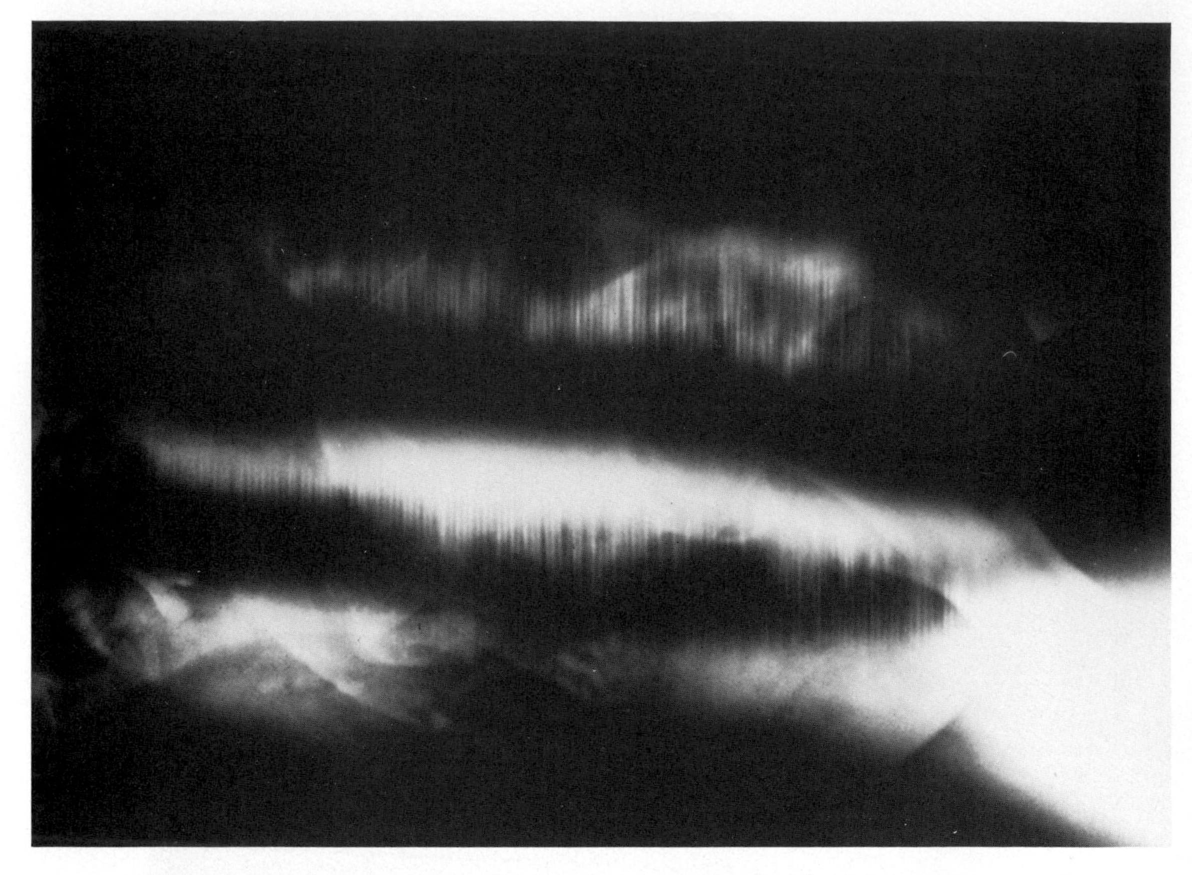

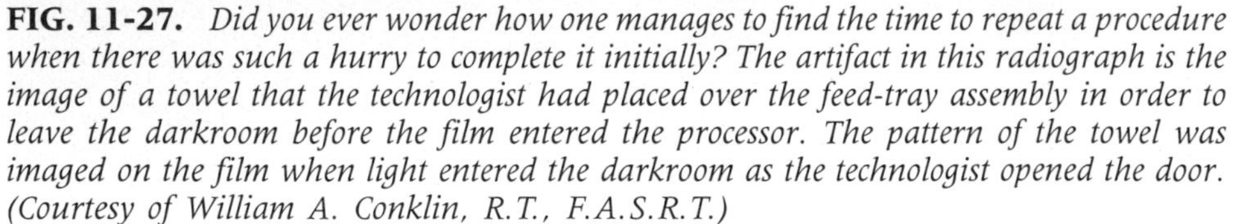

FIG. 11-27. *Did you ever wonder how one manages to find the time to repeat a procedure when there was such a hurry to complete it initially? The artifact in this radiograph is the image of a towel that the technologist had placed over the feed-tray assembly in order to leave the darkroom before the film entered the processor. The pattern of the towel was imaged on the film when light entered the darkroom as the technologist opened the door. (Courtesy of William A. Conklin, R.T., F.A.S.R.T.)*

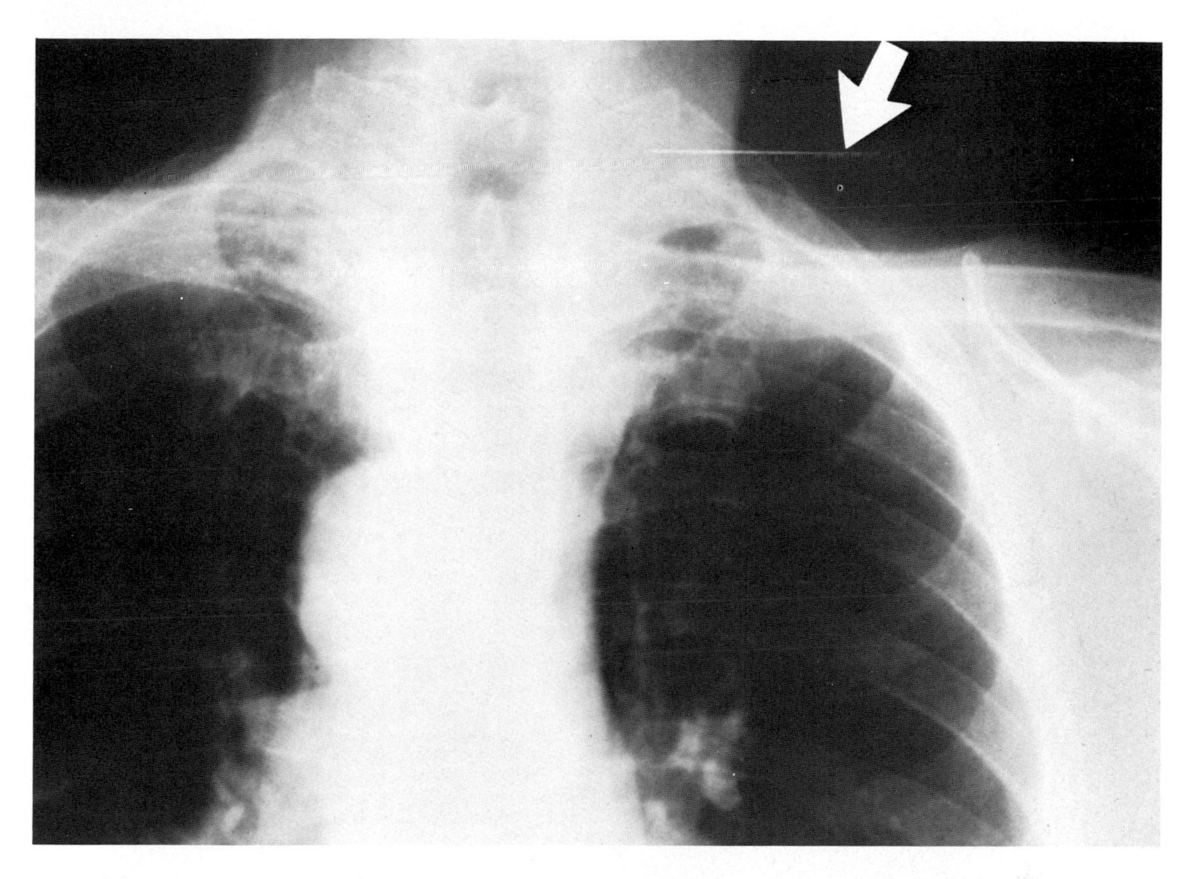

FIG. 11-28. *The artifact* (arrow) *in this radiograph is a scratch on the surface of the film. It occurred when the film was pulled up and out of the dryer section of the automatic processor. As a result, the soft film emulsion was scratched when it came into contact with a guide pin located on one of the air tubes. Such scratches can also occur if the dryer's air-tube pins or transport rollers are out of position.*

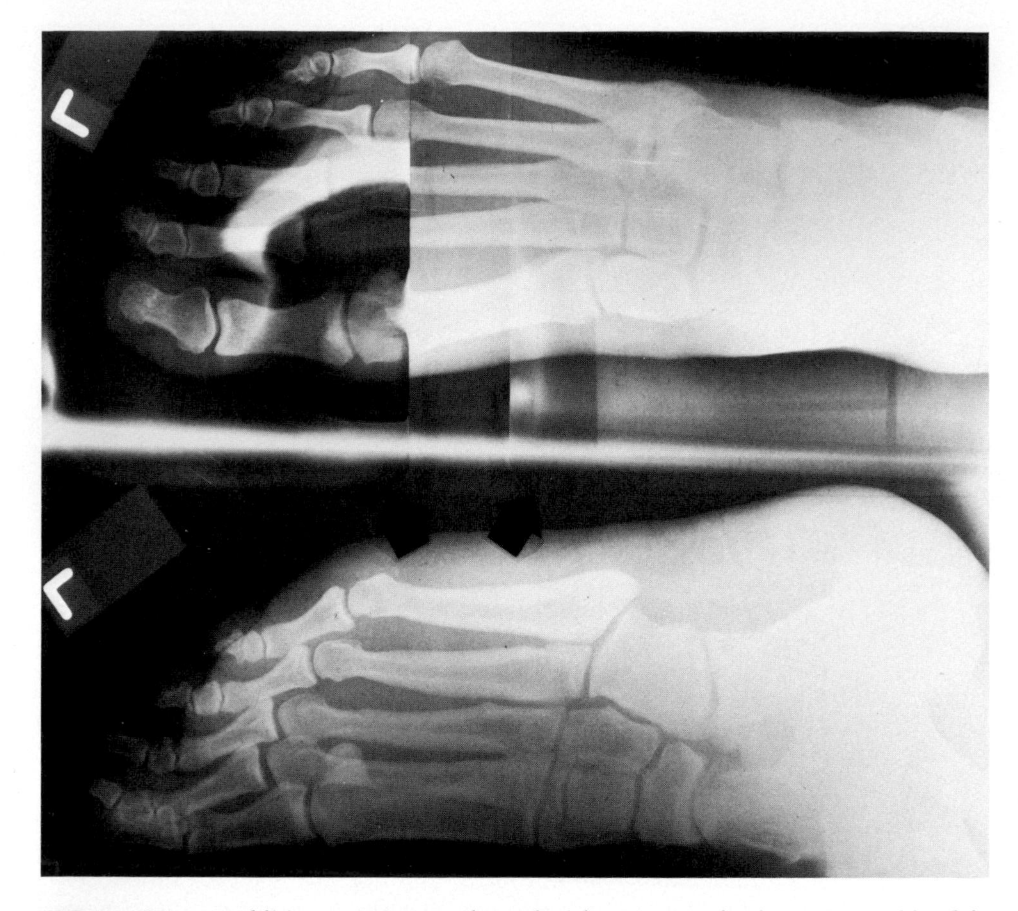

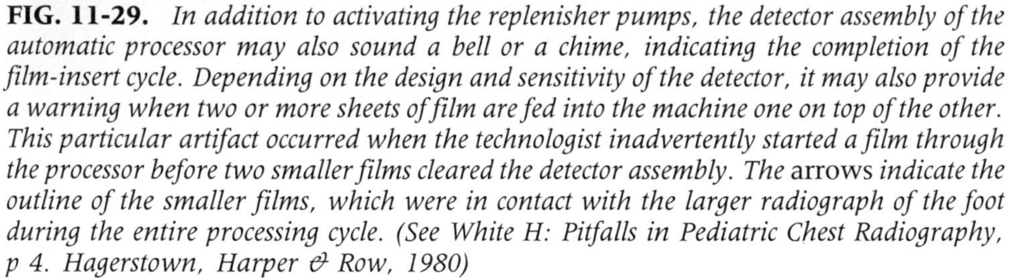

FIG. 11-29. *In addition to activating the replenisher pumps, the detector assembly of the automatic processor may also sound a bell or a chime, indicating the completion of the film-insert cycle. Depending on the design and sensitivity of the detector, it may also provide a warning when two or more sheets of film are fed into the machine one on top of the other. This particular artifact occurred when the technologist inadvertently started a film through the processor before two smaller films cleared the detector assembly. The* arrows *indicate the outline of the smaller films, which were in contact with the larger radiograph of the foot during the entire processing cycle. (See White H: Pitfalls in Pediatric Chest Radiography, p 4. Hagerstown, Harper & Row, 1980)*

Suggested Reading

Artifact Identification Program, Pub. No. 668066. Rochester, Eastman Kodak, 1977

Eastman TR: Radiographic Fundamentals and Technique Guide. St Louis, CV Mosby, 1979

Fischer HW: Radiology Departments: Planning, Operation and Management, pp 201–226. Ann Arbor, Edwards Brothers, 1982

Goldman L, Vucich JJ, Beech S, Murphy WL: Automatic processing quality assurance program: Impact on a radiology department. Radiology 125:591–595, 1977

Gray JE: Light fog on radiographic films: How to measure it properly. Radiology 115:225–227, 1975

Gray JE, Winkler NT, Stears J, Frank ED: Quality Control in Diagnostic Imaging, a, pp 35–41; b, pp 44–49. Baltimore, University Park Press, 1983

Hurtgen TP: Safelighting in the automated radiographic darkroom. Med Radiogr Photogr 54, No. 2:32–38, 1978

Lawrence DJ: A simple method of processor control. Med Radiogr Photogr 49, No. 1:2–6, 1973

Thompson TT: Cahoon's Formulating X-ray Techniques, 9th ed, pp 252–279. Durham, Duke University Press, 1979

Thompson TT, Kirby CC, McKinney WEJ: A Guide for Automatic Processing and Film Quality Control Chicago, American Society of Radiologic Technologists, 1975

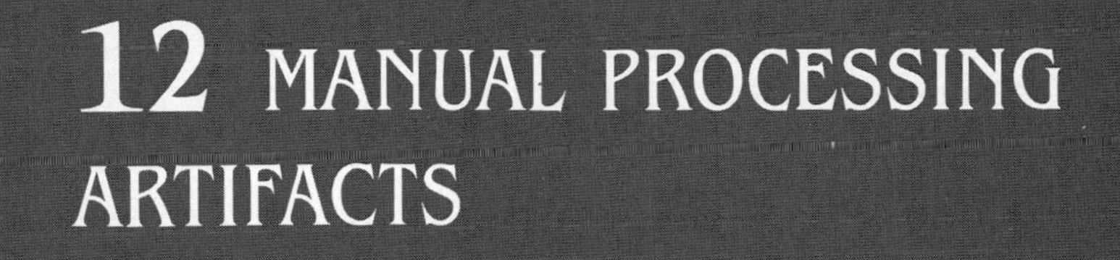

12 MANUAL PROCESSING ARTIFACTS

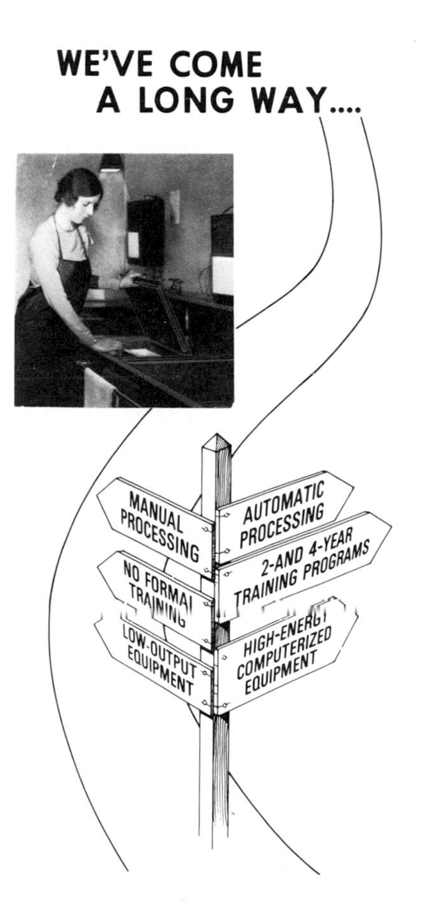

Manual processing has all but disappeared from the radiology department, and automatic systems have become the norm in both large and small radiography facilities. Automatic processing not only means more rapid processing of a large number of films, it ensures consistent results, provided that you follow the manufacturer's recommendations for the operation and maintenance of the unit. Automating the system, however, did not invalidate previous concepts of proper film processing. In fact, a great deal can be learned by reviewing some of the published material concerning film handling and processing in manual development.

Because some radiology departments still maintain manual processing facilities in various areas of the clinic or hospital, I have included the following figures—many of them accompanied by suggested reading—which illustrate a variety of artifacts that can occur in manual processing.

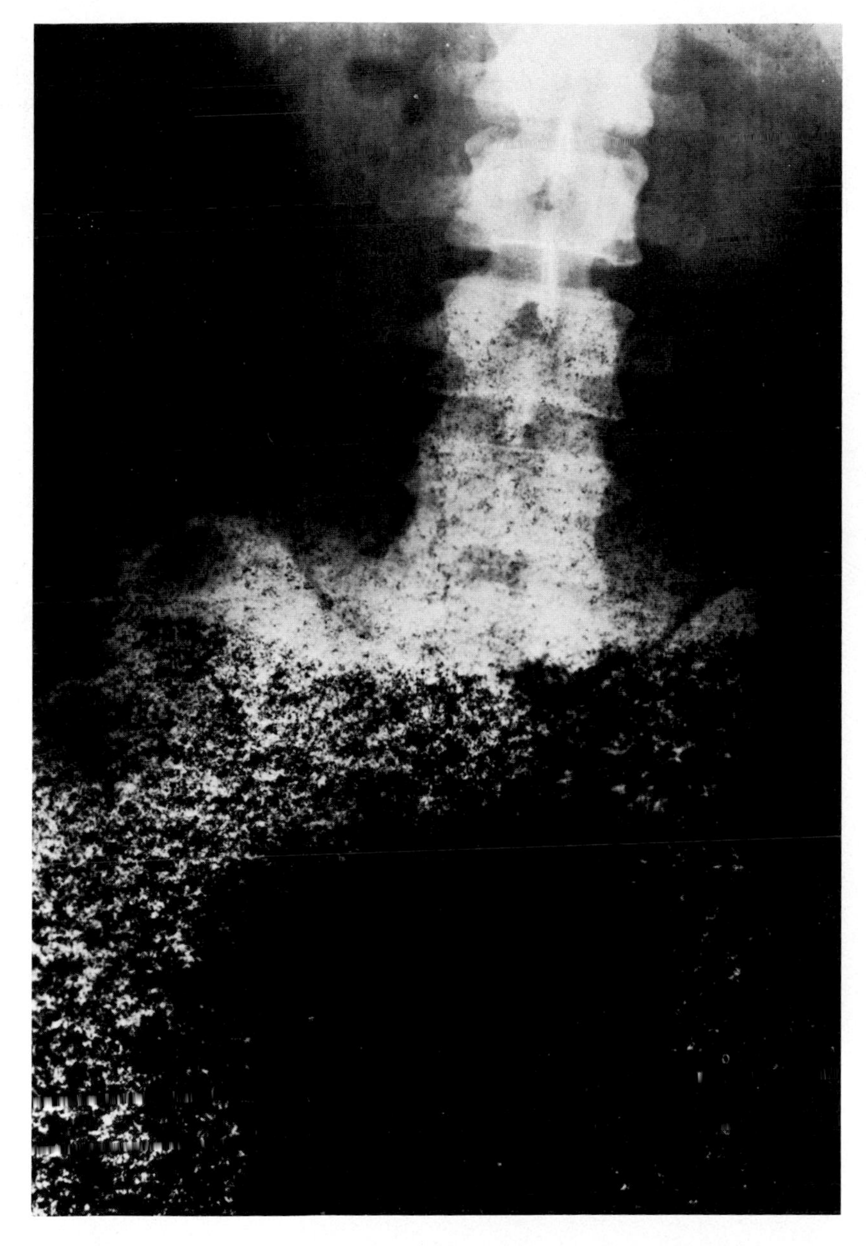

FIG. 12-1. *Interleaving paper, when struck by white light, results in a mottled appearance in the film. Notice that much of this film has been exposed and exhibits the mottled appearance. This image occurred when a sheet of interleaving paper was struck by white light as the film was being loaded into the cassette. (See Thompson TT: Cahoon's Formulating X-ray Techniques, 9th ed, p 33. Durham, Duke University Press, 1979)*

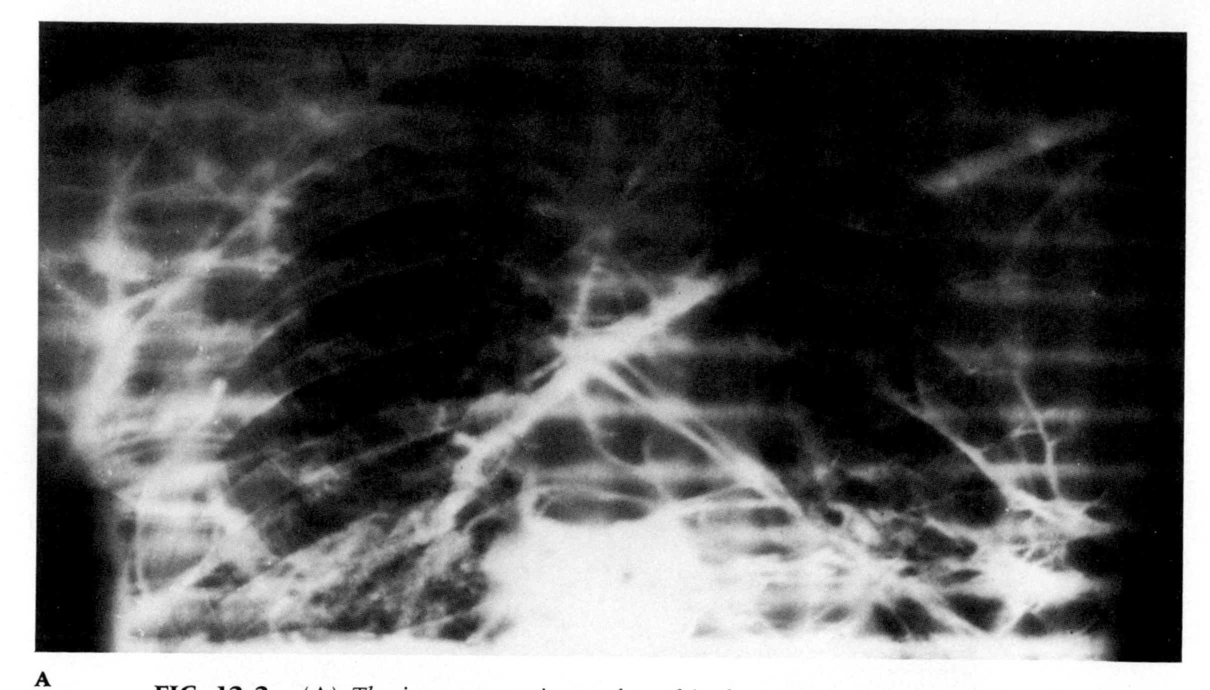

A

FIG. 12-2. (A) *The images superimposed on this chest radiograph are the branches of a tree that was outside the window of the darkroom. The artifact was caused by the camera obscura effect when light entered the processing room through a small hole in the blackened window of the darkroom* (B). *(See Camera obscura exposure of x-ray film. Med Radiogr Photogr 25, No. 2:47–48, 1949; Neblette GB: Photography: Its Principles and Practice, 4th ed. New York, Van Nostrand Reinhold, 1942)* (A *and* B, *from Camera obscura exposure of x-ray film. Med Radiogr Photogr 25, No. 2, 1949. Copyright © by Eastman Kodak Company, Rochester, NY)*

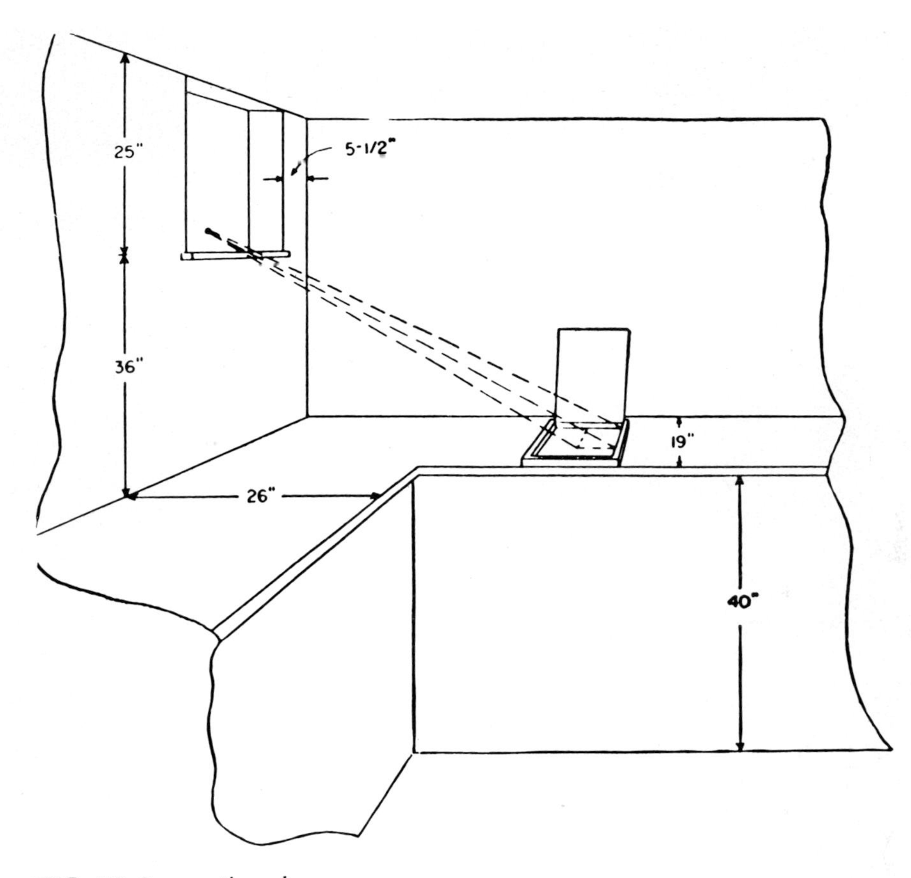

FIG. 12-2. *continued*

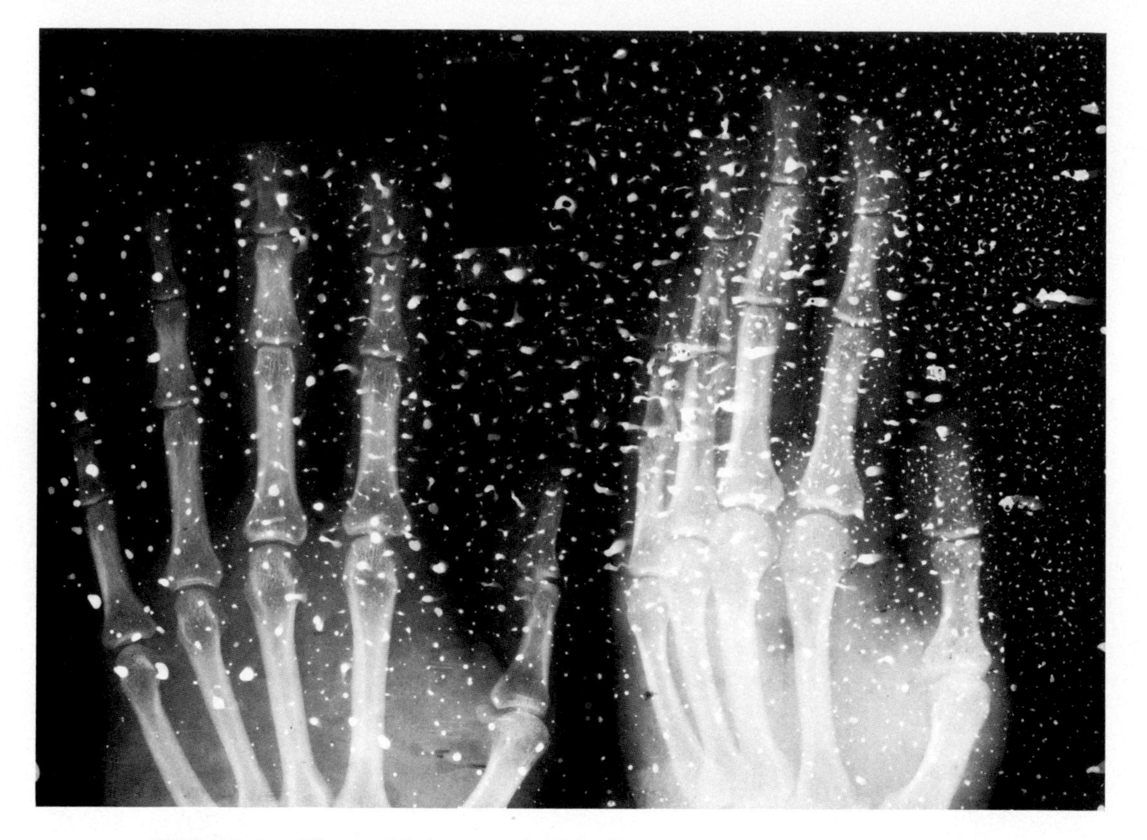

FIG. 12-3. *The speckled pattern in this direct-exposure radiograph of the hand occurred when both the interleaving paper and the film were clipped to the processing hanger and inserted into the developer solution. (Courtesy of Robert A. Short, R.T. [R])*

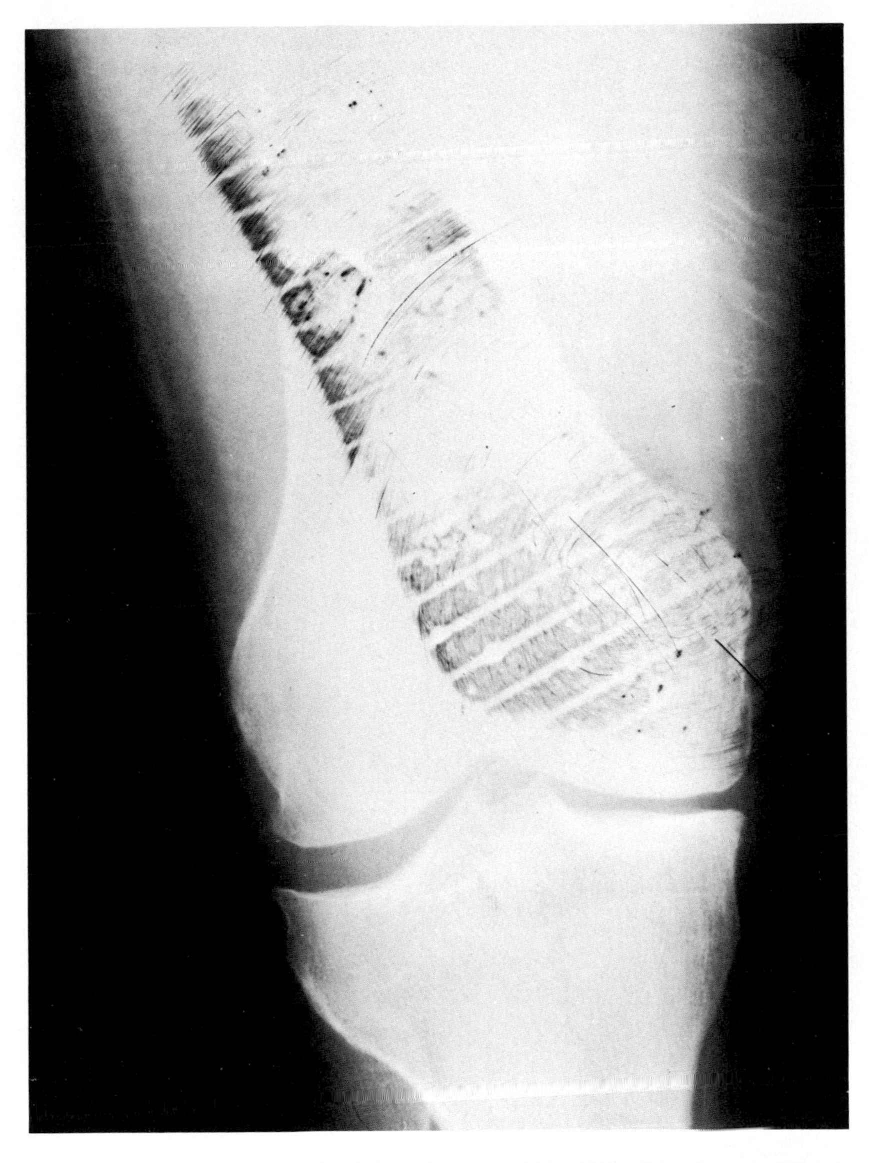

FIG. 12-4. *Probably the best name for this artifact would be "the old soft-shoe defect." This artifact occurred when the technologist managed to step on the film after dropping it onto the floor.*

FIG. 12-5. (A) *This radiographic image of the chest is inverted in order to demonstrate a so-called cathedral artifact. Although the exact origin of the term is uncertain, it seems quite appropriate when you compare the spired appearance of the artifact with the architecture of a cathedral* (B). *This artifact occurred during manual processing when the film was accidentally placed in the fixer bath and, upon discovery, rapidly transferred to the developer.*

B

FIG. 12-5. *continued*

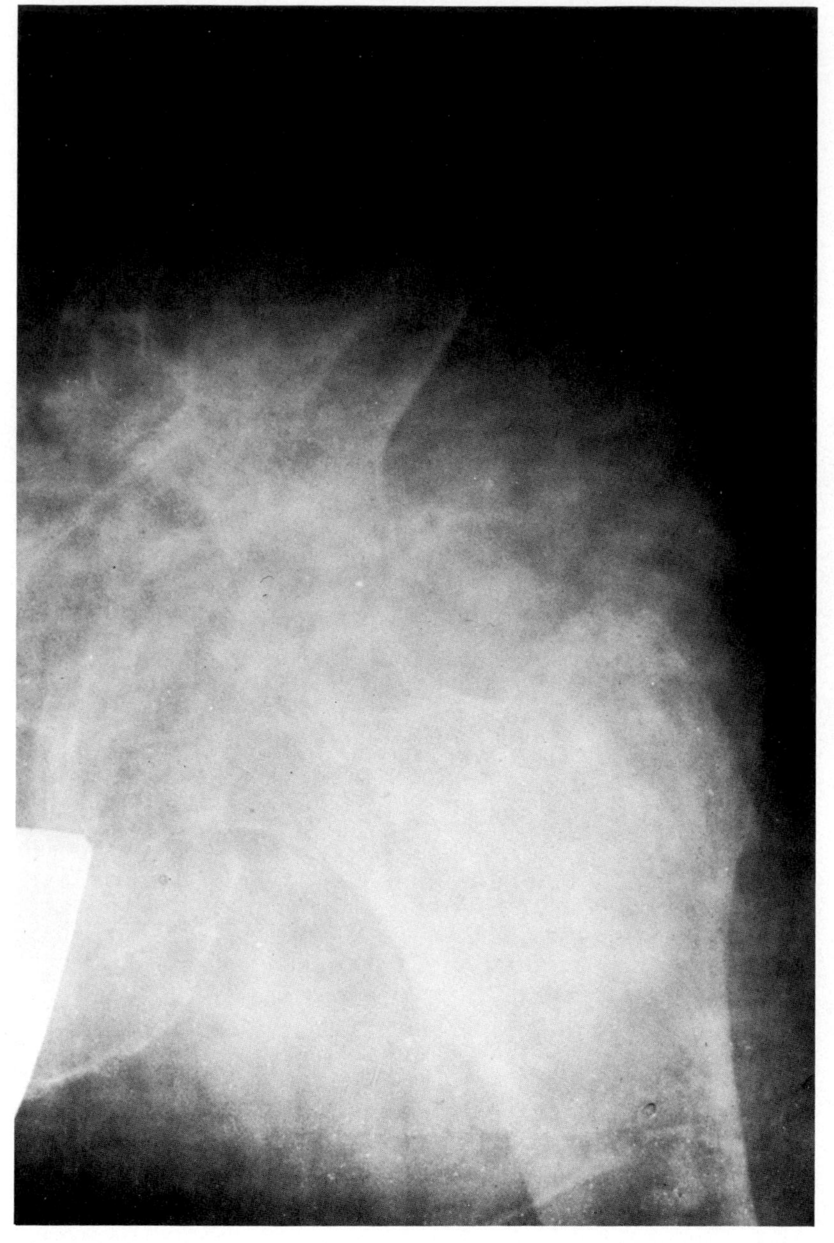

A

FIG. 12-6. *(A and B) The lack of contrast and density in these radiographs resulted from overexposure and underdevelopment. Failing to follow the prescribed methods of radiographic technique selection and film processing are responsible for such disastrous results.*

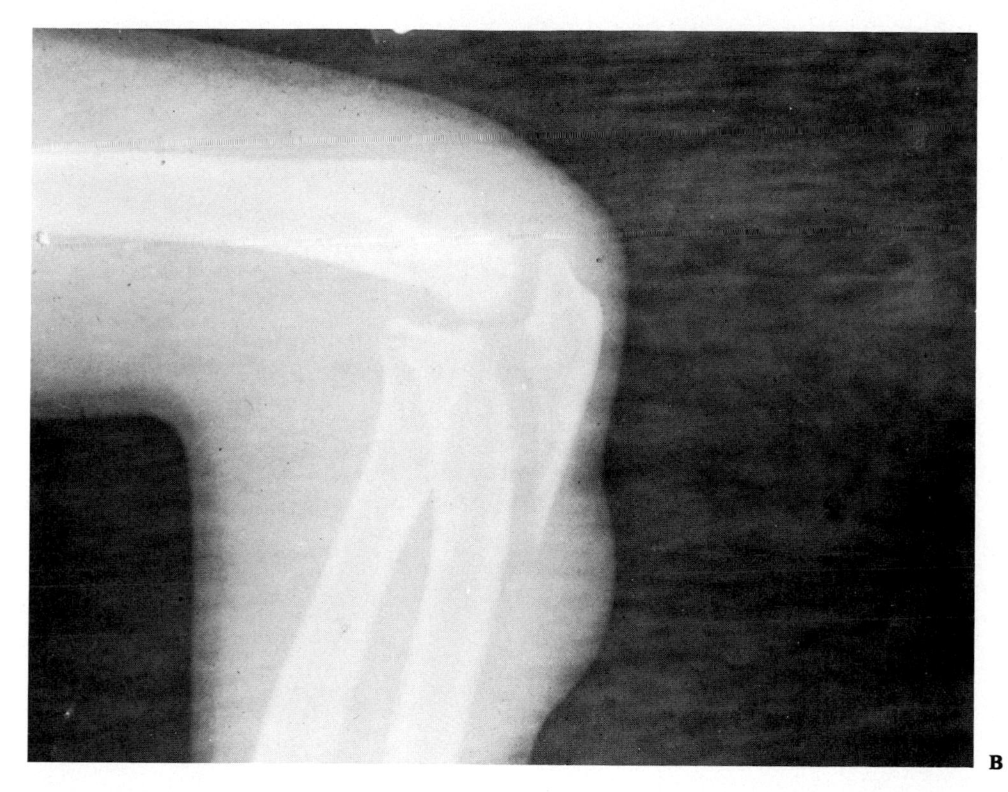

FIG. 12-6. *continued*

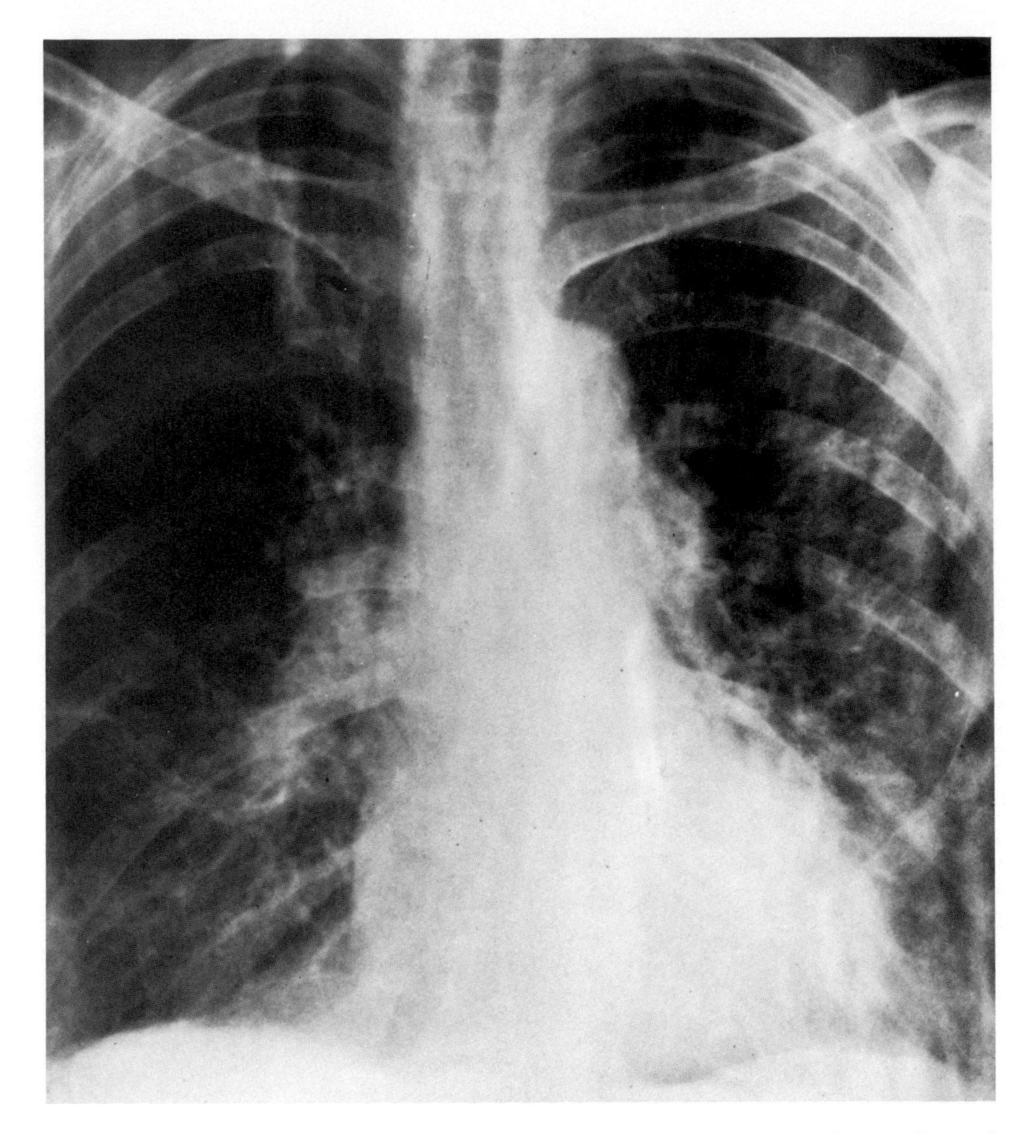

FIG. 12-7. *The lack of contrast and the streaking of the image in this chest radiograph indicate that the developer solution has become oxidized and should be changed. The artifacts could also be attributed to stratification of the developer chemicals resulting from failing to stir the solution prior to use. (See Selman J: The Fundamentals of X-ray and Radium Physics, 6th ed, p 318. Springfield, IL, Charles C Thomas, 1979)*

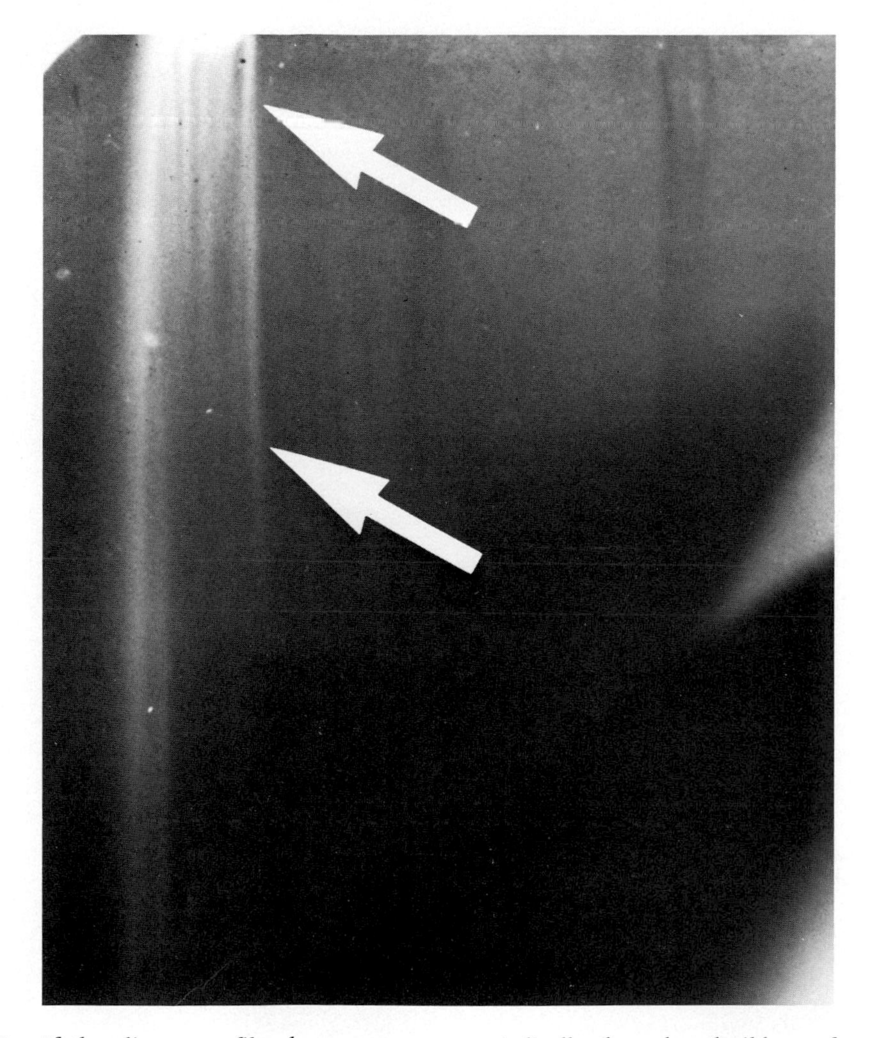

FIG 12-8 *If the clips on a film hanger are not periodically cleaned, a build-up of chemicals on their surface can be responsible for the streaklike artifact* (arrows) *seen in this radiograph. (See Cleaning hanger clips. DuPont X-ray News, No 6. Wilmington, EI du Pont de Nemours & Co; Darkroom Technique for Better Radiographs Processed Manually or Automatically, Pub. No. A-61109, p 3. Wilmington, EI du Pont de Nemours & Co; Lyons NJ: Care of processing hangers. DuPont X-ray News, No. 52. Wilmington, EI du Pont de Nemours & Co)*

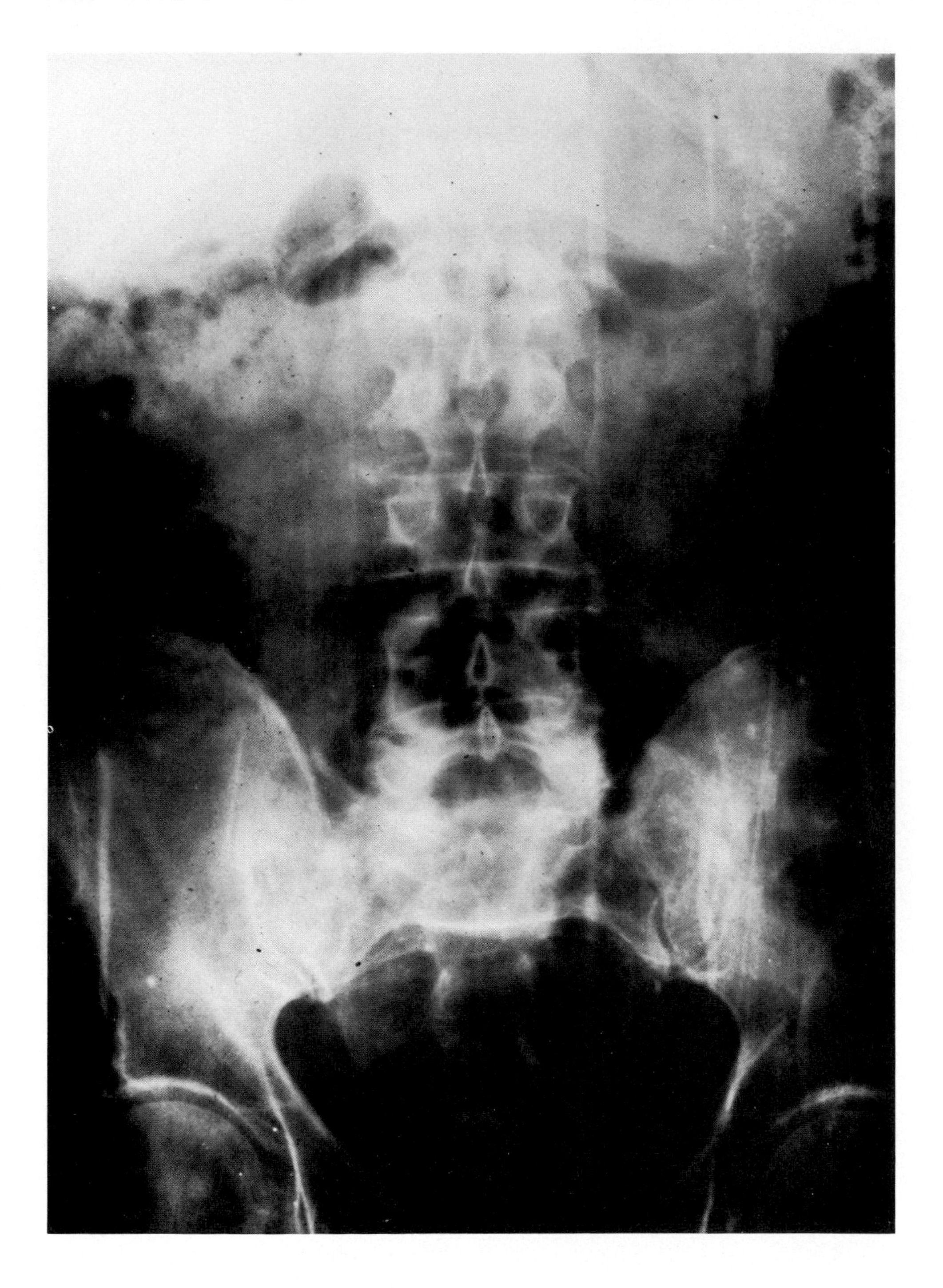

◁ **FIG. 12-9.** *The vertical streaks seen in the abdominal radiograph occurred when developer was spilled on the surface of the intensifying screen. The stains inhibited the light from the screen from reaching the film, producing the areas of minus density. Although it may seem unusual that such a condition could occur today with automatic processing, just think how many times you have seen cassettes left open in an area where processing chemicals were being mixed or where the processor was being dismantled for service or cleaning.*

FIG. 12-10. *This mammographic study was conducted with direct-exposure film and processed by hand. The streaked and light-fogged regions of the radiograph resulted from repeatedly viewing the film under safelight illumination during development. Fortunately, this practice has been eliminated by the use of a film that is automatically processed.*

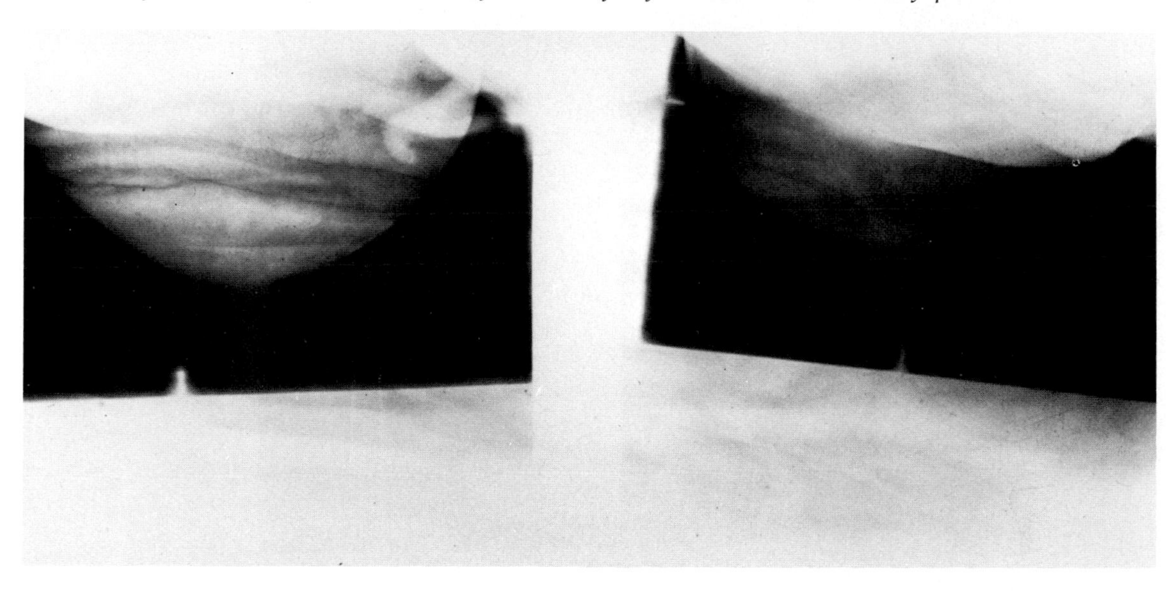

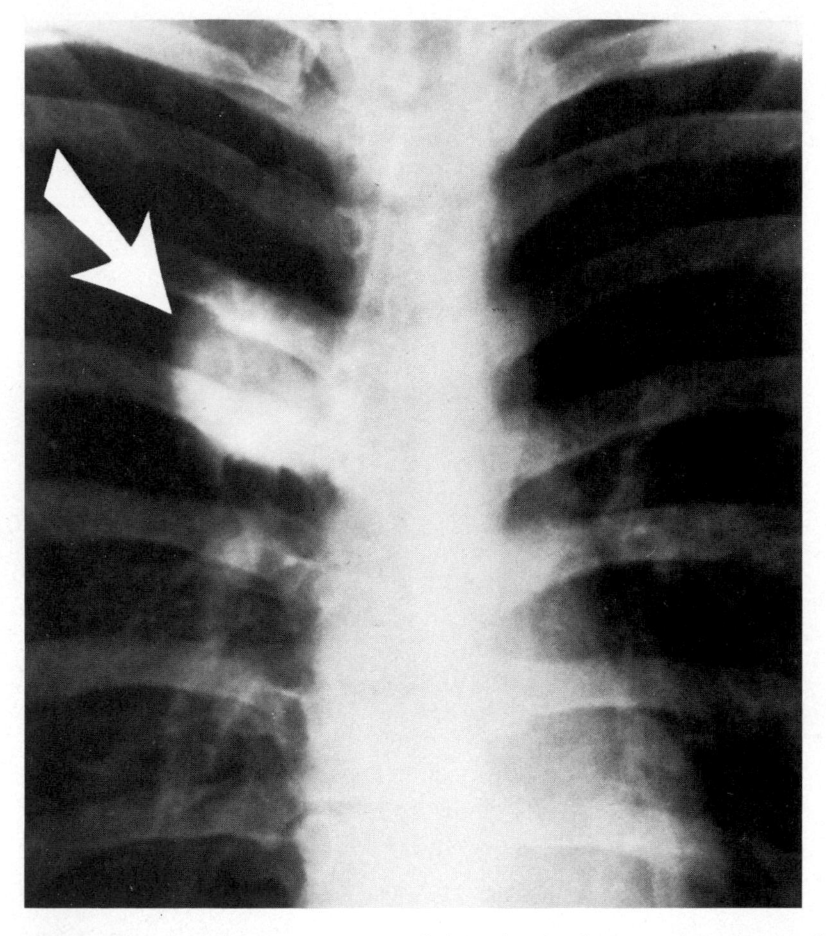

FIG. 12-11. *If films are overcrowded in the developer tank and are allowed to touch during processing, the result will be an area of nondevelopment. This is referred to as a* kissing-type artifact *and may also occur if the film comes into contact with the side of the developer tank during processing. (See Cahoon JB: Formulating X-Ray Techniques 5th ed, p 18. Durham, Duke University Press, 1961; Meschan I: Radiographic Positioning and Related Anatomy, p 40. Philadelphia, WB Saunders, 1978)*

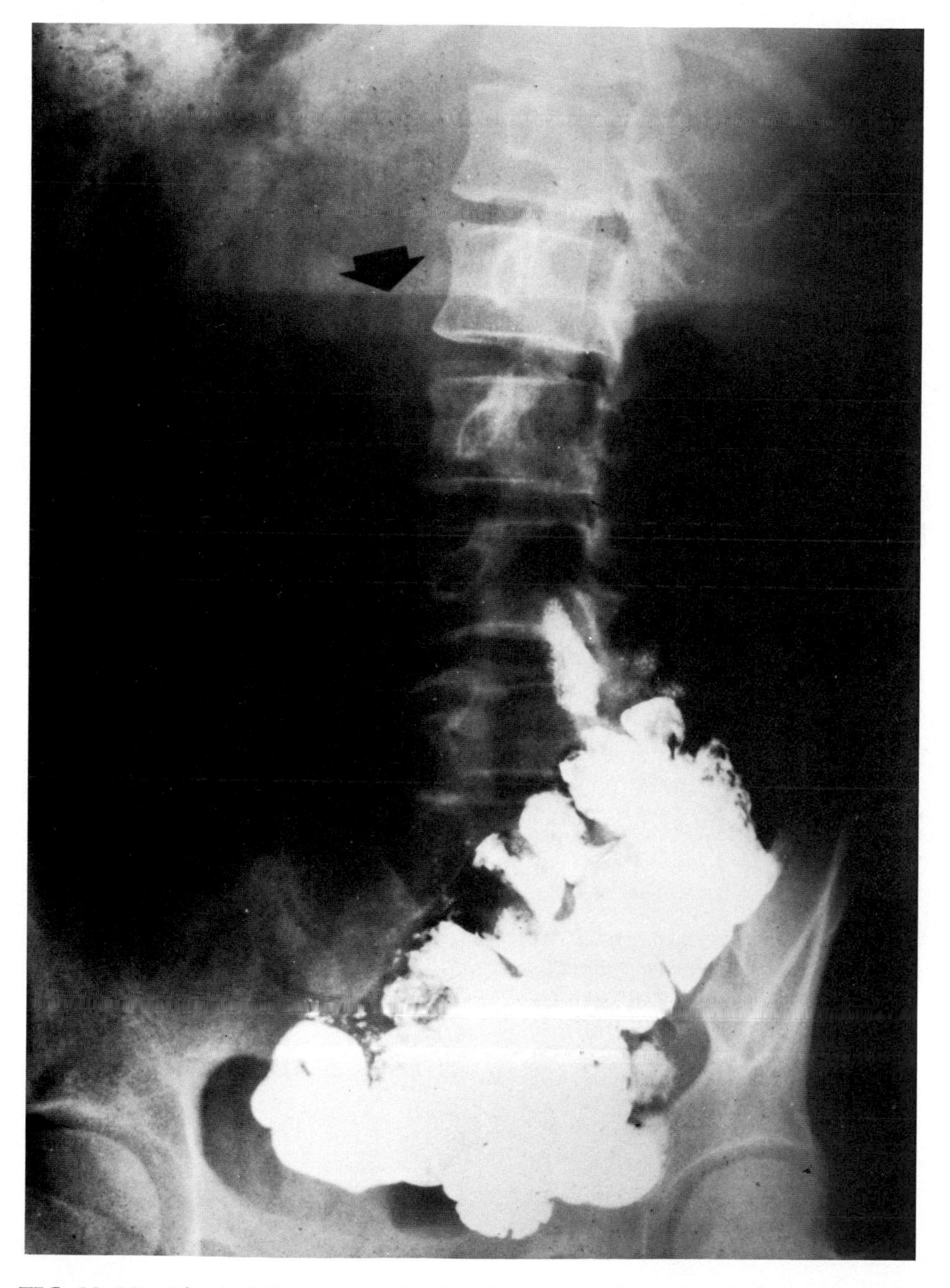

FIG. 12-12. *The dark line* (arrow) *in this abdominal radiograph is the point where this film was deliberately immersed in the fixer solution in order to demonstrate the effect of prolonged fixing on the appearance of the image.*

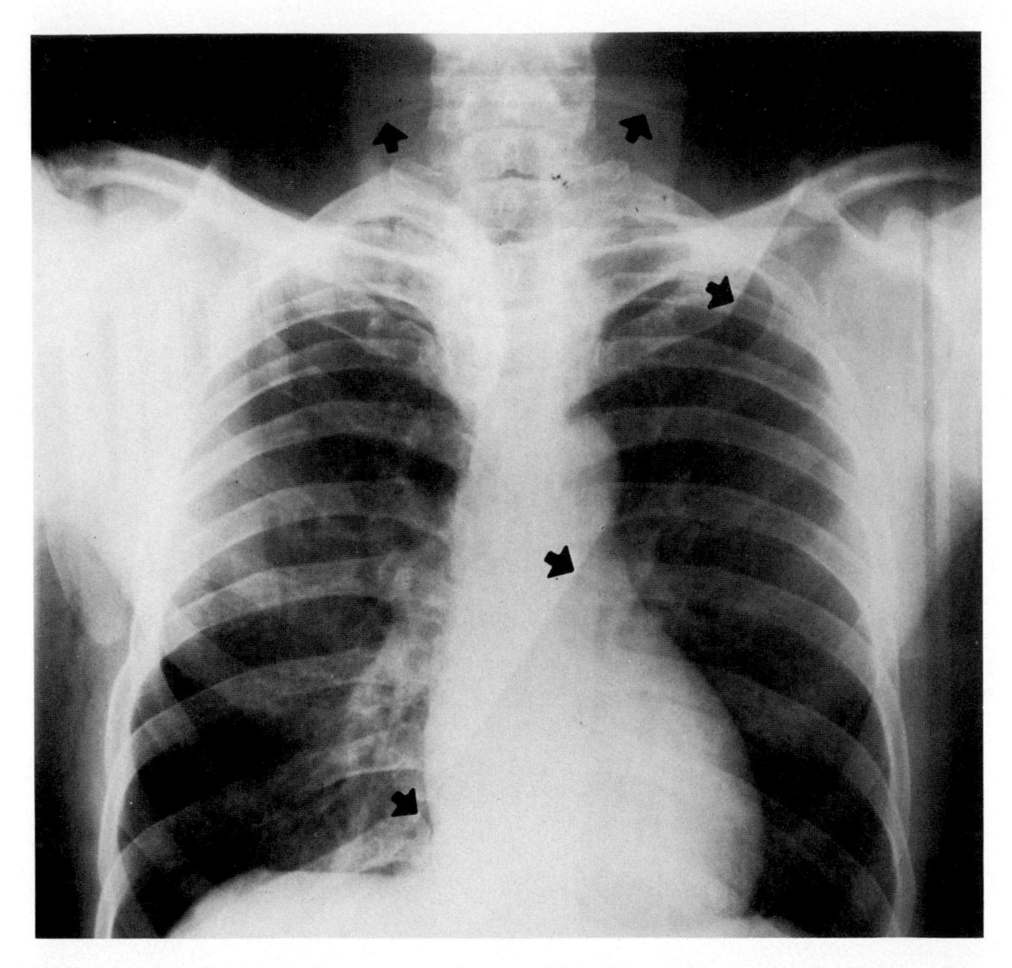

FIG. 12-13. *The* vertical arrows *at the top of this chest radiograph indicate the outline of a film hanger. The image running diagonally* (horizontal arrows) *is the edge of another film. These artifacts occurred when the film was exposed to prolonged safelight illumination while stacked in a pile on the loading bench. (Courtesy of Ralph Coates, R.T. [R])*

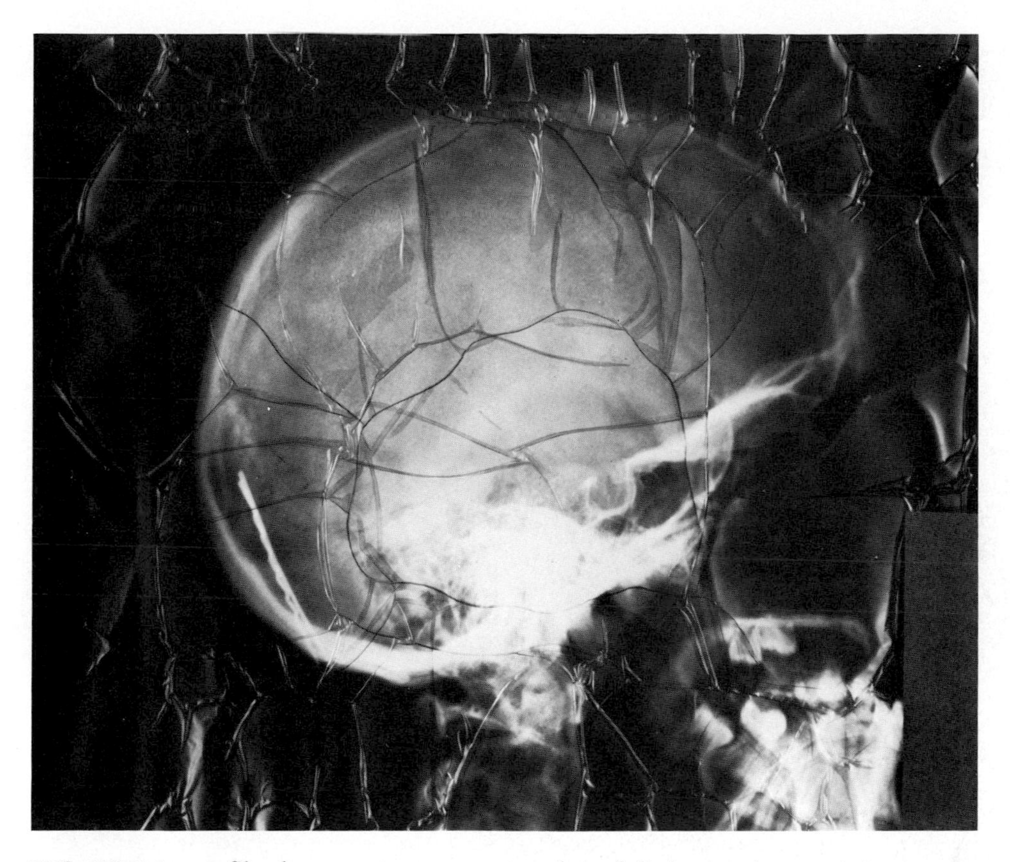

FIG. 12-14. *A film base must possess a number of distinctive characteristics, among which is its dimensional stability—that is, the shape and size of the base must not change during development or during the stored life of the film. This radiograph demonstrates a wrinkling of the film emulsion due to the shrinkage of the film base over a period of 27 years. (See Christensen EE, Curry TS, Dowdey JE: An Introduction to the Physics of Diagnostic Radiology, pp 127–128. Philadelphia, Lea & Febiger, 1978)*

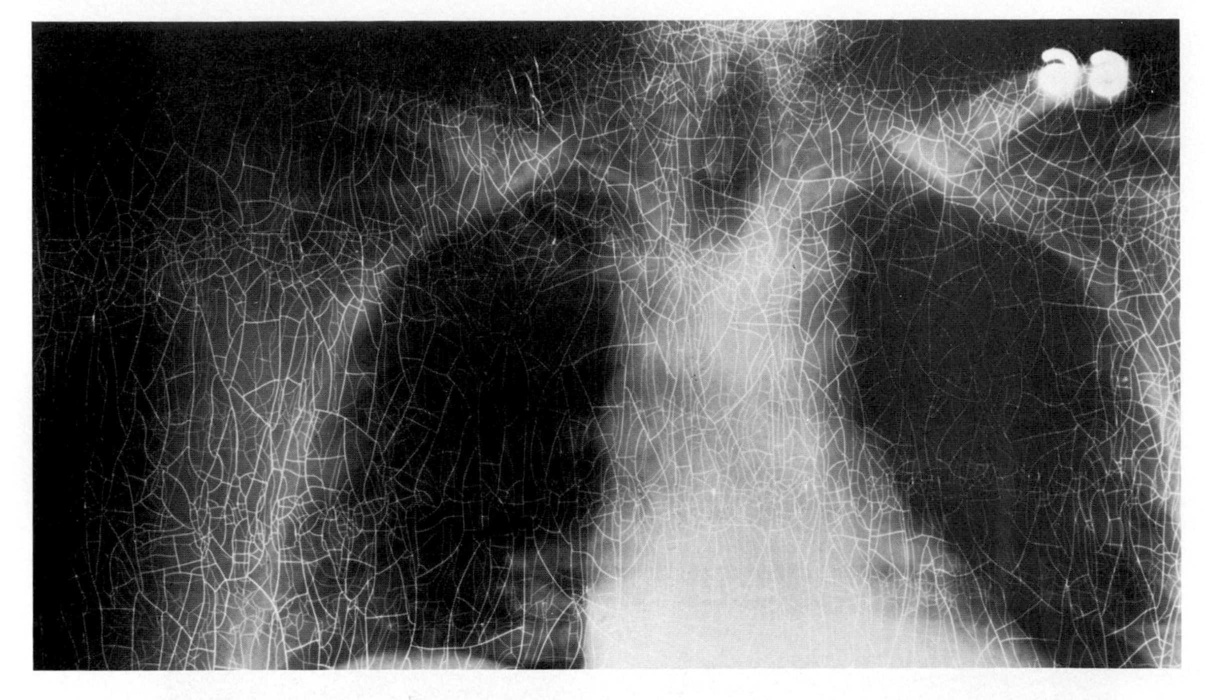

FIG. 12-15. *Two of the prime characteristics of the gelatin used for film emulsion is that it swells in solution, allowing the processing chemicals to permeate, and that it contracts to its original thickness during drying. The fine cobweb appearance in the surface of this radiograph is known as* reticulation, *or shrinking of the emulsion. It occurred when the film was subjected to wide variations in temperature during processing. (See Selman J: The Fundamentals of X-ray and Radium Physics, 6th ed, p 318. Springfield, IL, Charles C Thomas, 1979; Thompson TT: Cahoon's Formulating X-ray Techniques, 9th ed, pp 18–20. Durham, Duke University Press, 1979)*

FIG. 12-16. *It appears in this radiograph as if something was spilled on the surface of the film, but these artifacts are water marks caused by placing the film in a dryer cabinet in which the temperature was too high. As a result, the water on the surface of the film was dried almost on contact, causing this streaking pattern. Such problems can be eliminated by reducing the temperature of the dryer and by immersing the film in a wetting agent before placing it into the dryer. (See Selman J: The Fundamentals of X-ray and Radium Physics, 6th ed, p 316. Springfield, IL, Charles C Thomas, 1979)* ▷

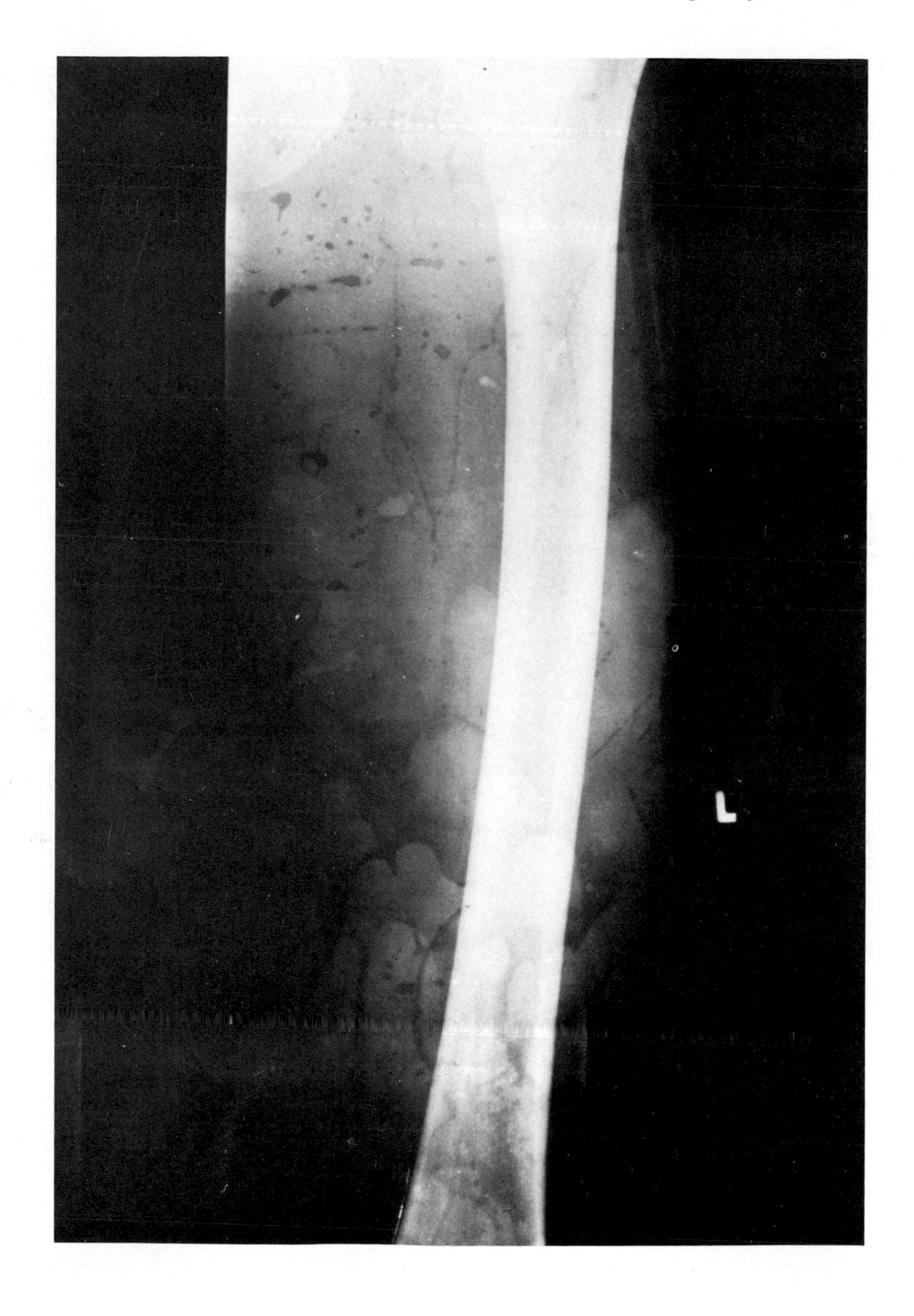
L

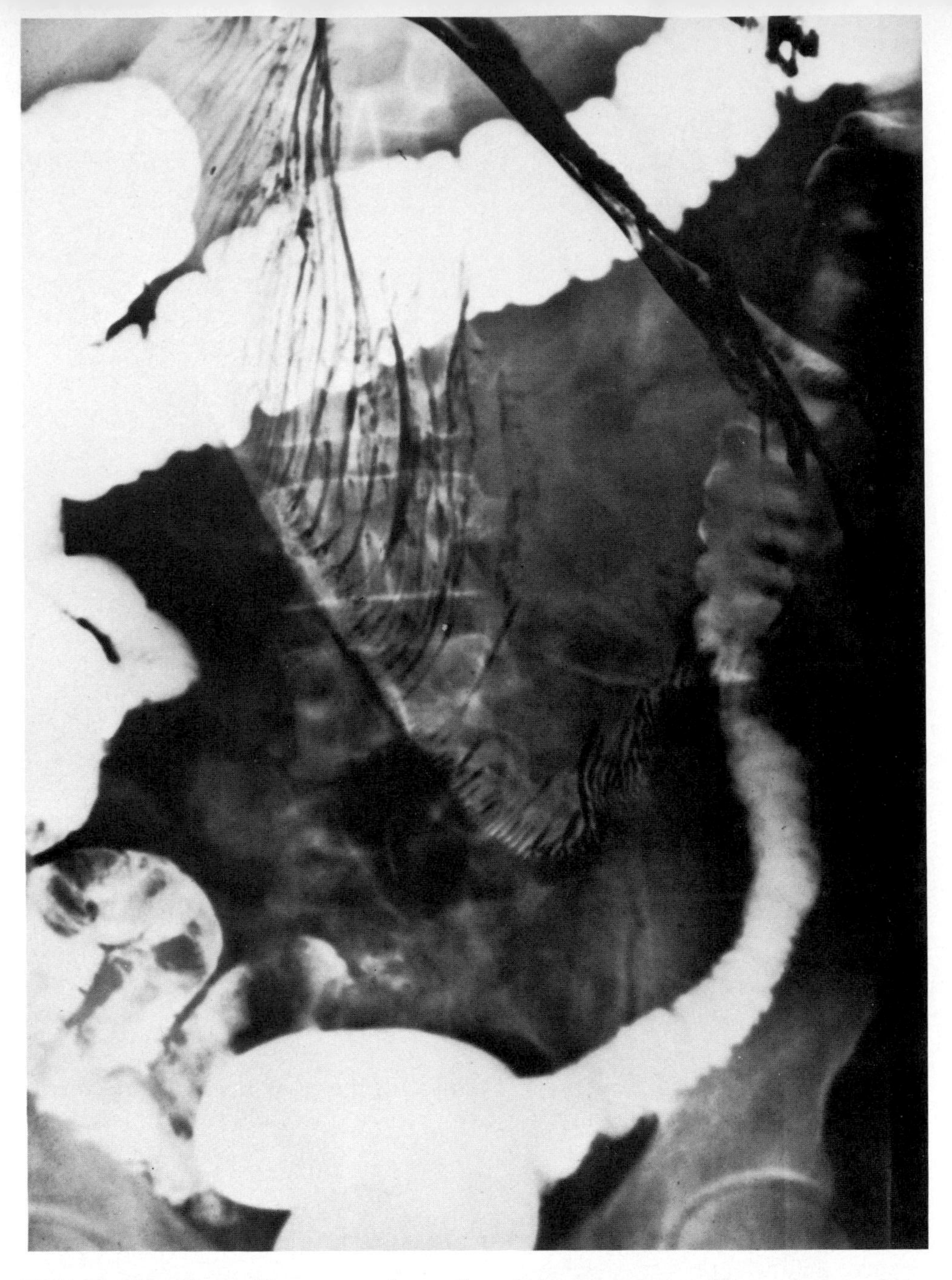

FIG. 12-17. *If wet films are stuck together, it is almost impossible to separate them without damaging the emulsion. Although this problem frequently occurs during manual processing when films are not properly spaced in the dryer cabinet, it can also occur in automatic processing if film becomes jammed in the automatic processor or if the radiographs are water damaged while in storage in the file room. (See Hurdle CC, Bowers JP, Rogers LF: Flood in the file room: Water damage to file areas. Appl Radiol 8, No. 5: 43–46, 1979; Separating stuck radiographs. Med Radiogr Photogr 25, No. 2:57, 1949)*

13 ARTIFACTS OF MISCELLANEOUS ORIGIN

There is something reassuring about investigating the appearance of various artifacts: if you collect enough information, you might eventually succeed in identifying the problem. Very often, an immediate follow-up after the artifact appears will provide the necessary clues to determine the probable cause. However, in some instances the condition responsible for the artifact may *never* be determined. This may be because the condition was not evaluated thoroughly, or following an unsuccessful attempt to determine the cause, the investigation is concluded. In addition, some artifacts may not be seen during the initial inspection of the radiograph because we are not looking for them or because we have become so accustomed to seeing them that they are ignored. For example, study Figures 13-1 and 13-2 and try to identify the artifact in each illustration. After completing this exercise, turn to pages 365 through 367 for the answers. Figures 13-3 through 13-11 illustrate a variety of bizarre artifacts that are difficult to categorize.

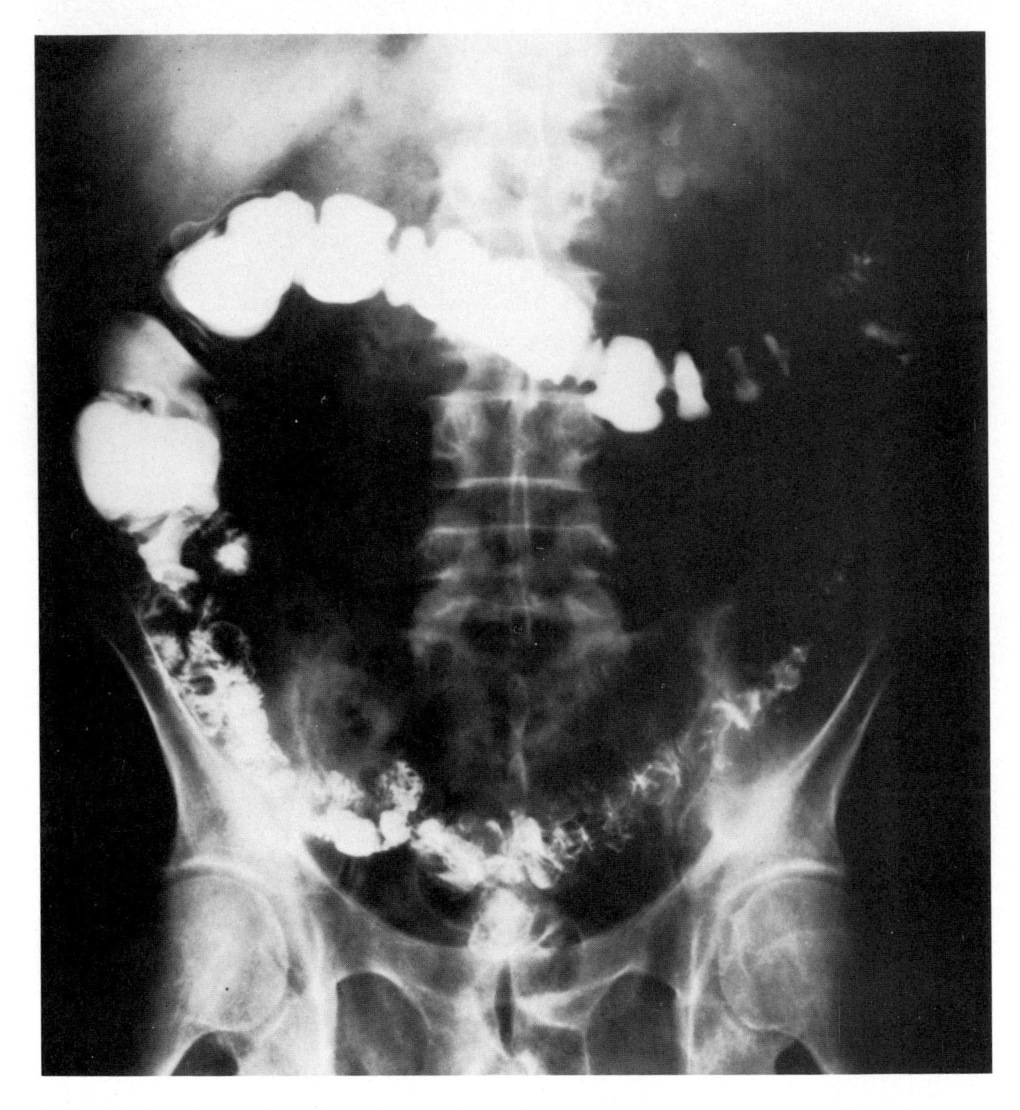

FIG. 13-1. *Somewhere in this gastrointestinal study, there are two small artifacts. See whether you can locate them and determine the reason for their appearance. (Answer on pp. 365–366) (Radiograph courtesy of Dr. Paul O'Connor)*

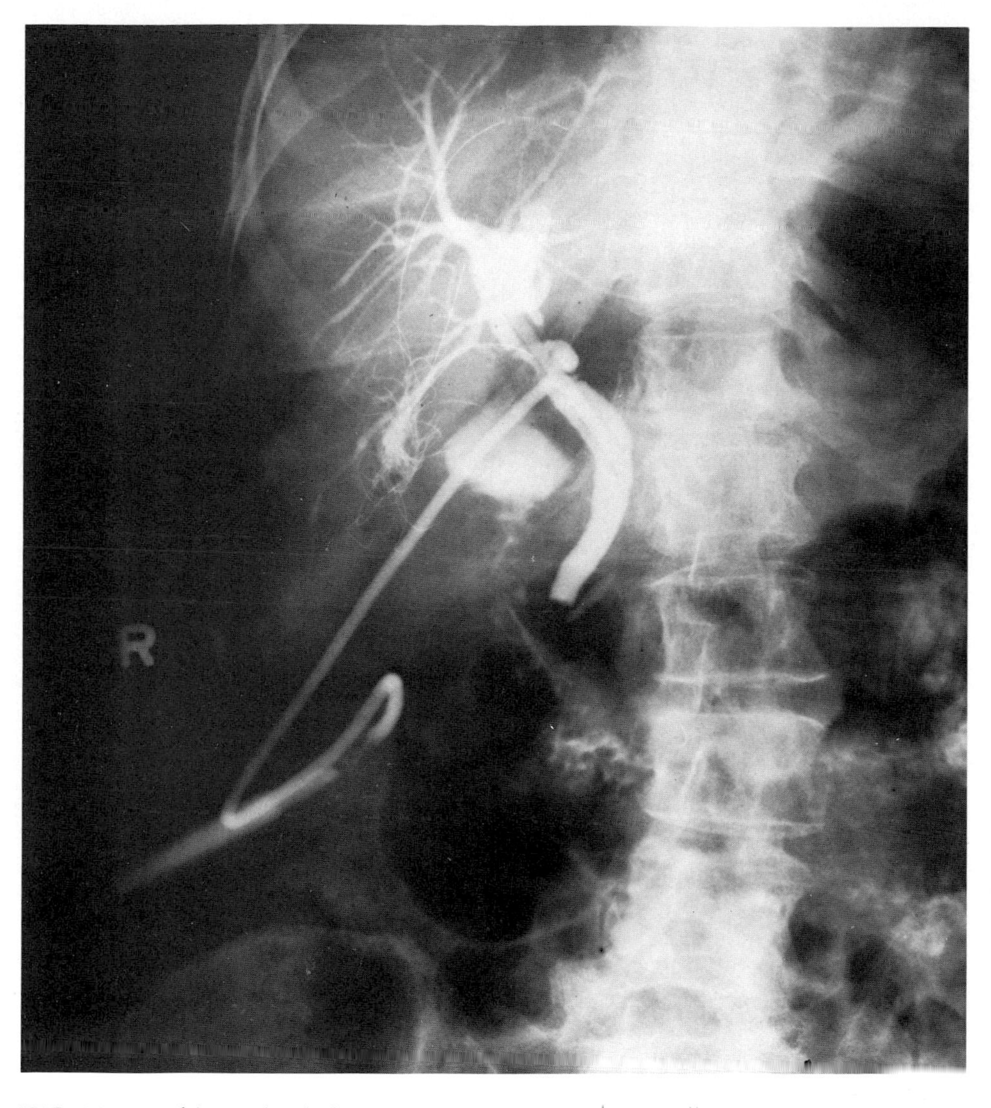

FIG. 13-2. *This T-tube cholangiogram exhibits a rather peculiar artifact. Can you identify the imaging problem? (Answer on p. 367)*

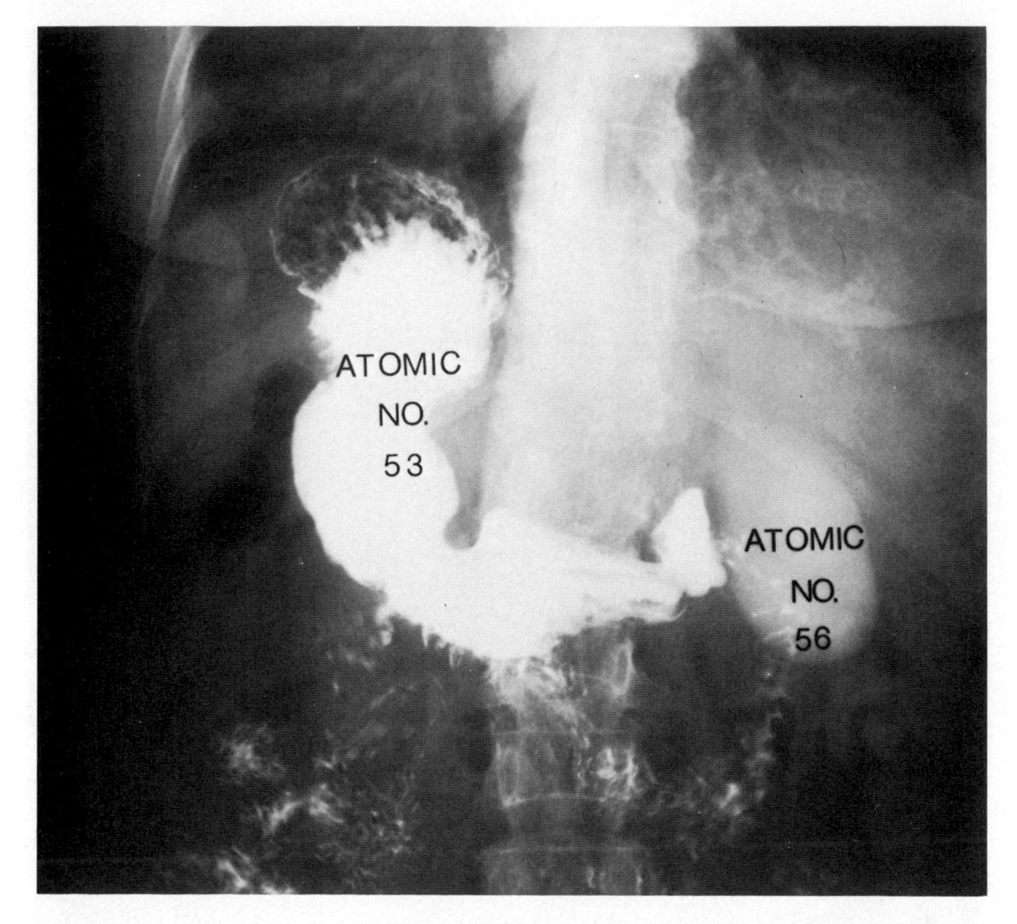

FIG. 13-3. *Because barium sulfate and iodinated compounds have a high atomic number, they are suitable as radiographic contrast agents. However, if such substances are spilled or absorbed into the positioning sponges, table pads, and so forth, various artifacts will result.*

FIG. 13-4. (A) *A radiopaque sponge used during operative procedures. The opaque material is not wire or lead; it consists of approximately 65% barium sulfate covered with a synthetic substance.* B *demonstrates the variation in the radiographic appearance of the opaque material used in three types of sponges.* ▷

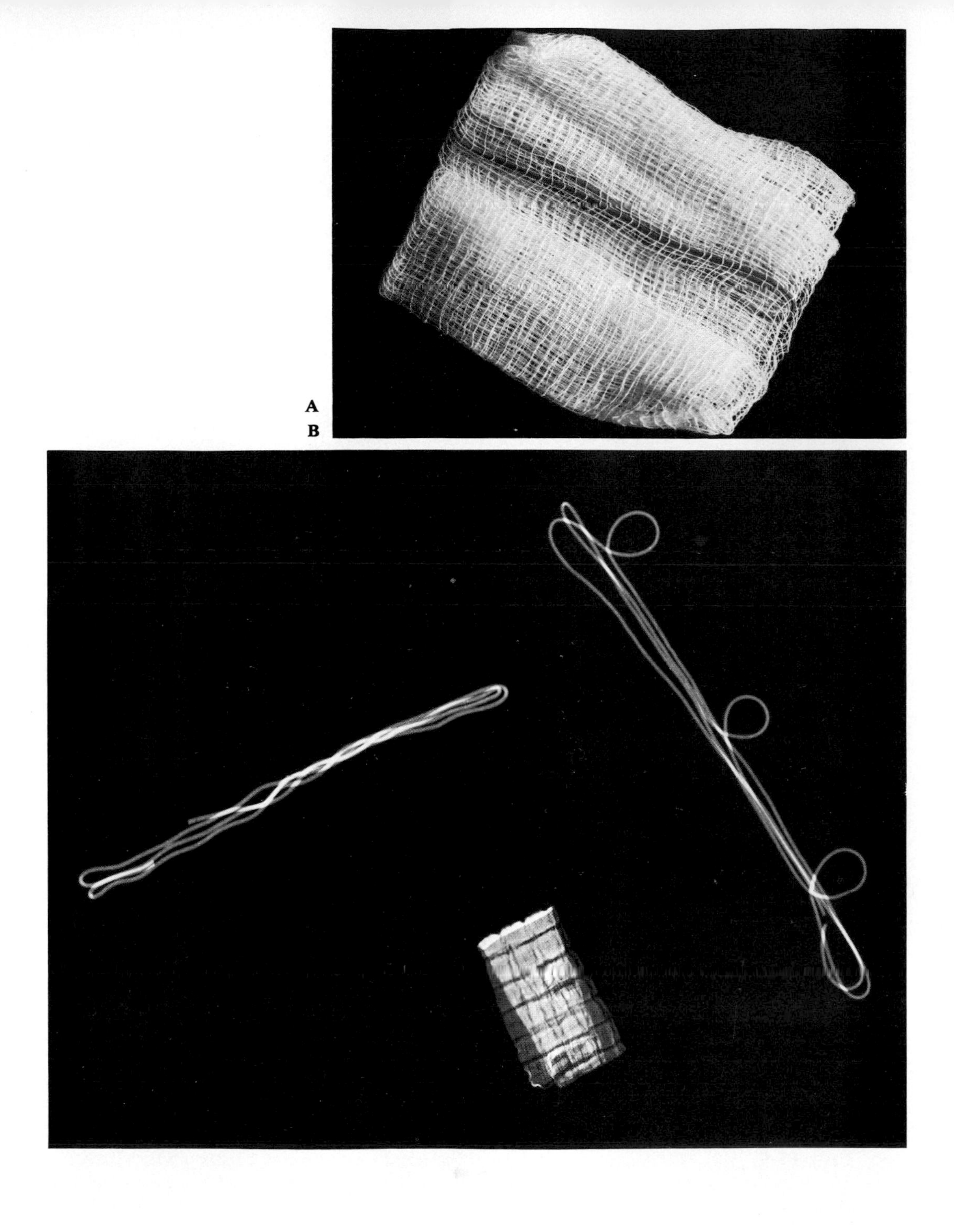
A
B

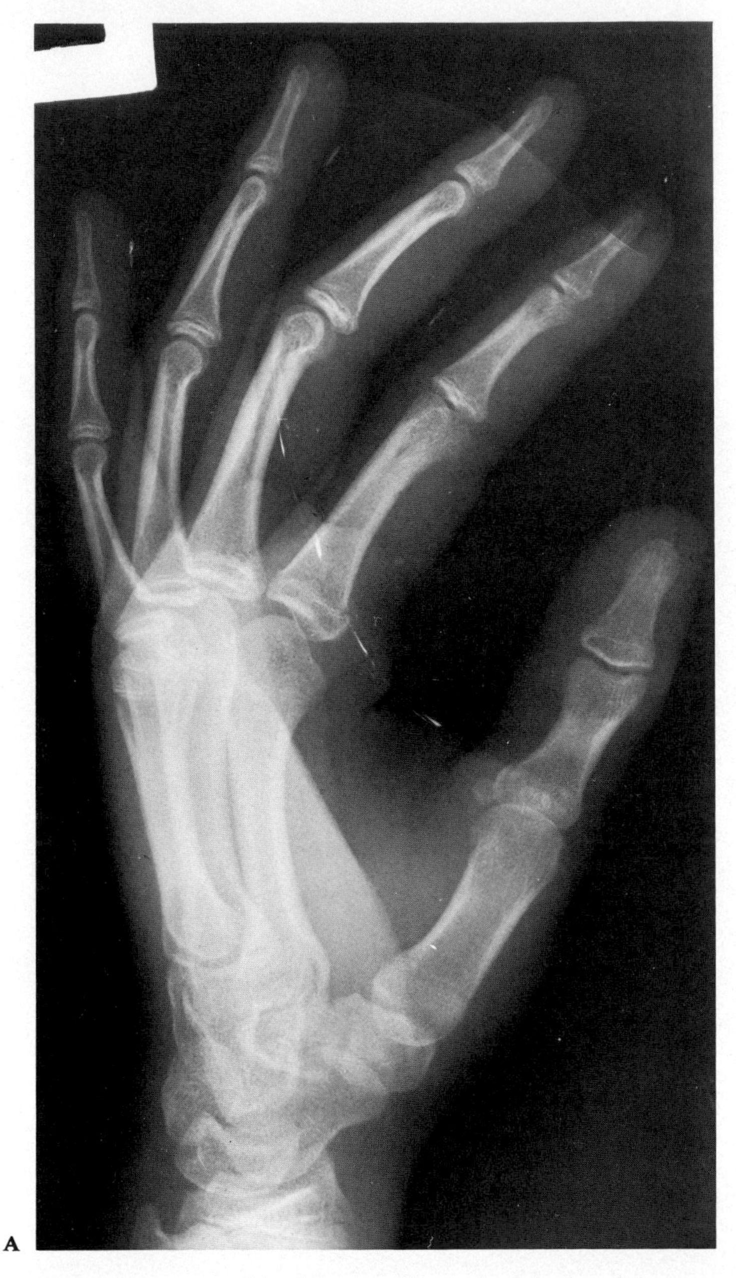

FIG. 13-5. (A) *This oblique radiograph of the hand demonstrates a semicircular streak of minus density. The artifact resulted from the absorption of an iodinated contrast substance that was in the sponge seen in* B. (C) *A radiograph of the sponge verifies the presence of the contrast medium.*

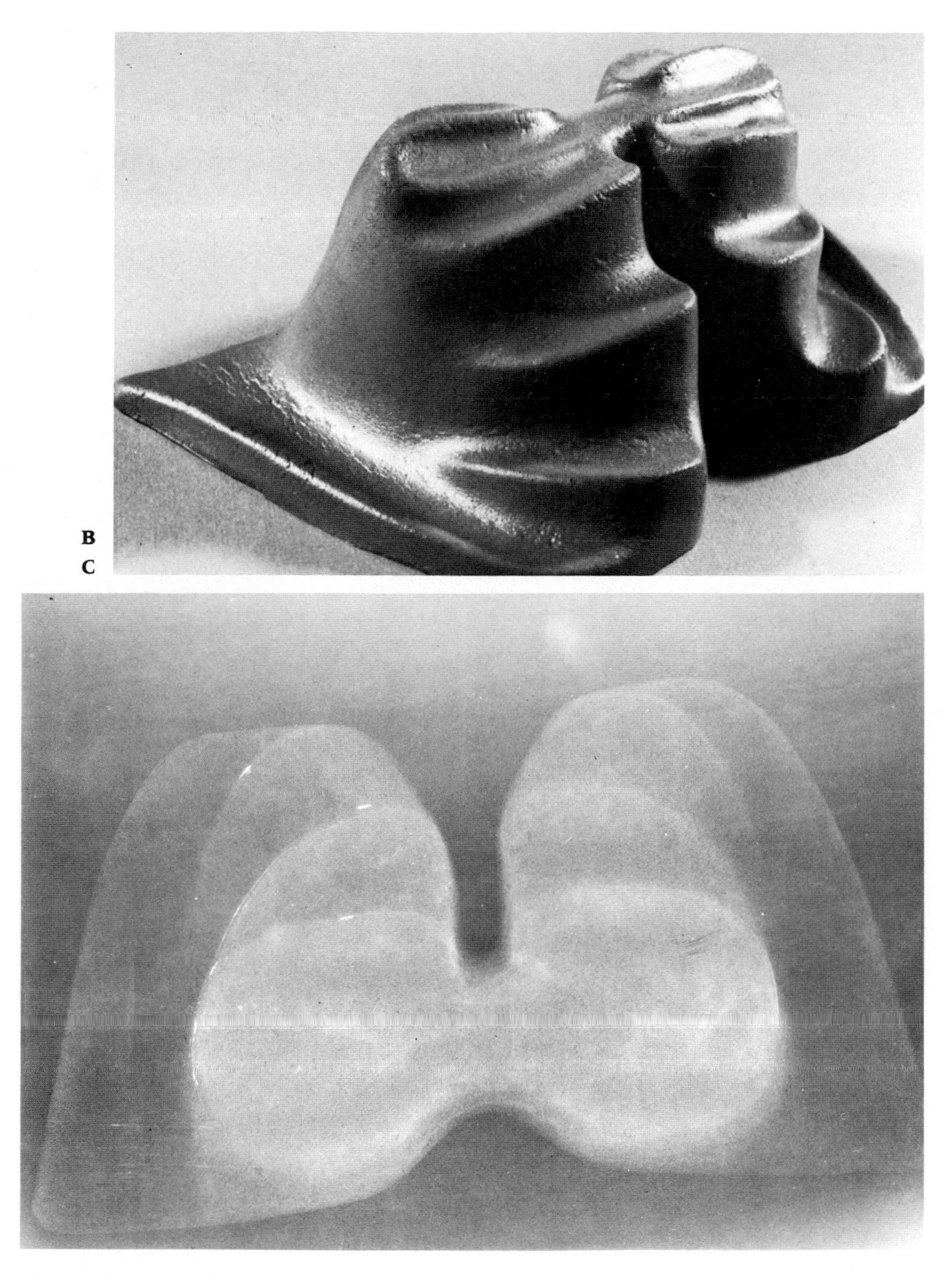

FIG. 13-5. *continued*

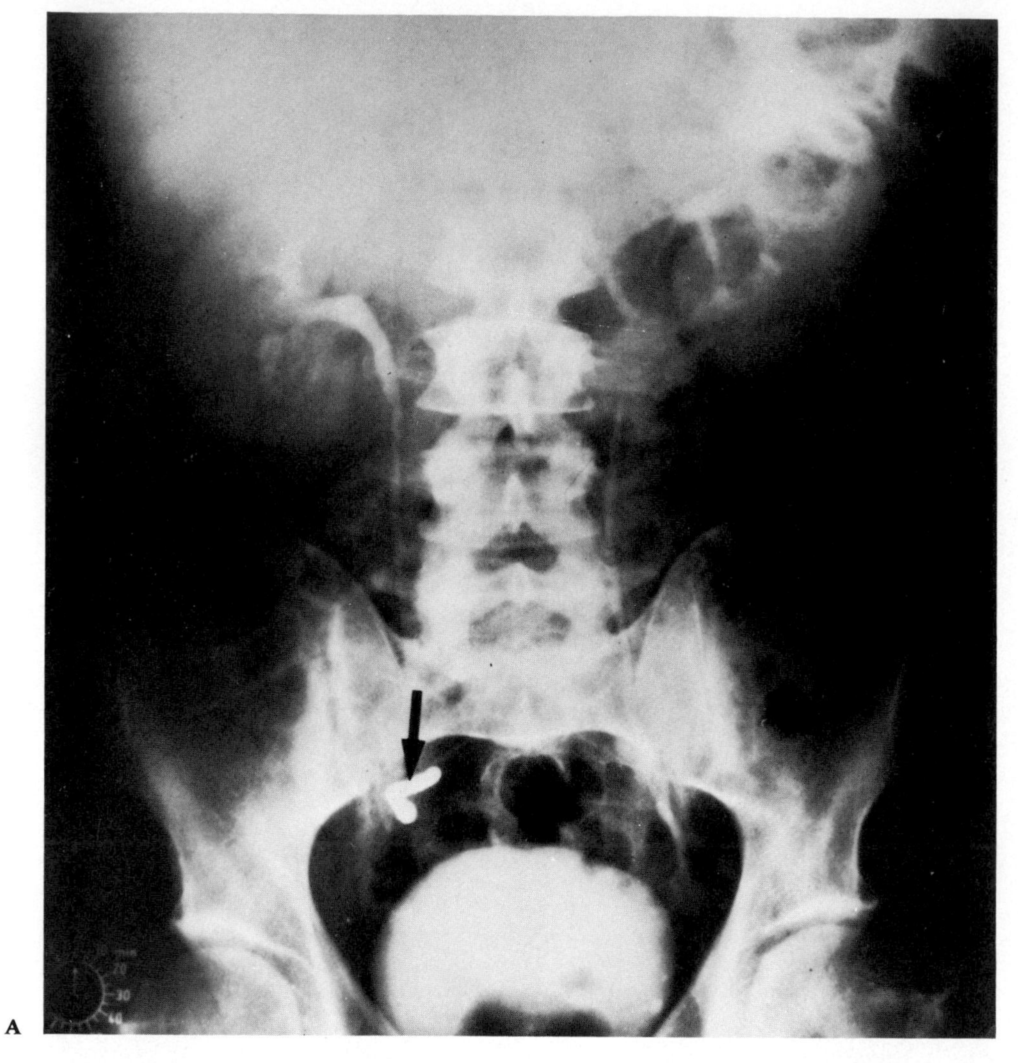

A

FIG. 13-6. (A) *The L-shaped artifact* (arrow) *in this abdominal radiograph could easily be mistaken for an image of a ''left'' marker incorrectly positioned over the right border of the cassette. However, this happens to be residual barium in the patient's appendix.* (B) *A later film shows the image* (arrow) *altered owing to the peristaltic movement of the appendix. (From Sweeney RJ: On the Technical Side. Radiol Technol 50, No. 3:312, 1978)*

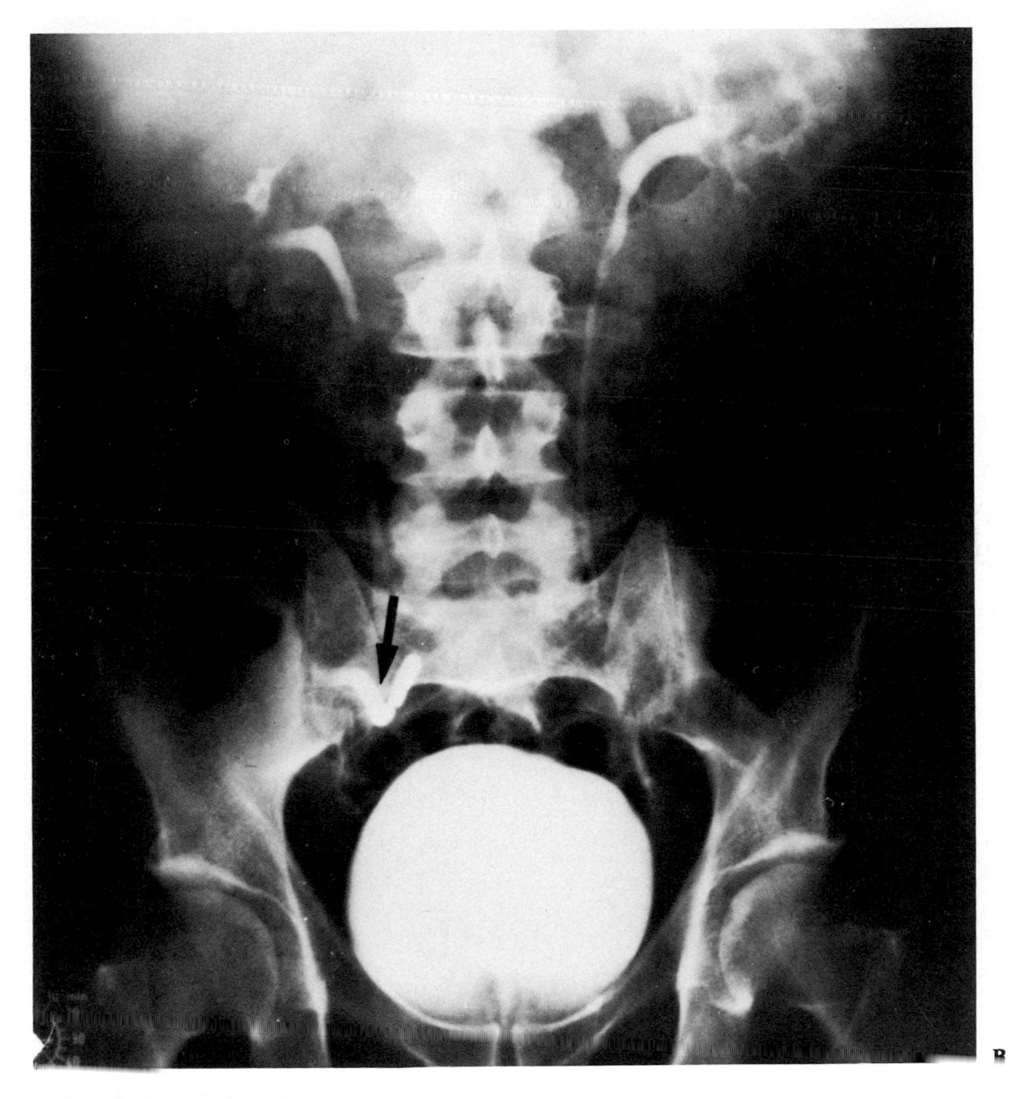

FIG. 13-6. *continued*

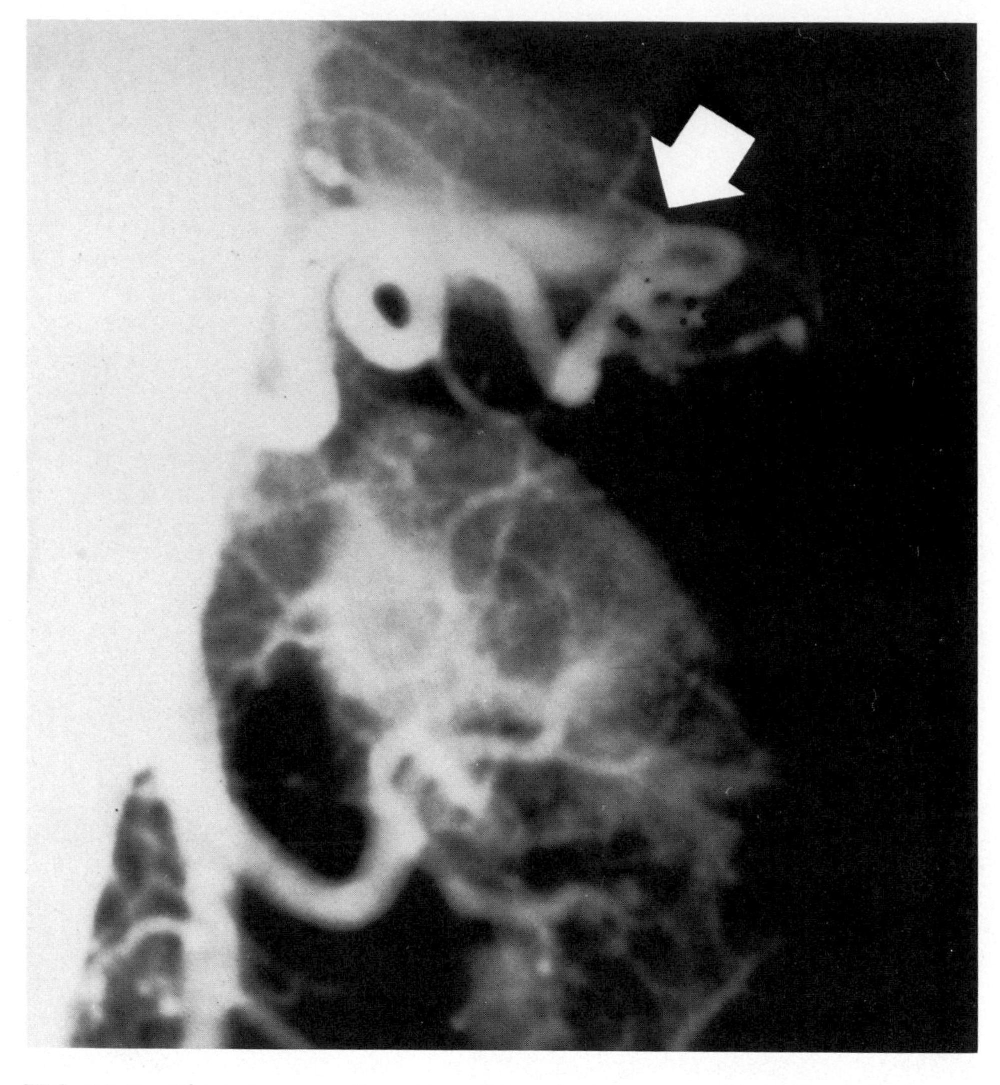

FIG. 13-7. *The tortuous configuration of the splenic artery in this radiograph seems to form the word* love. *This in itself is unusual, even more so when one considers that the expression* to vent one's spleen *is associated with hostility or anger. (Courtesy of William A. Conklin, R.T., F.A.S.R.T.)*

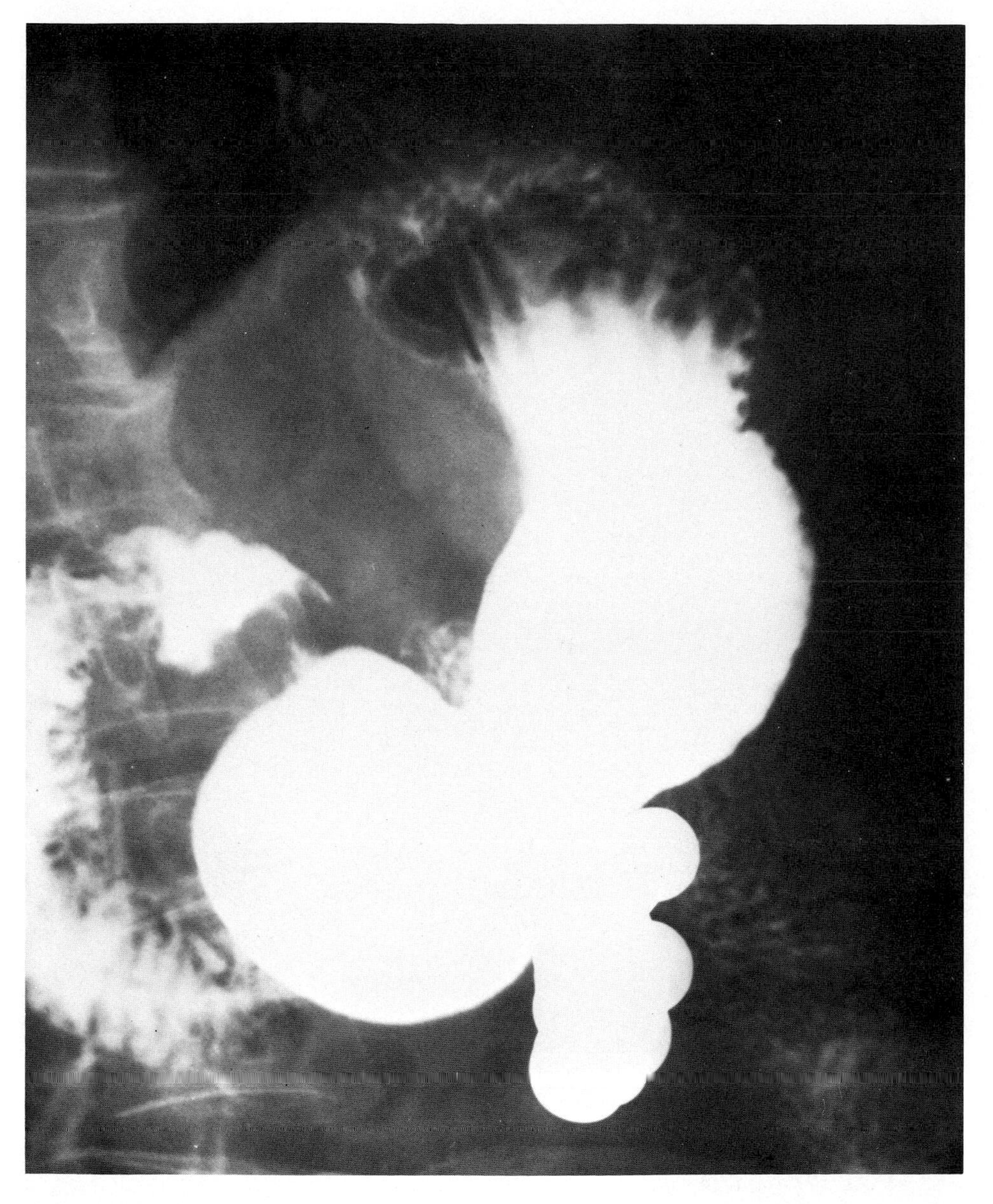

FIG. 13-8. *The rather unusual appearance of this barium-filled stomach was caused by the presence of a number of coins in the patient's pajama pocket. (Courtesy of Ralph Coates, R.T. [R])*

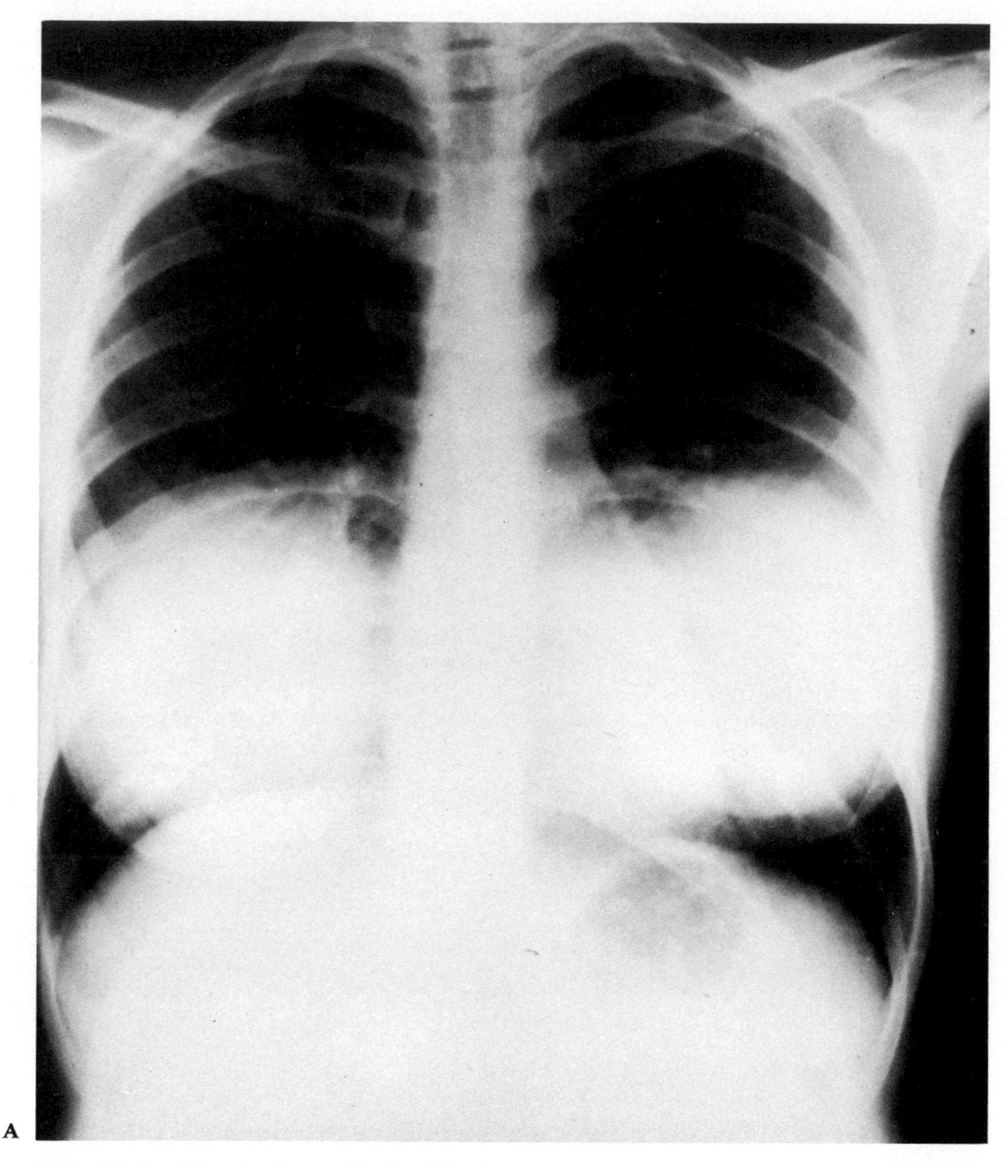

FIG. 13-9. *The shadows of both of the breasts appear extremely dense in these anterior* (A) *and lateral* (B) *views of the chest because of the presence of a silicone implant in each breast.* (A *and* B, *courtesy of William A. Conklin, R.T., F.A.S.R.T.)*

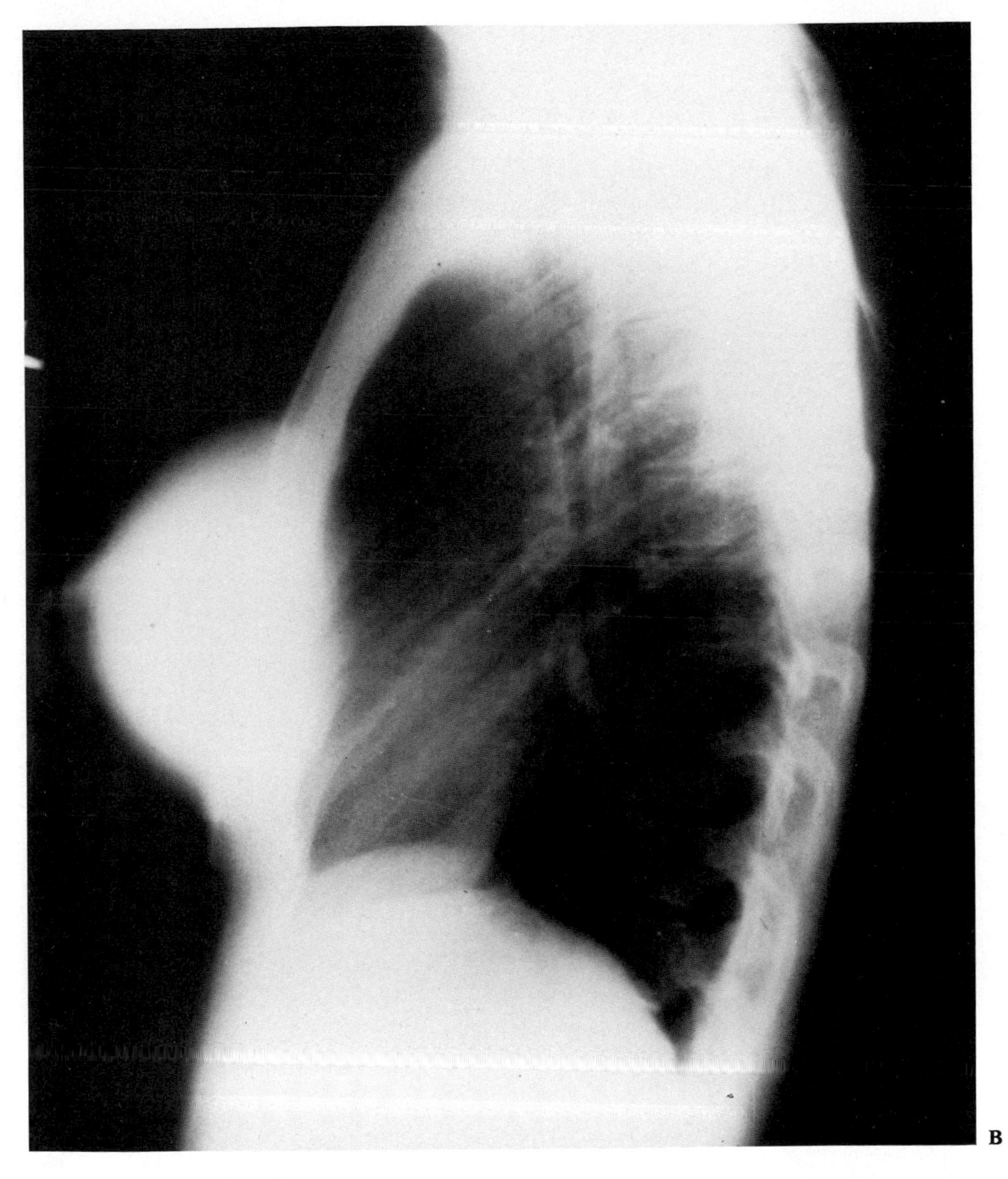

FIG. 13-9. *continued*

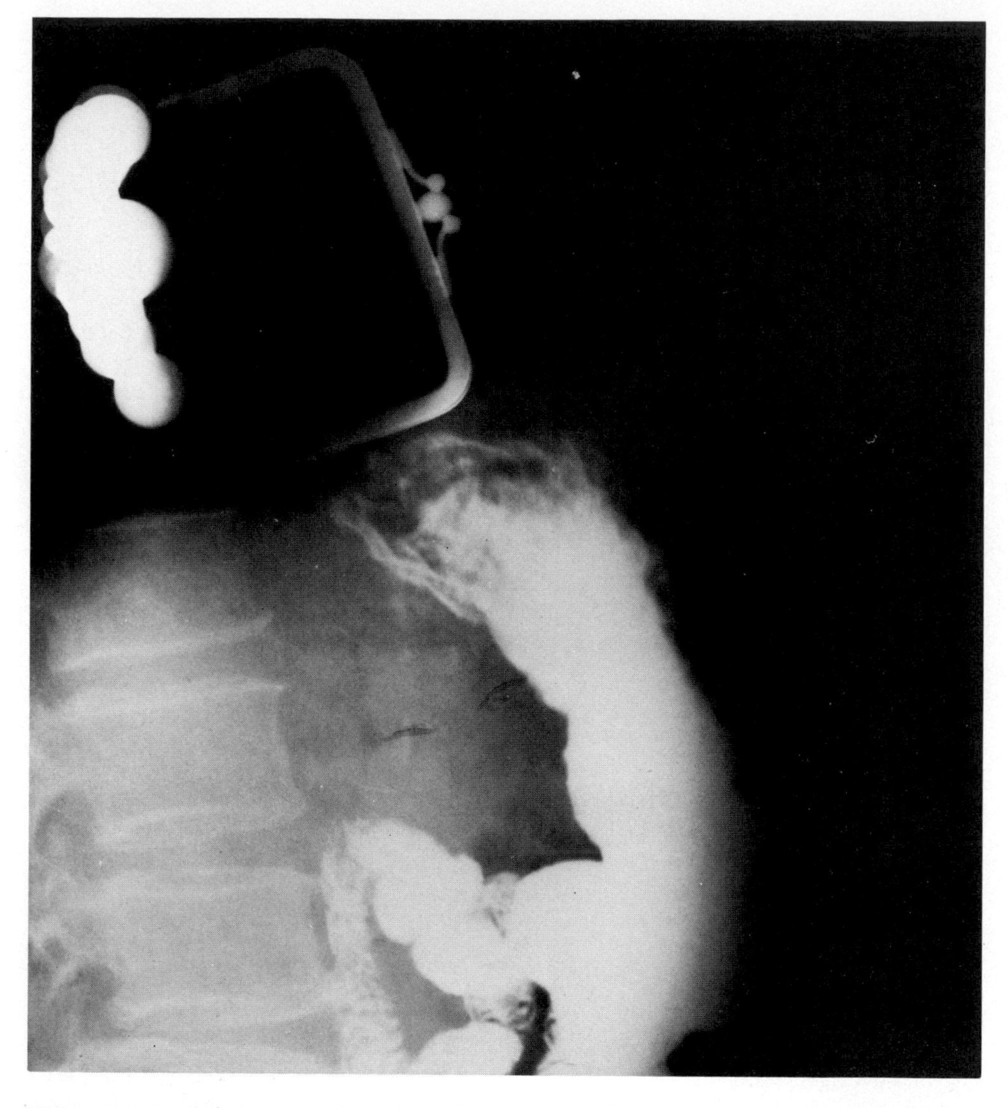

FIG. 13-10. *While the outline of a coin purse and its contents is obvious in this radiograph, compare the image of the artifact with that seen in Figure 13-8. (Courtesy of Ralph Coates, R.T. [R])*

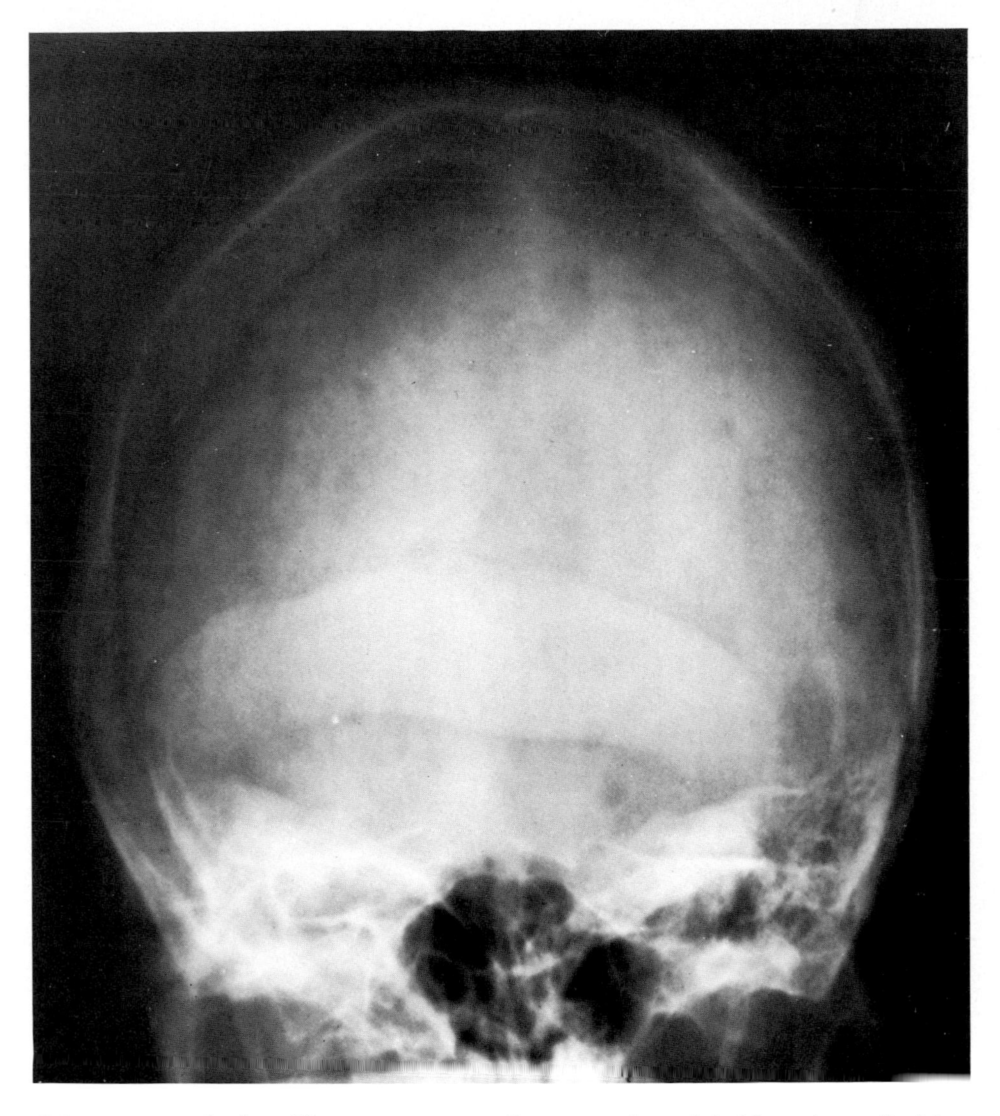

FIG. 13-11. *The bandlike structure extending across the occipital bone is a roll of fatty tissue located in the lower cervical region of an obese patient. The angle of the x-ray tube was responsible for imaging this structure in the occipital area.*

Phantom Renal Calculus

For a change, the radiograph in Figure 13-12 does not exhibit any artifacts, and the radiologist's report confirms the presence of bilateral renal calculi. Notice the granular appearance of the larger stone in the left kidney, and compare it with the artifact in Figure 13-13. The artifact occurred when an iodinated contrast medium was absorbed into the table pad on which the patient was positioned for the radiographic examination. The granular appearance is typical of an iodine-based substance that is allowed to dry after being spilled or absorbed into positioning sponges, table pads, and the like. Unfortunately, as with this example, most artifacts usually appear within or near the anatomic region of interest. Consequently, the somewhat suspicious appearance of such artifacts could lead to troublesome situations if not immediately identified and corrected.

Incidentally, the difference in appearance and radiodensity of the iodinated spot on the table pad (Fig. 13-13, *left*) and the artifact in the region of the kidney (Fig. 13-13, *right*) is due to the weight of the patient, which compressed the table pad during the exposure.

Figures 13-14 through 13-30 illustrate some additional artifacts of miscellaneous origin.

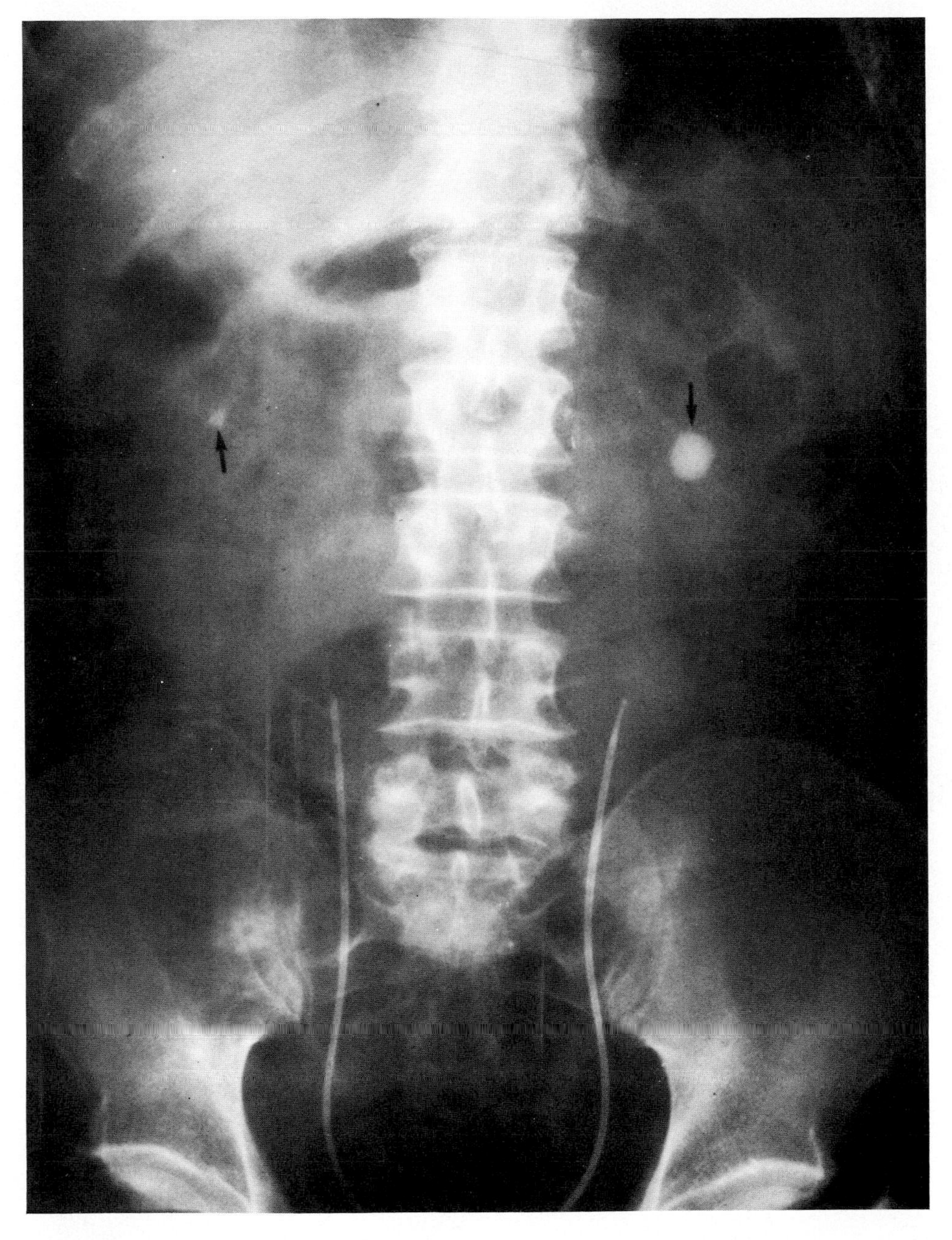

FIG. 13-12. *This scout film obtained during a retrograde pyelogram demonstrates the presence of bilateral renal calculi* (arrows).

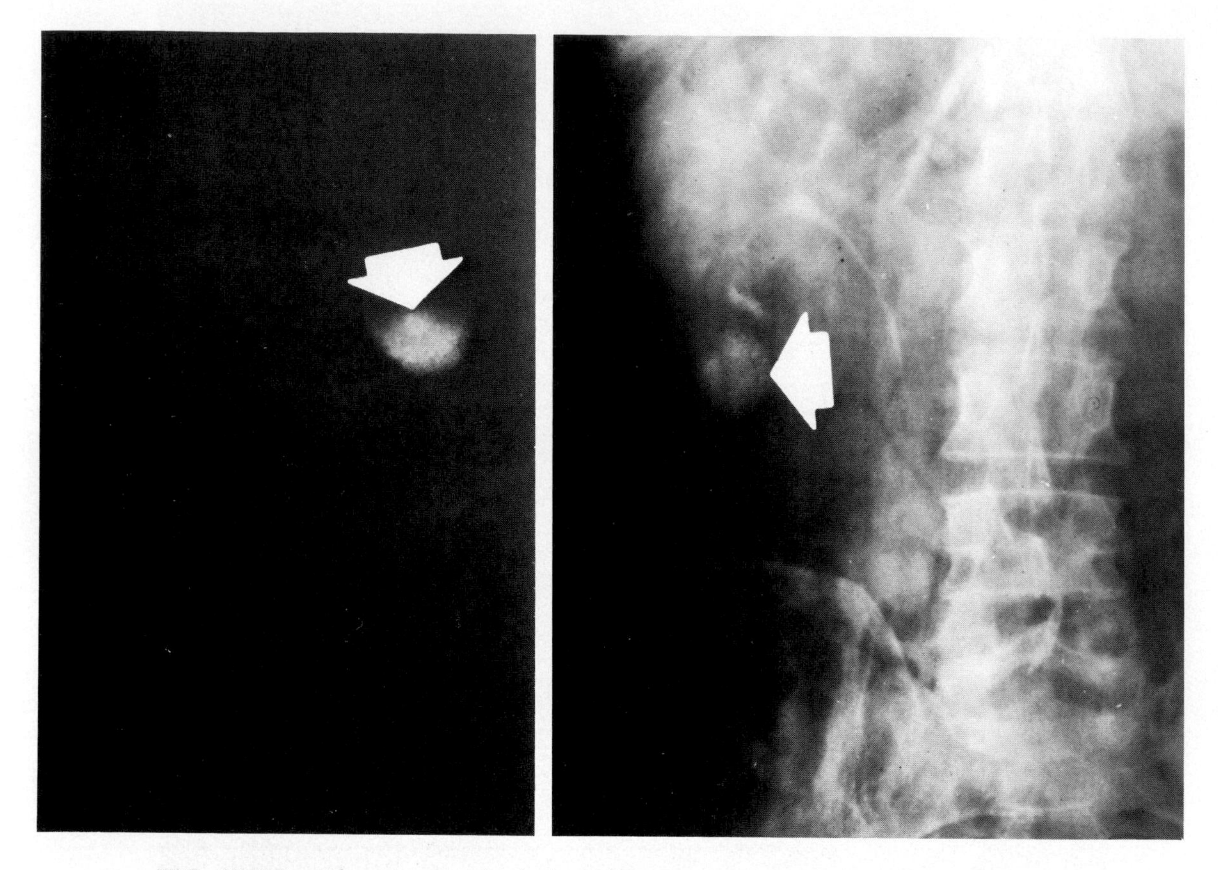

FIG. 13-13. *Phantom renal calculus* (right, arrow), *resulting from the absorption of contrast medium within the pad on the radiographic table* (left, arrow).

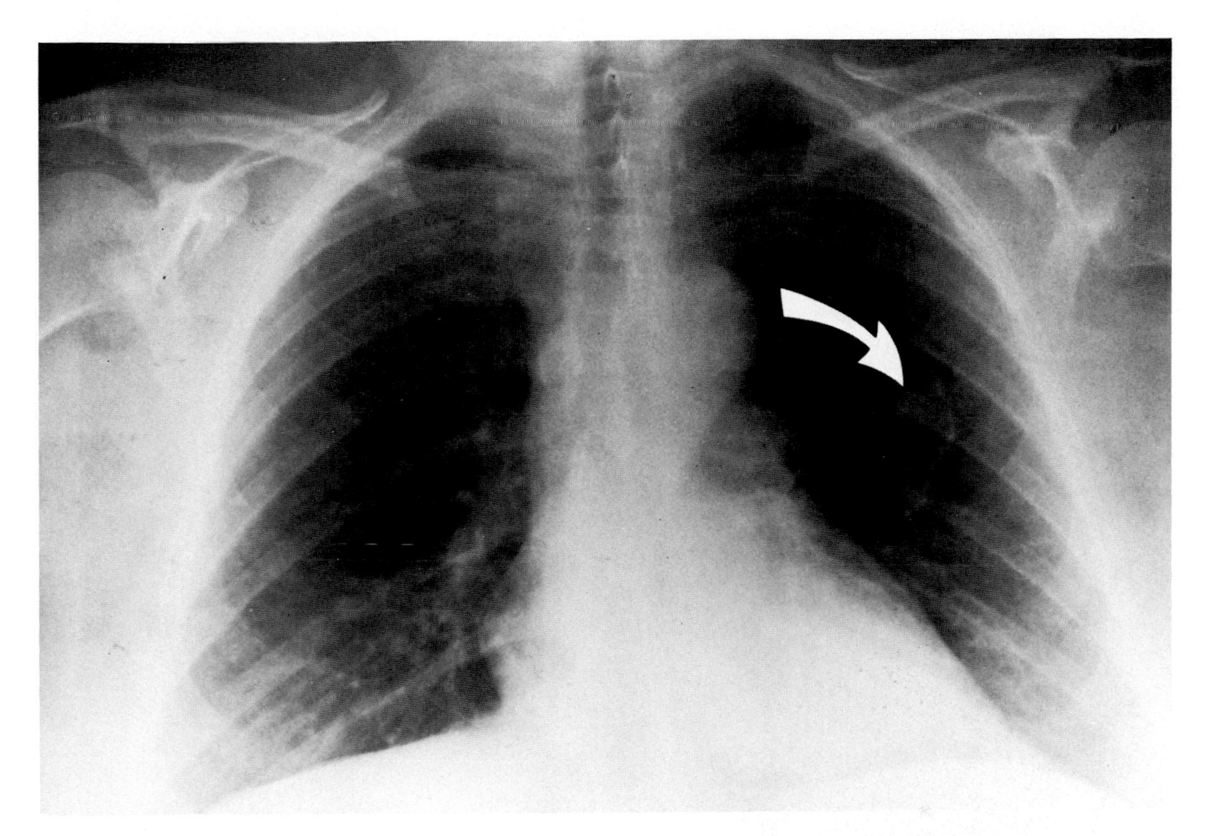

FIG. 13-14. *This radiograph of the chest was obtained following a gastrointestinal series. The artifact* (arrow) *is a barium stain located on the front of the patient's gown.*

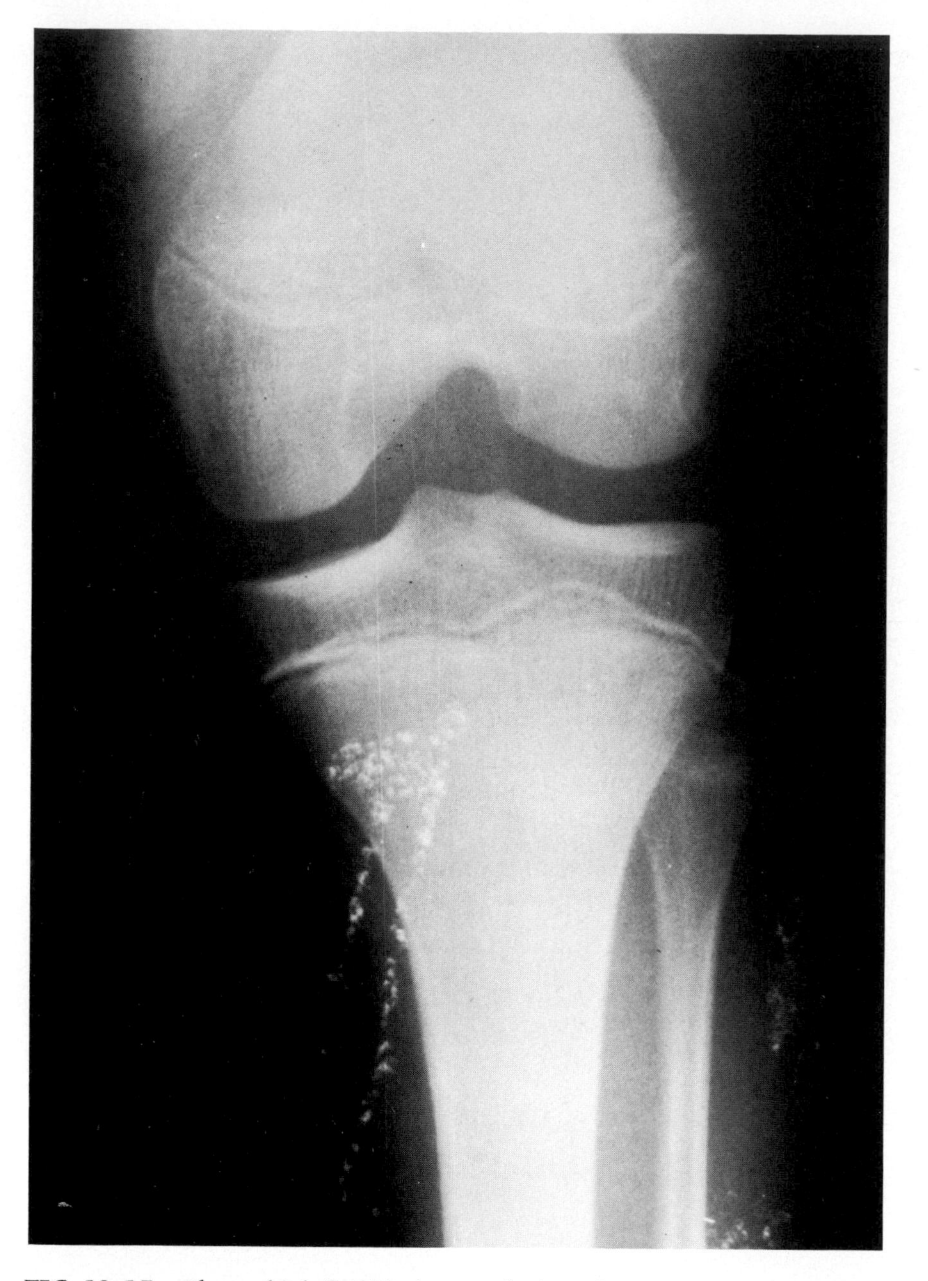

FIG. 13-15. *The multiple flecklike images of minus density in this radiograph of the knee were caused by the absorption of a contrast medium that was in the positioning sponge used to support the patient's knee.*

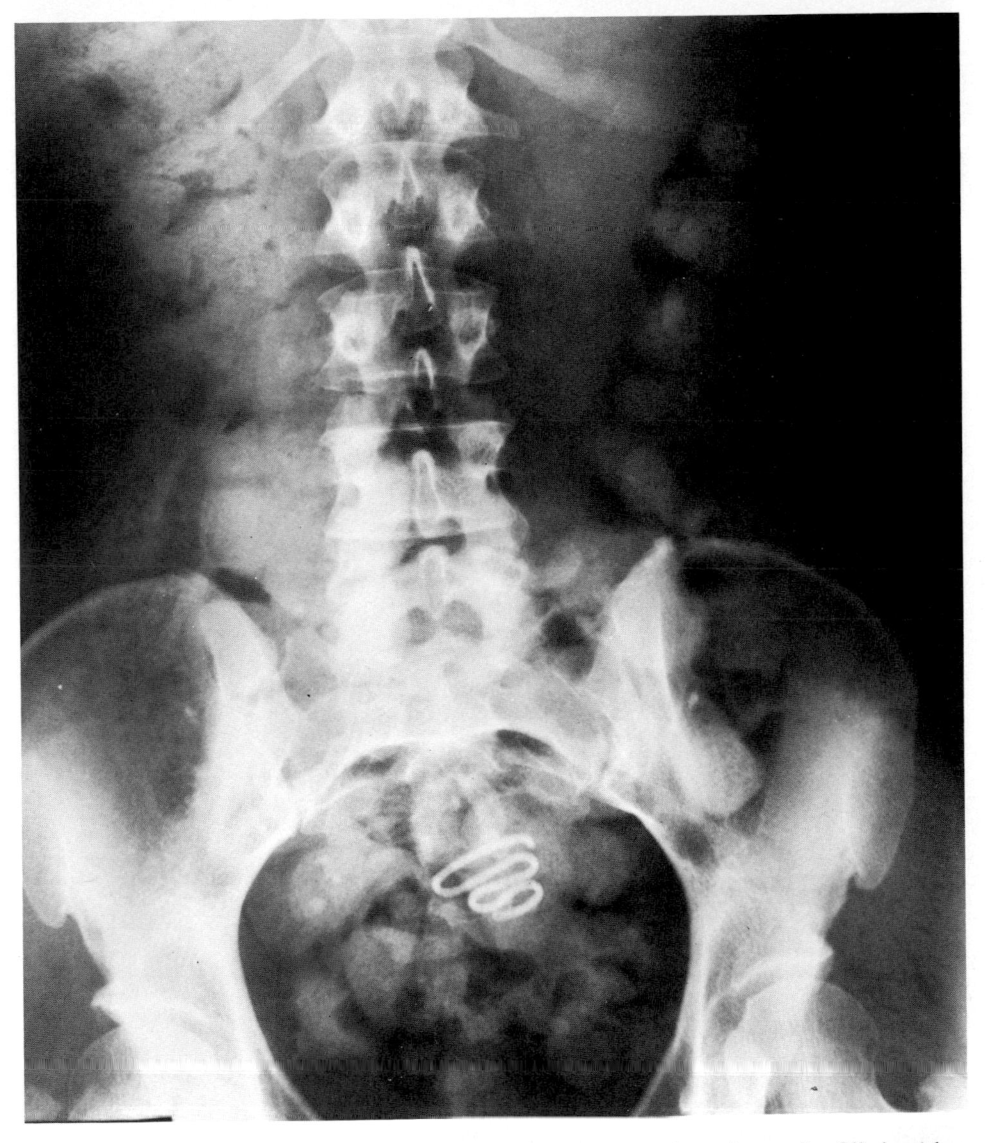

FIG. 13-16. *This radiograph of the abdomen revealed the colon to be filled with a radiopaque substance. The patient denied having ingested any medicine or having undergone any recent radiologic procedure. After persistent questioning by the emergency-room physician, the patient finally admitted that she consumed large amounts of clay daily. Although it is not generally known that geophagia (dirt eating) is practiced by some adults, the literature suggests that such practice should be suspected whenever there is an unexplained increase in the radiopacity of the fecal content of the colon. (Courtesy of Dr. Paul O'Connor. See Clayton RS, Goodman PH. The roentgenographic diagnosis of geophagia [dirt eating]. Am J Roentgenol Rad Ther Nucl Med 73, No. 2: 203–207, 1955)*

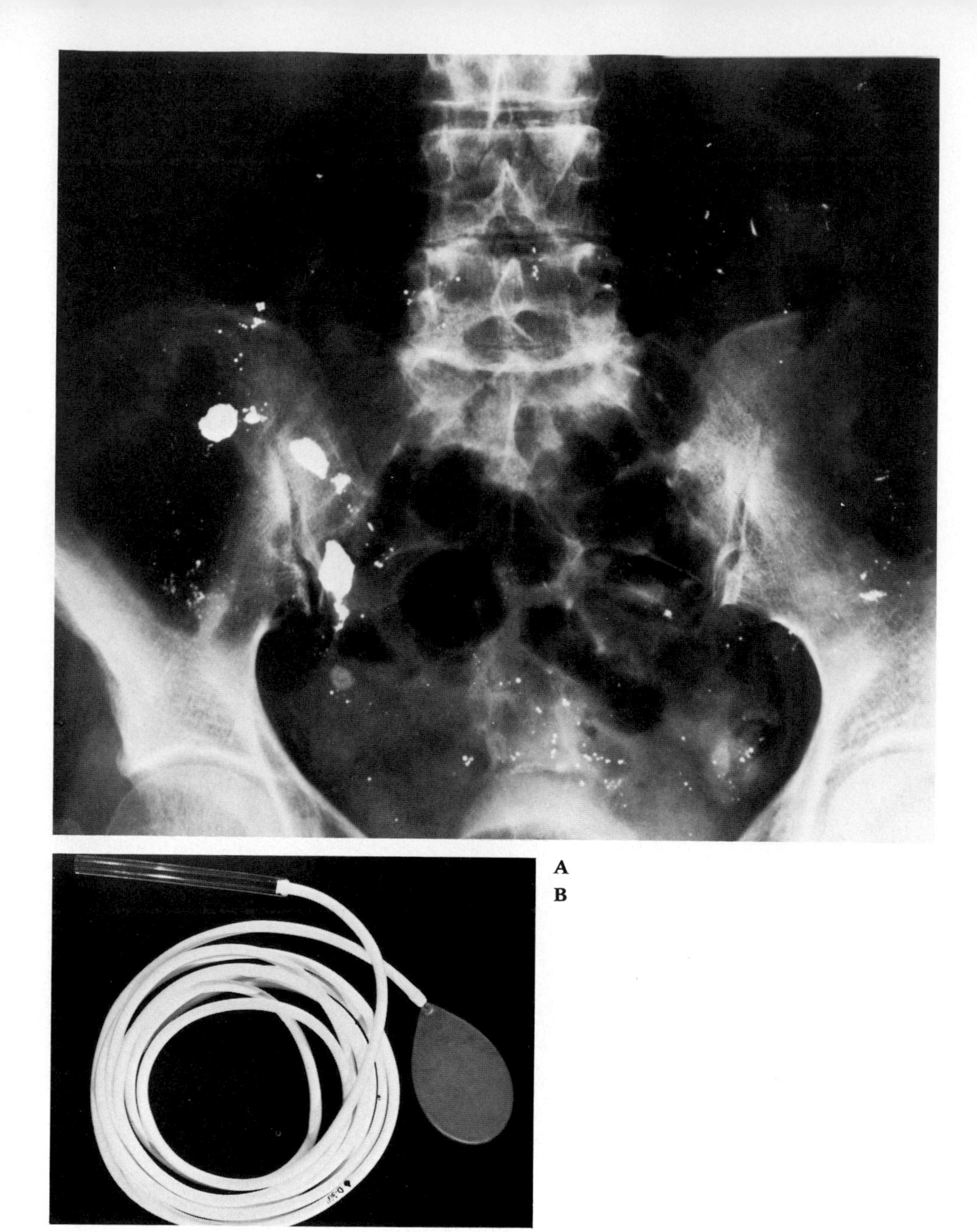

FIG. 13-17. (A) *The artifacts of minus density in this abdominal study appear to be residual traces of barium. However, this is mercury that escaped into the intestine when a bag containing 4 ml of mercury at the end of a Cantor tube broke while it was being removed from the patient.* (B) *The bag at the end of the Cantor tube is often injected with mercury to facilitate its passage through the stomach and intestine. (See Torres LS, Morrill CM: Basic Medical Techniques and Patient Care for Radiologic Technologists, 2nd ed, pp 92–93. Philadelphia, JB Lippincott, 1983)*

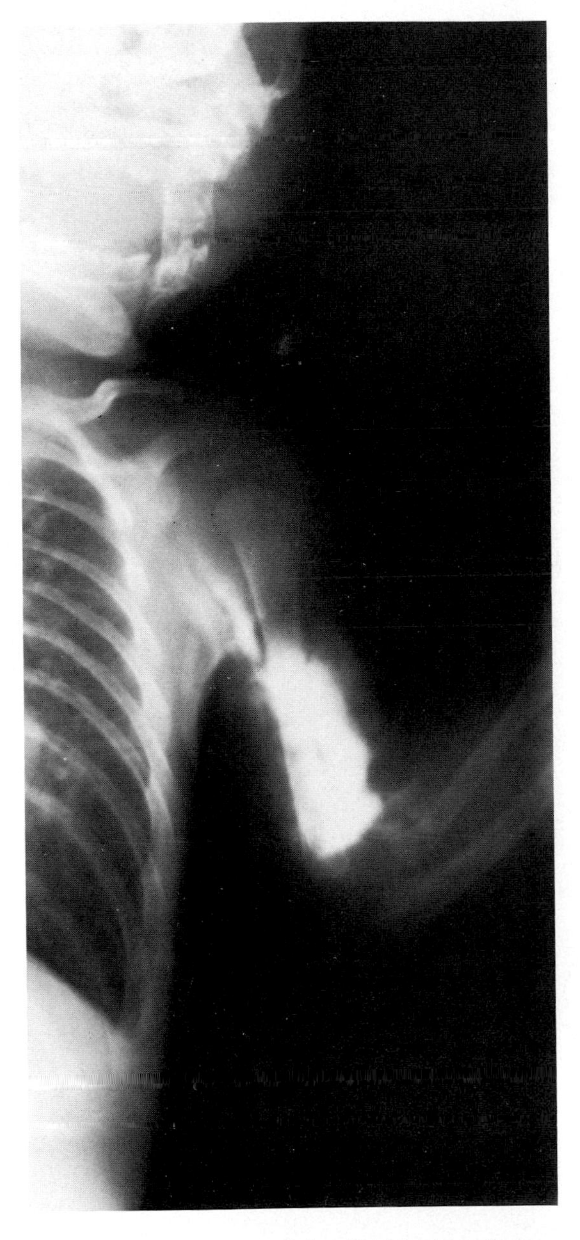

FIG. 13-18. *The opaque image in this radiograph of the arm is localized in the soft-tissue region of the humerus. The radiograph was obtained after the urinary tract failed to opacify following the injection of a contrast medium. This radiograph demonstrates that the contrast substance had been injected subcutaneously. (See Caffey J: Pediatric X-ray Diagnosis, Vol. 2, p 1582. Chicago, Year Book Medical Publishers, 1973)*

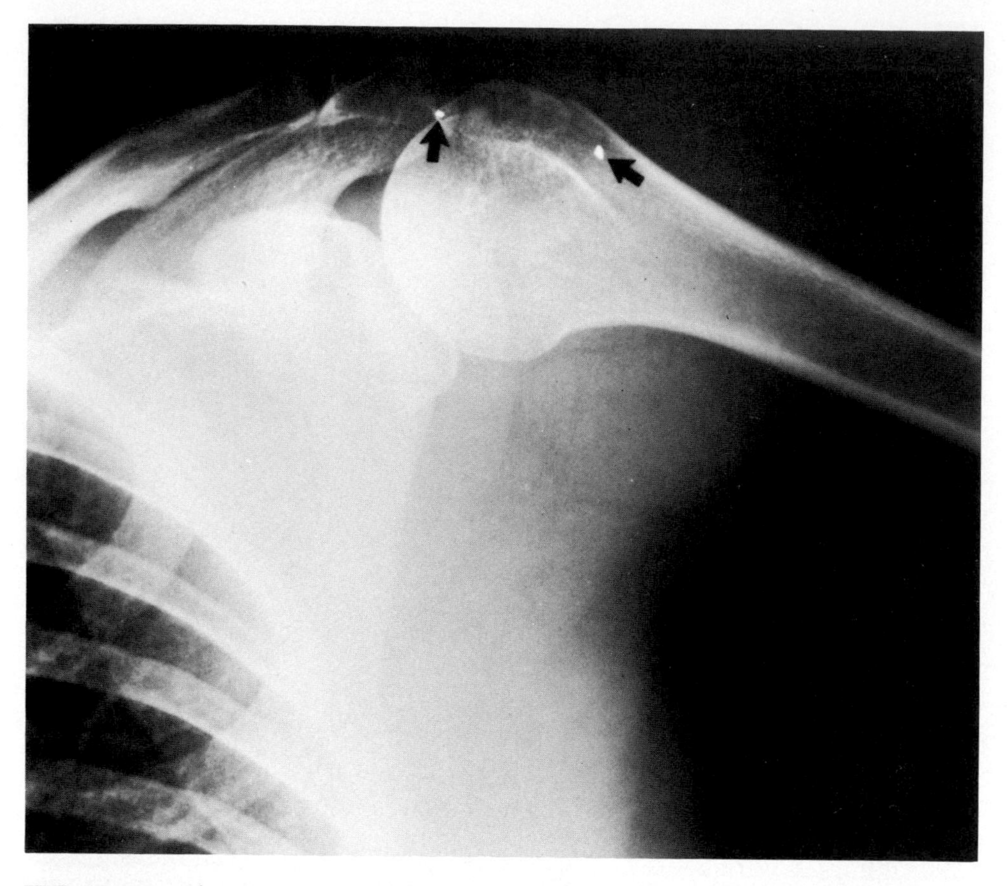

FIG. 13-19. *The two artifactual images* (arrows) *in this radiograph of the shoulder were caused by dust particles on the surface of the intensifying screen. Compare the similar appearance of these images with that of the pellets in Figures 13–20 and 13–21.*

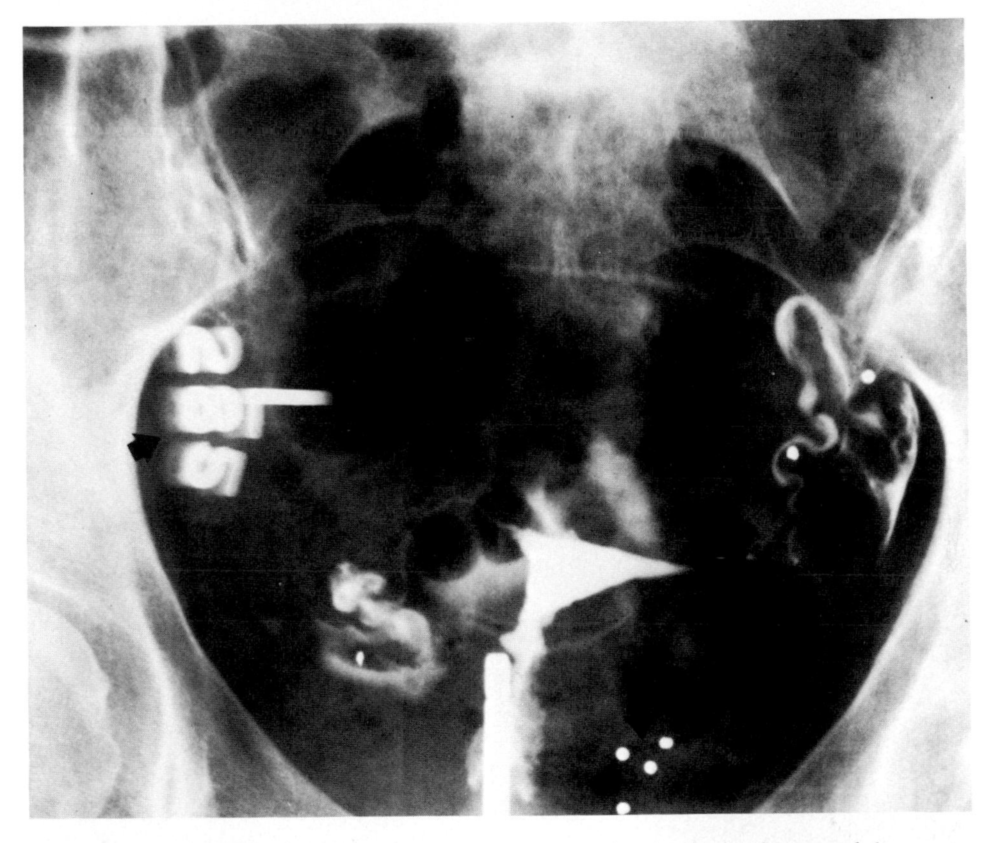

FIG. 13-20. *The multiple circular images of minus density in this hysterosalpingogram are BB pellets in the intestine. The patient ingested the pellets the evening before the radiographic examination, when she ate some wild duck that her husband had shot and prepared for dinner. The lead identification marker* (arrow) *appears unsharp, but the adjacent structures are well defined. The unsharpness resulted when the lead marker became caught and was dragged to this point while located on the underside of a moving-type grid. (Courtesy of Dave Sack, R.T. [R])*

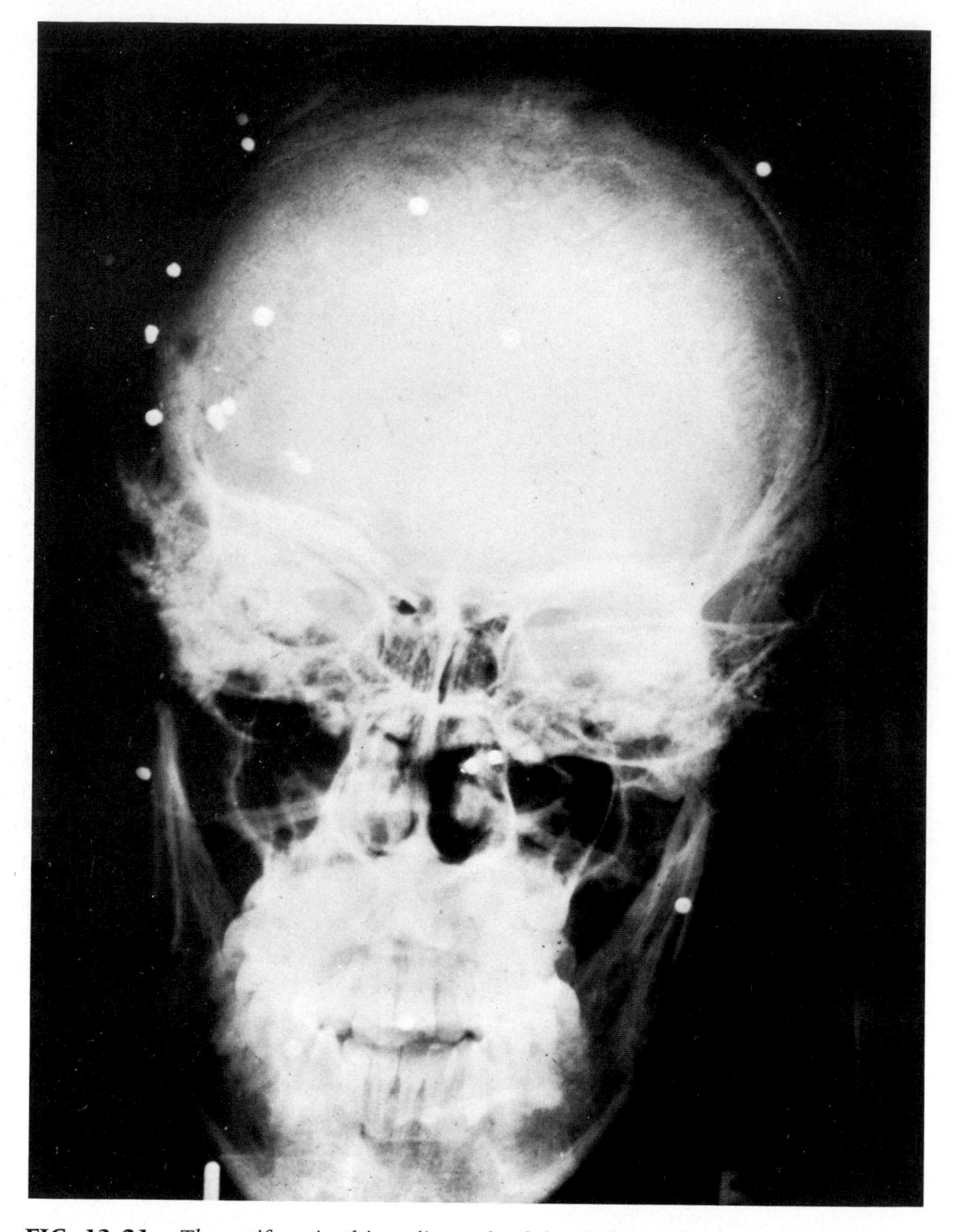

FIG. 13-21. *The artifacts in this radiograph of the skull are shotgun pellets in the soft tissue and bone. The patient had been shot during a robbery 5 years prior to having this radiograph of his skull made following an automobile accident.*

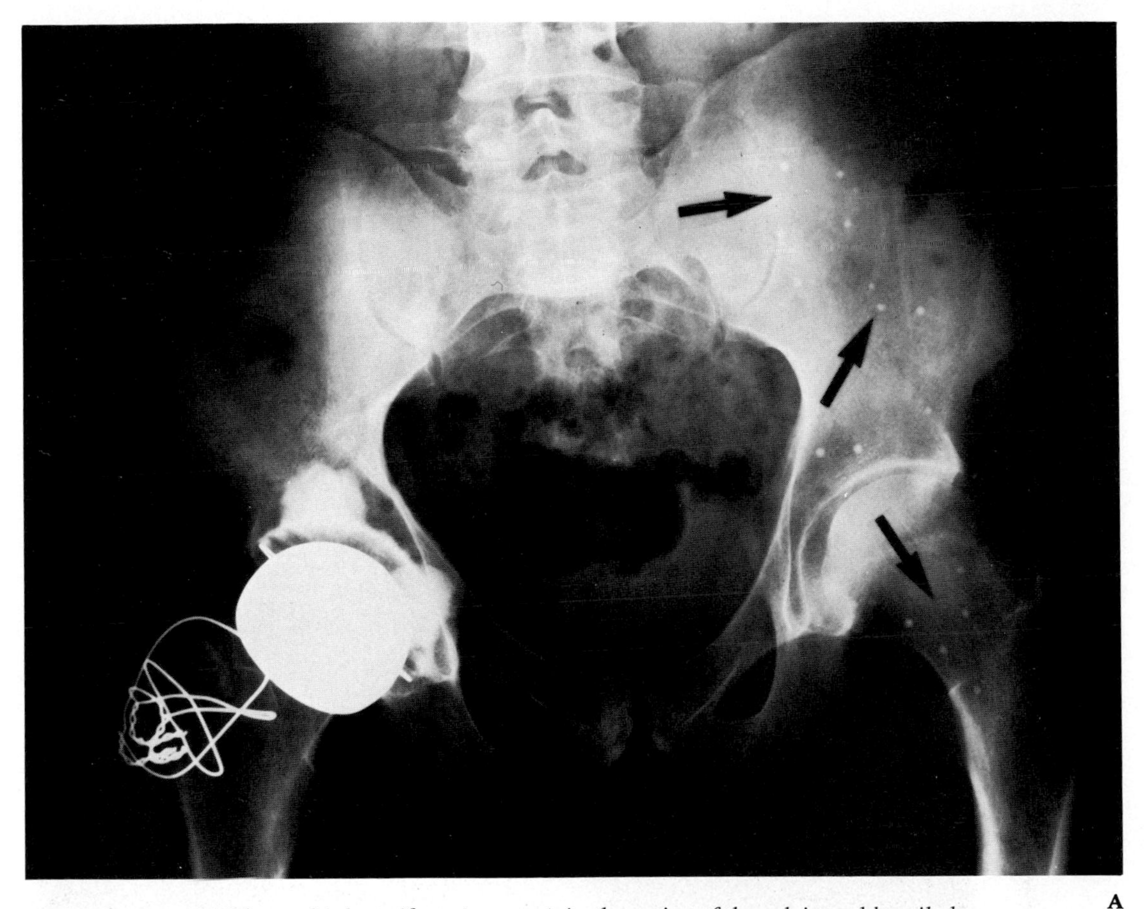

A

FIG. 13-22. (A) *The multiple artifacts* (arrows) *in the region of the pelvis could easily be mistaken for shotgun pellets. However, the images appeared in additional radiographic studies made in the same x-ray room* (B, arrows). (C) *After an investigation of the condition of the cassettes and intensifying screens for the possible cause of the artifacts, the x-ray tabletop was removed. It revealed that the artifacts were caused by the presence of barium located on the underside of the table.* *(Fig. 13-22 continues.)*

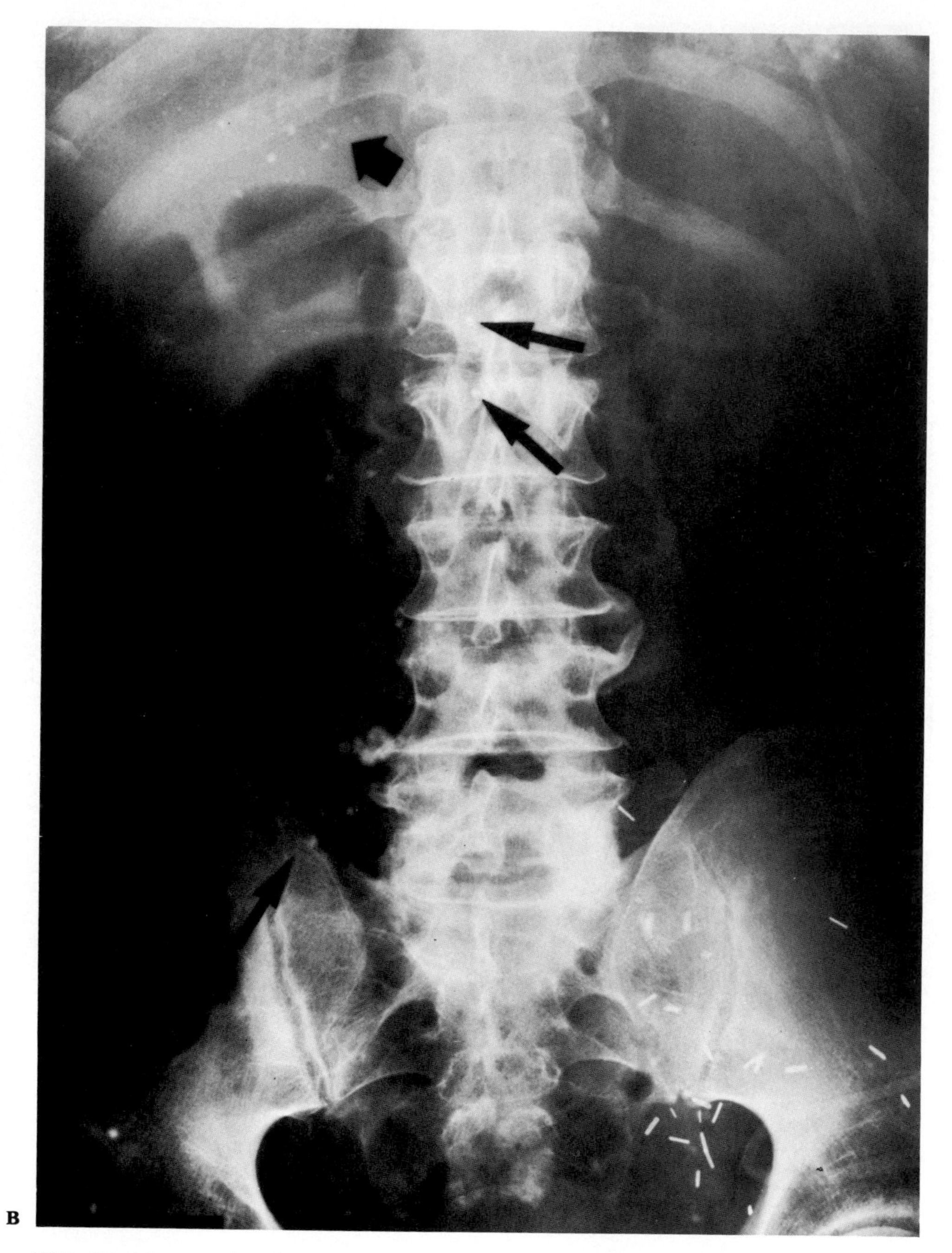

B

FIG. 13-22. *continued*

FIG. 13-22. *continued*

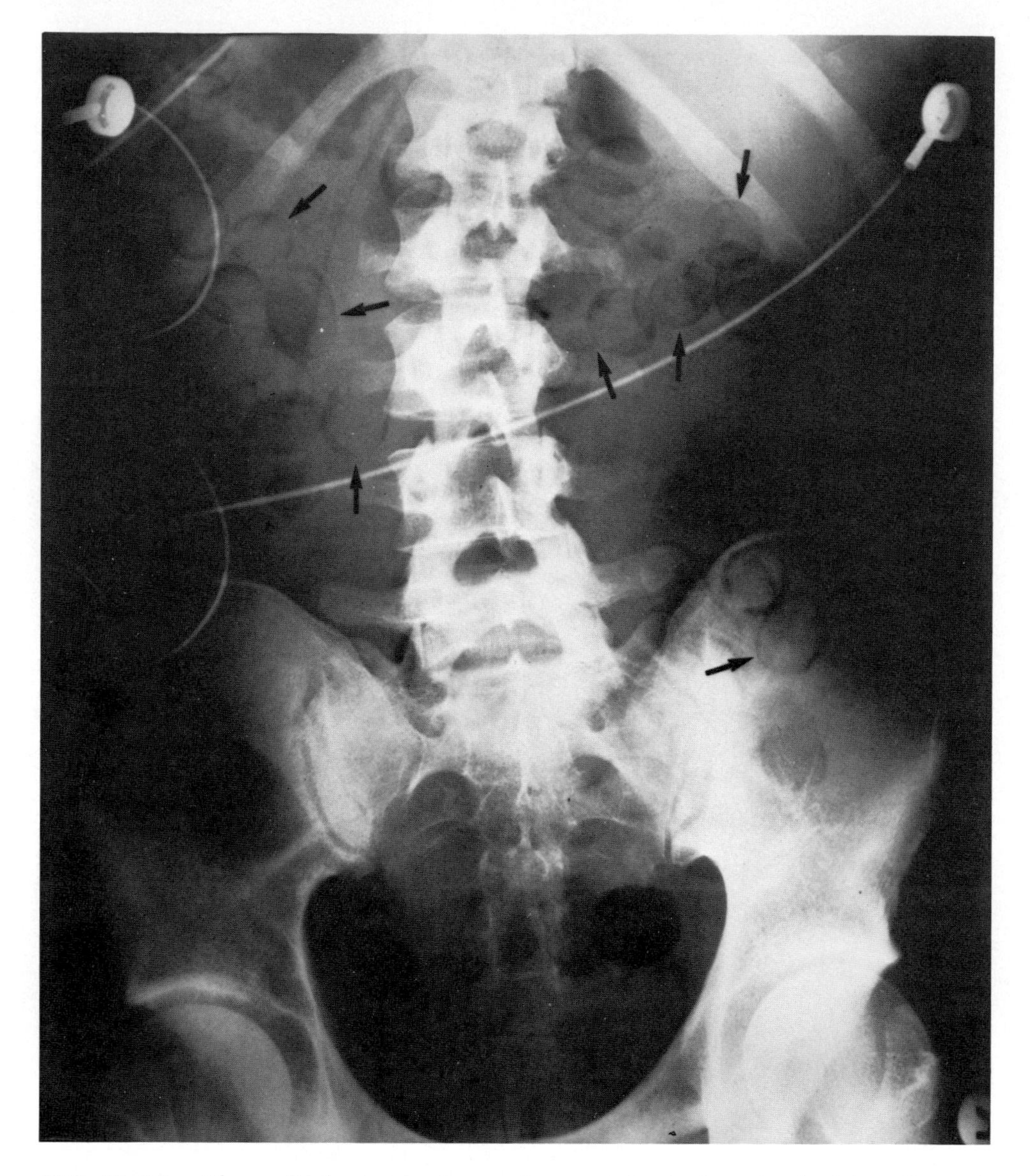

FIG. 13-23. *The multiple circular images* (arrows) *in the intestine in this radiograph are condoms filled with cocaine. The patient admitted himself to the hospital complaining of abdominal cramps. A radiographic study revealed the presence of a foreign material (no pun intended) in his intestine. After the physician confronted the patient with his findings, the patient admitted that while overseas he had swallowed the cocaine-filled condoms in order to smuggle the drugs into the country. Approximately 230 g of cocaine was contained within the various ingested condoms, which were removed surgically. (Courtesy of Dave Sack, R.T. [R]. See Freed TA, Sweet LN, Gauder PJ: Balloon obturation bowel obstruction: A hazard of drug smuggling. AJR 127:1033–1034, 1976)*

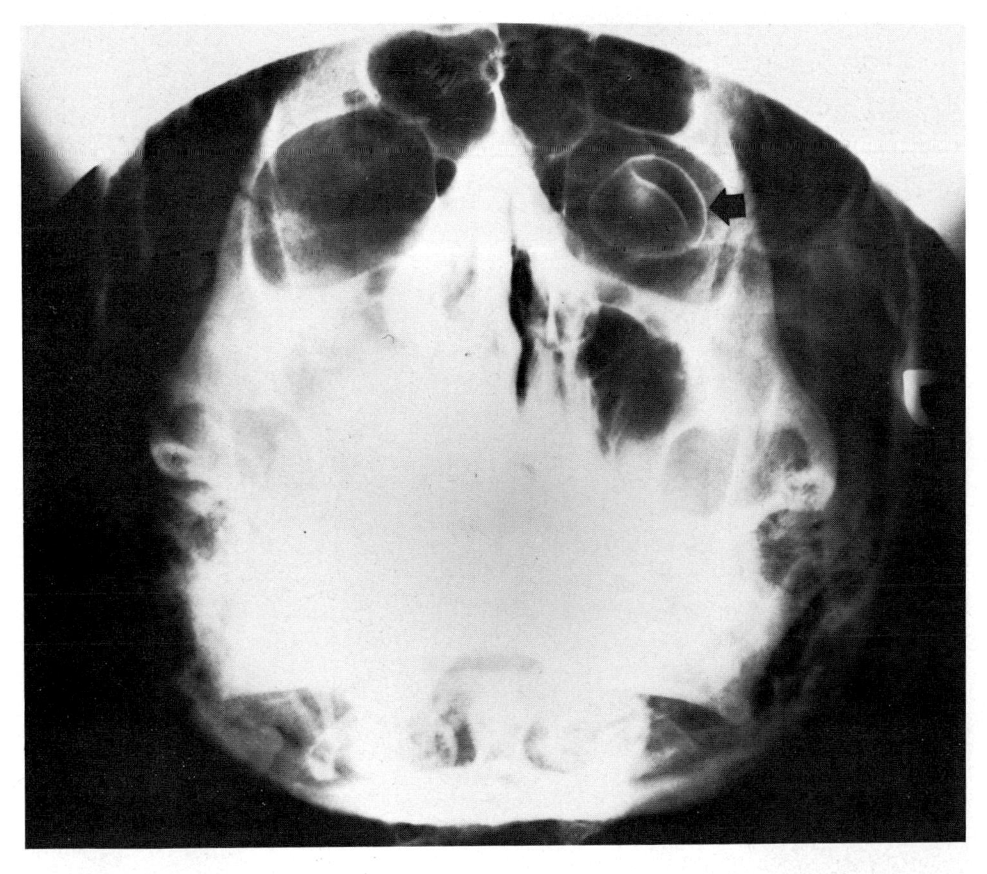

FIG. 13-24. *The well-defined artifactual image* (arrow) *in the orbit is a glass eye.*

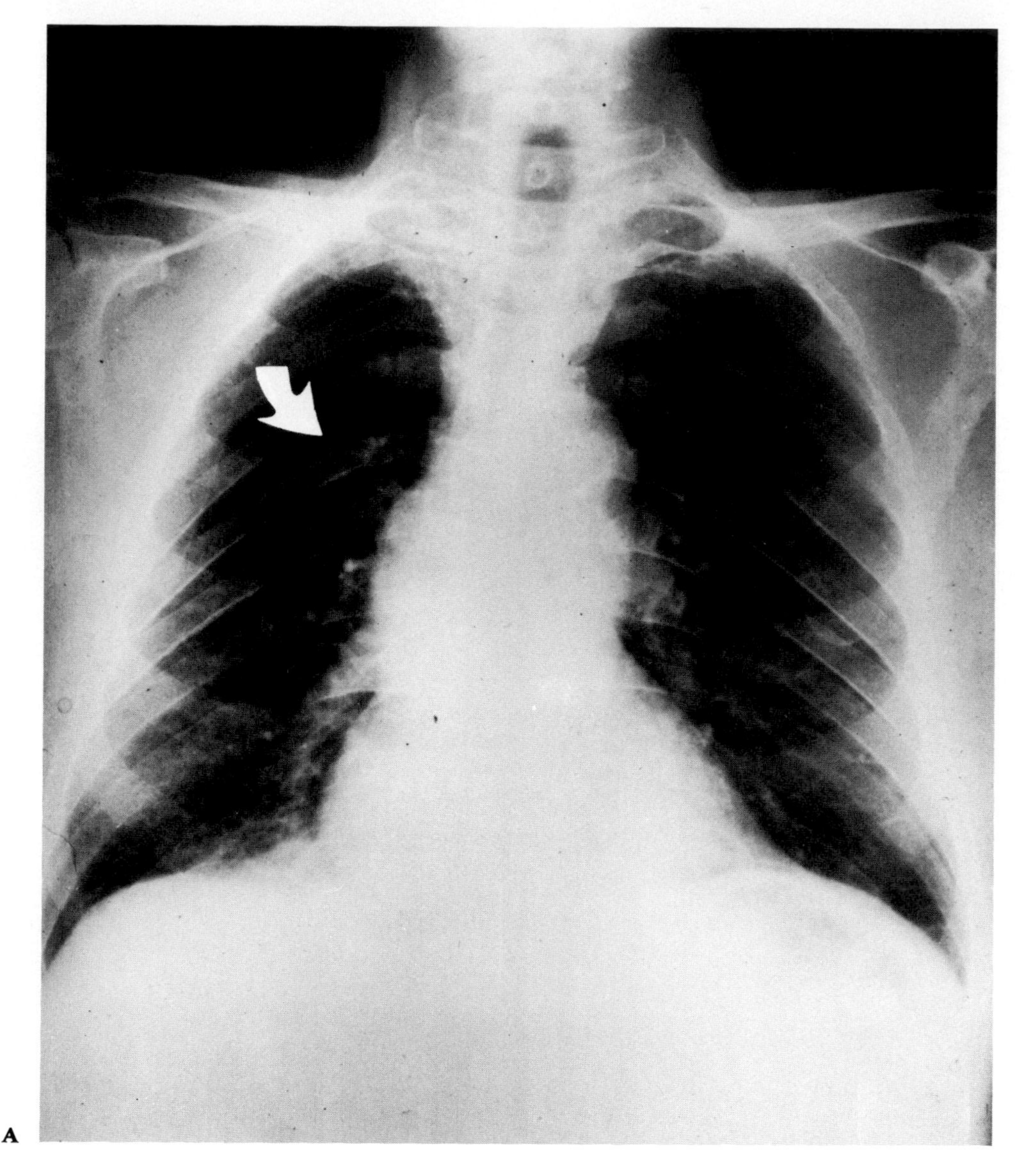

FIG. 13-25. (A) *This radiograph, made in a mobile chest x-ray screening facility, indicated the presence of a lesion in the right lung, according to the report. The patient was sent to a thoracic surgeon for a more extensive evaluation. While examining the patient, the surgeon discovered a large melanoma that was responsible for the appearance of the image* (arrow) *seen in the radiograph. Notice in* B *the similarity in appearance and location of the large melanoma to those of the image in the lung seen in* A. *It is obvious why such a finding was reported.*

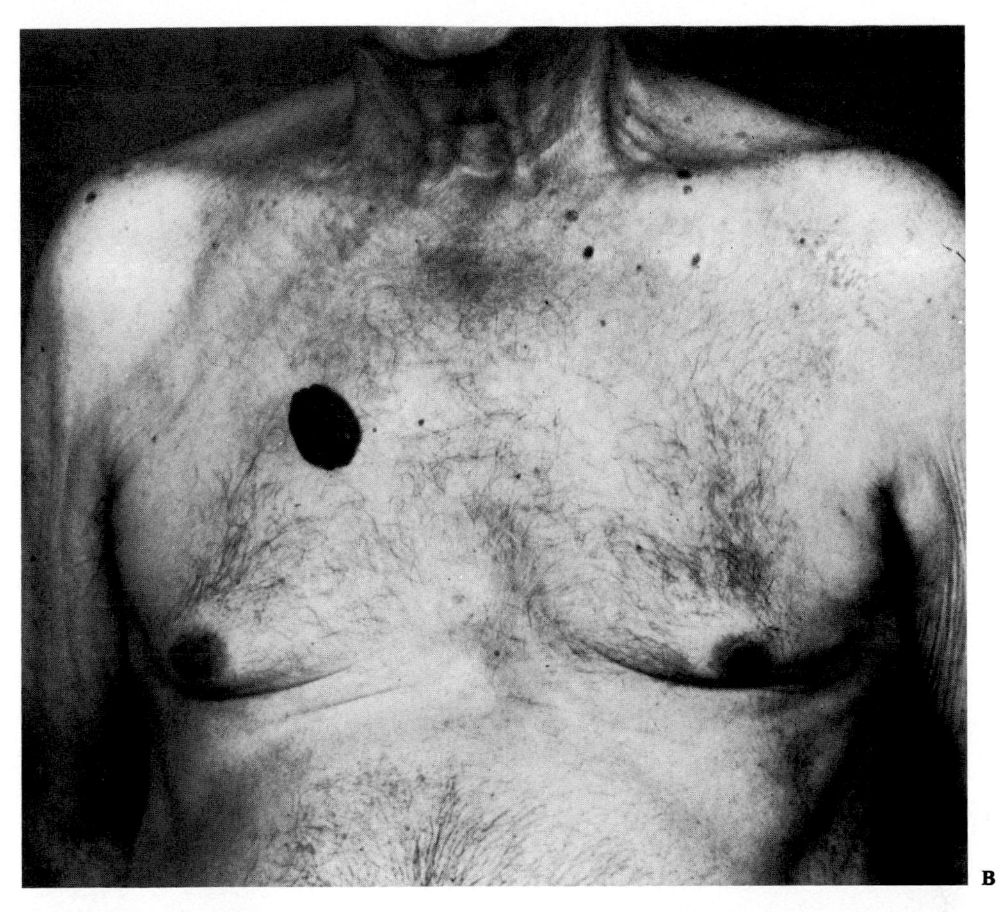

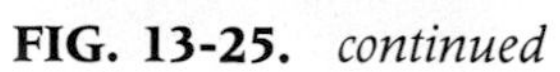
FIG. 13-25. *continued*

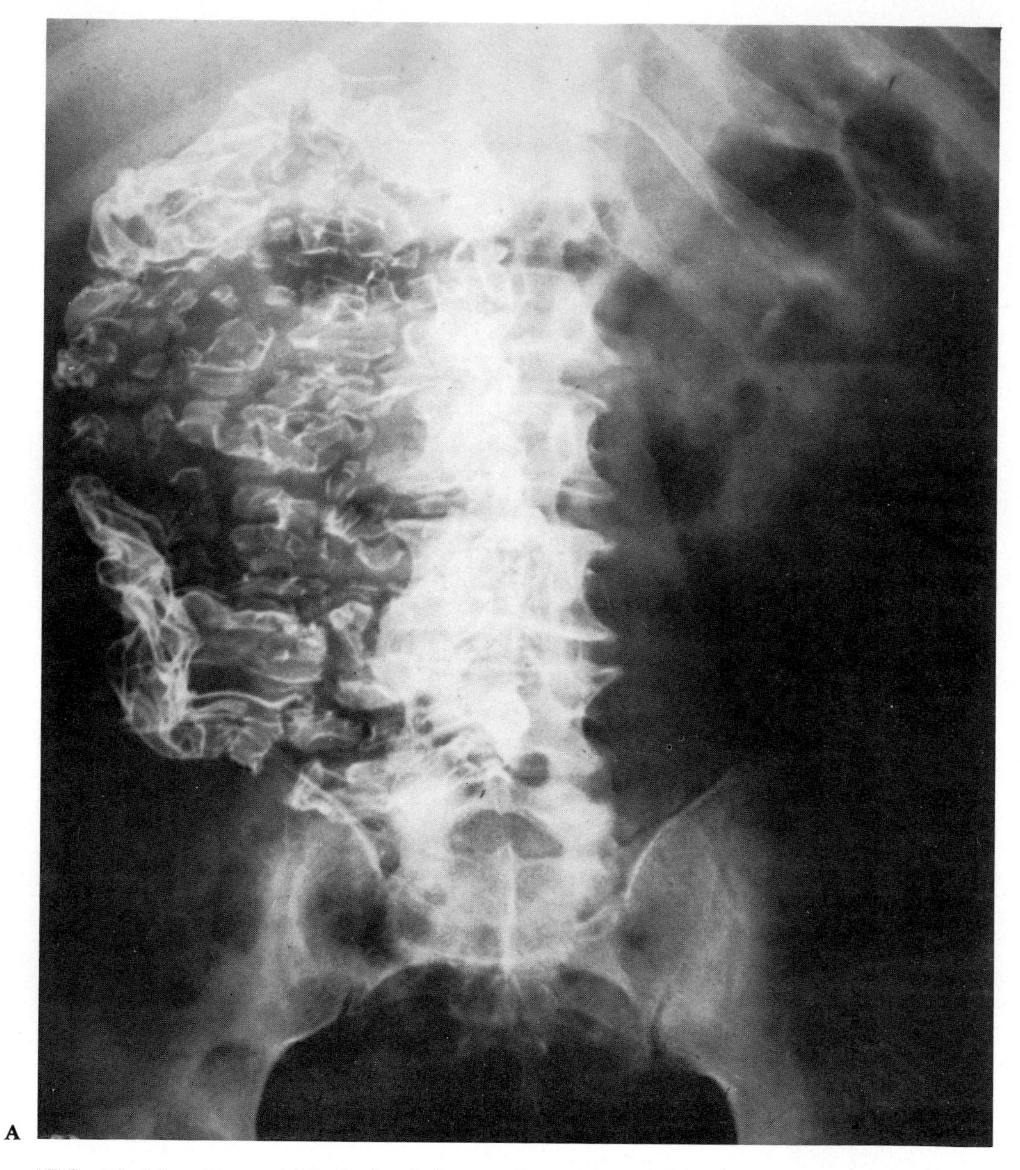

A

FIG. 13-26. (A *and* B) *Both of these radiographs exhibit the presence of an opaque substance known as tantalum, a noncorrosive, malleable metal that is used for plates or disks to replace cranial defects due to wounds and for making prosthetic appliances. In these instances, the tantalum was used to repair abdominal* (A) *and inguinal* (B) *hernias.*

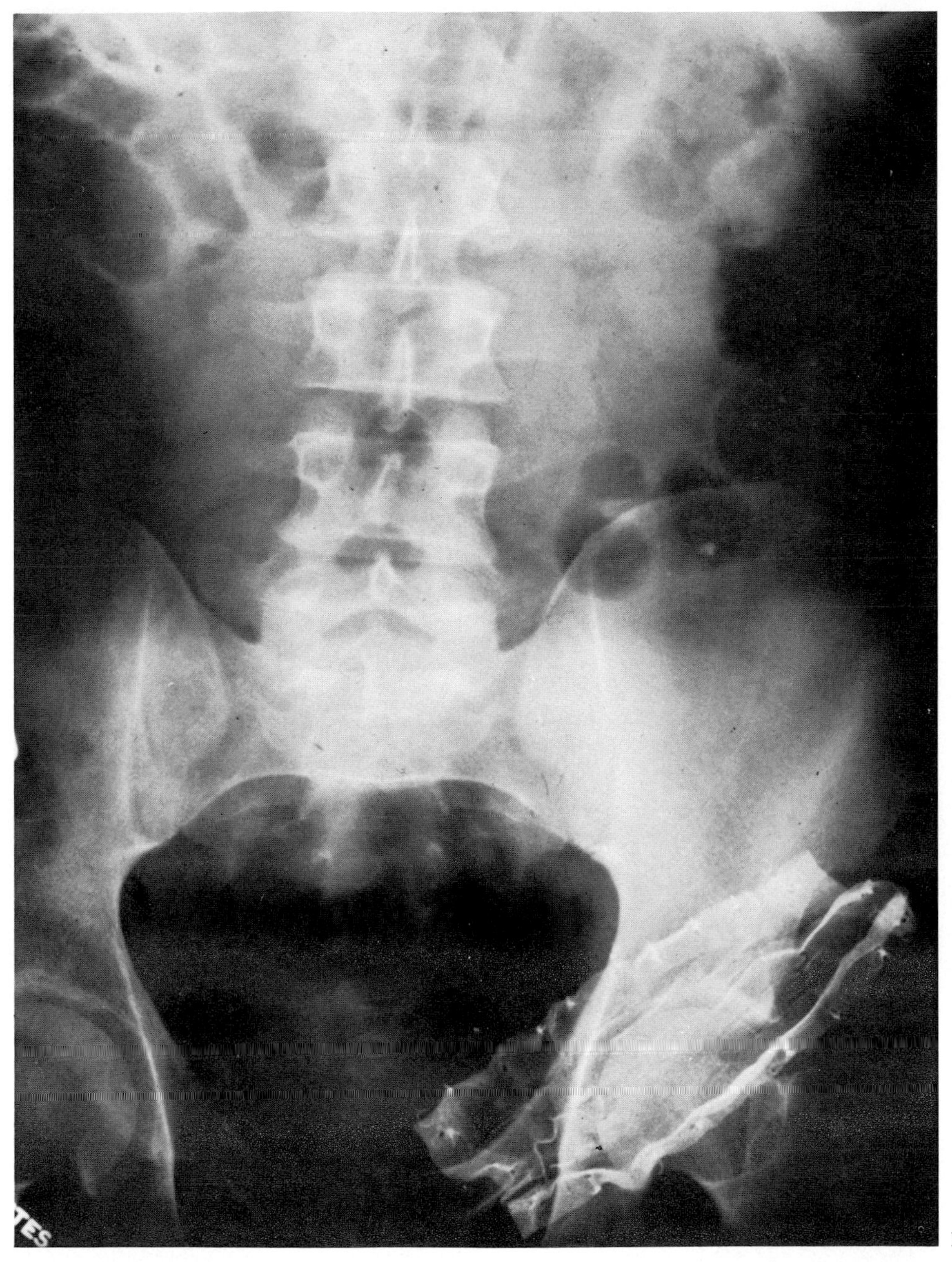

B

FIG. 13-26. *continued*

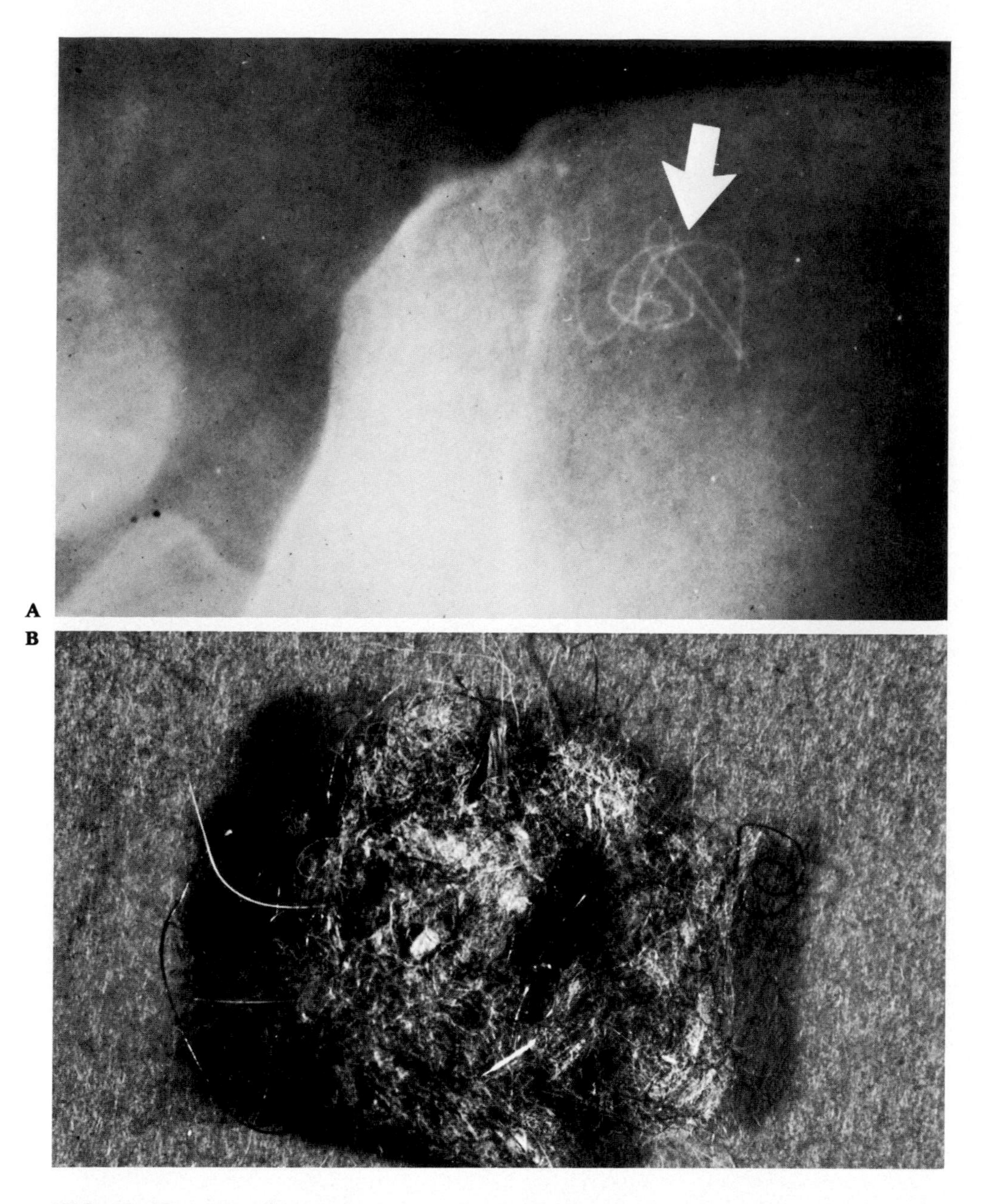

FIG. 13-27. (A) *This portion of a radiograph of the pelvis demonstrates a coil of wire* (arrow) *that appears to be suture material, but the patient had never had surgery.* (B) *The substance responsible for the artifact was a small tuft of dust composed of many things, including some small strands of copper wire. This material was found to have been on the surface of the plastic drape located on the underside of a moving radiographic tabletop. The movement of the table and the tabletop was responsible for the erratic imaging pattern of the artifact.*

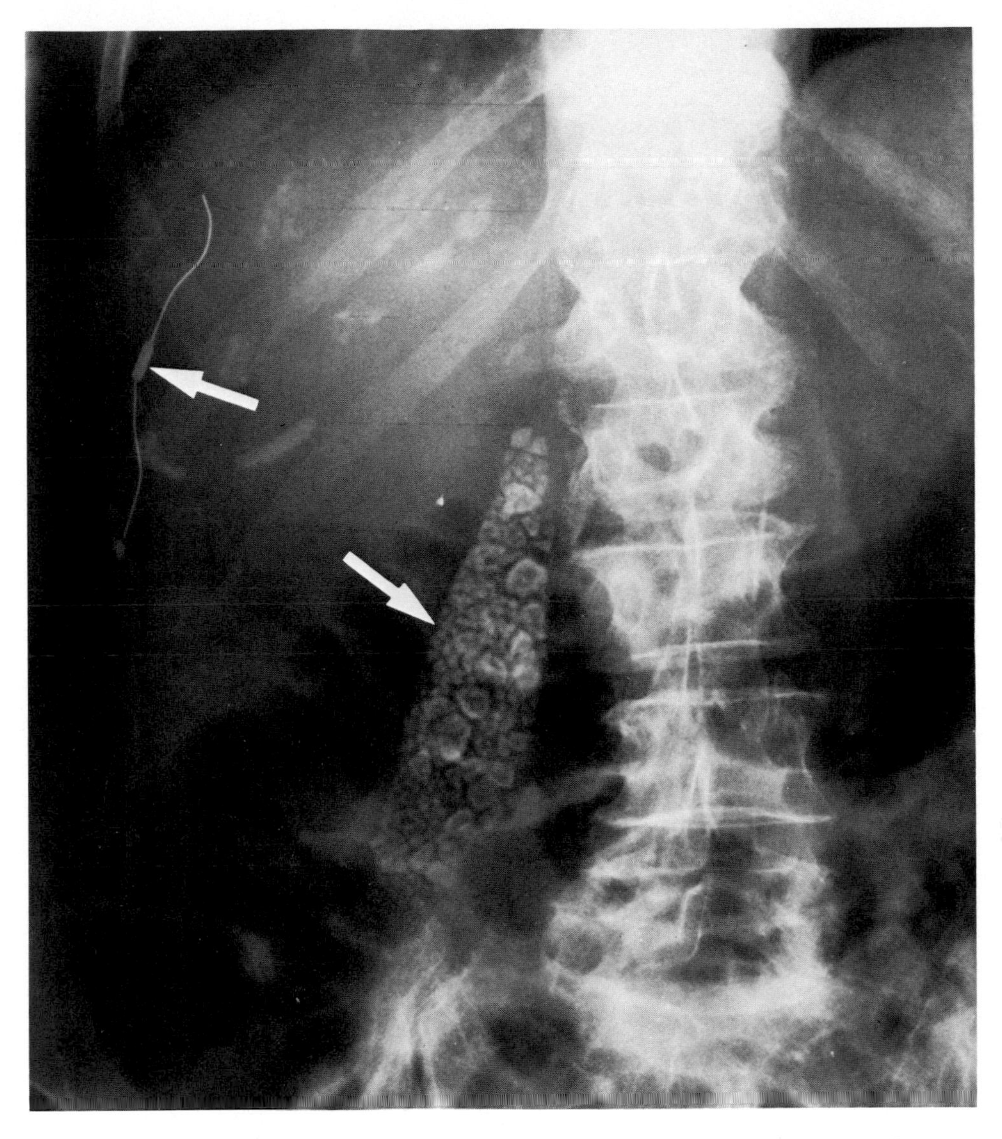

FIG. 13-28. *The radiologist's report on this radiograph indicates the presence of numerous radiopaque stones* (bottom arrow) *in the gallbladder, as well as a metallic object* (top arrow) *in the region of the right lower ribs. The latter is the image of a capacitor that was located under a sheet on the radiographic tabletop. It is reasonable to assume that this component was accidentally left on the surface of the table when the unit was being repaired.*

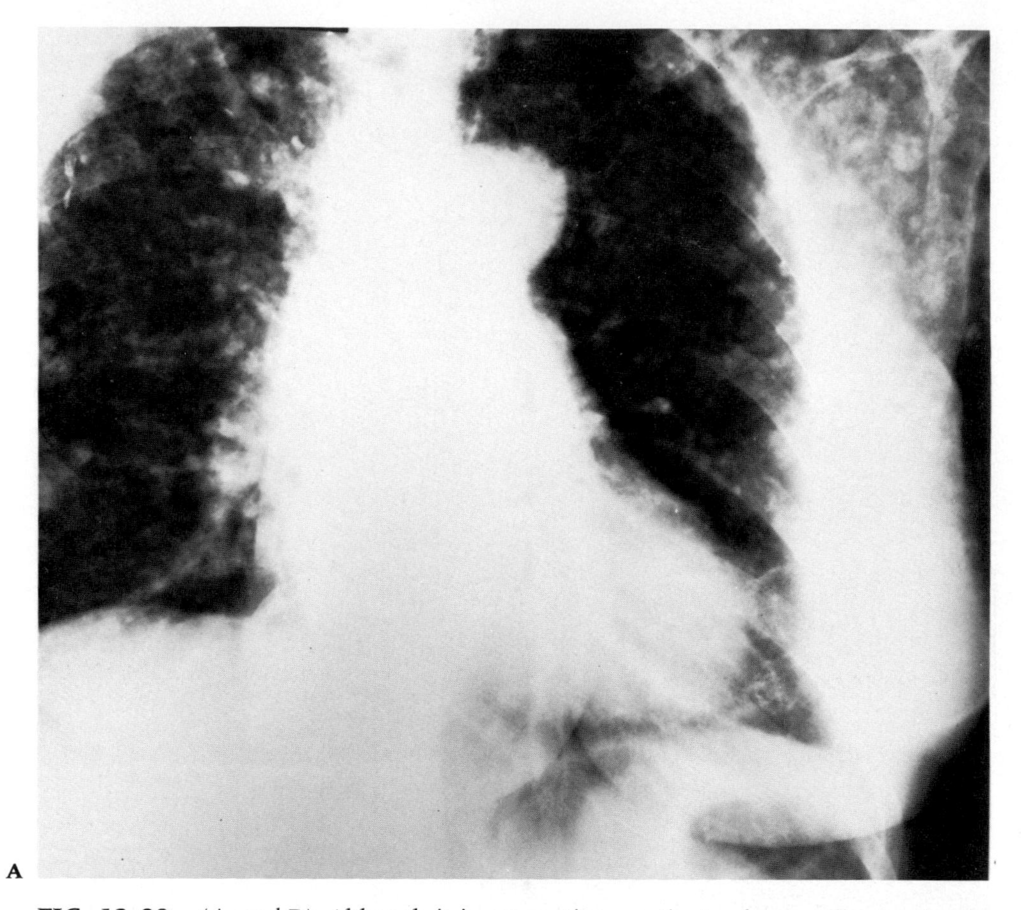

FIG. 13-29. (A *and* B) *Although it is not routine practice to place a pillow under the anatomic region of interest, it is quite possible that a pillow that is supposed to be nonradiopaque will somehow appear in a radiograph, as it does in these two films. Most manufacturers indicate that the filler in their pillows is nonradiopaque, but I am sure that if you inspect the various pillows that have found their way into your department over the years, you will find that several contain radiopaque materials. In order to eliminate such artifacts, all new pillows should routinely be radiographed or viewed under the image intensifier. (See Nemgar W: On the Technical Side. Radiol Technol 51:666–668, 1980)*

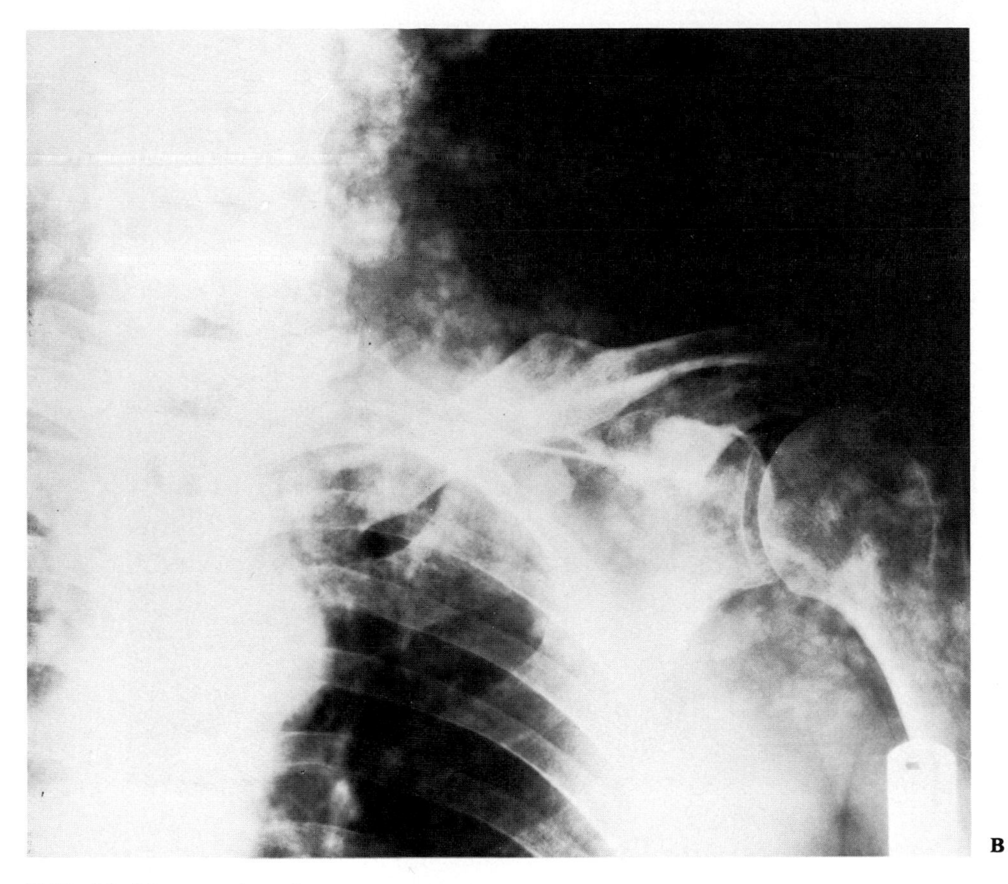

FIG. 13-29. *continued*

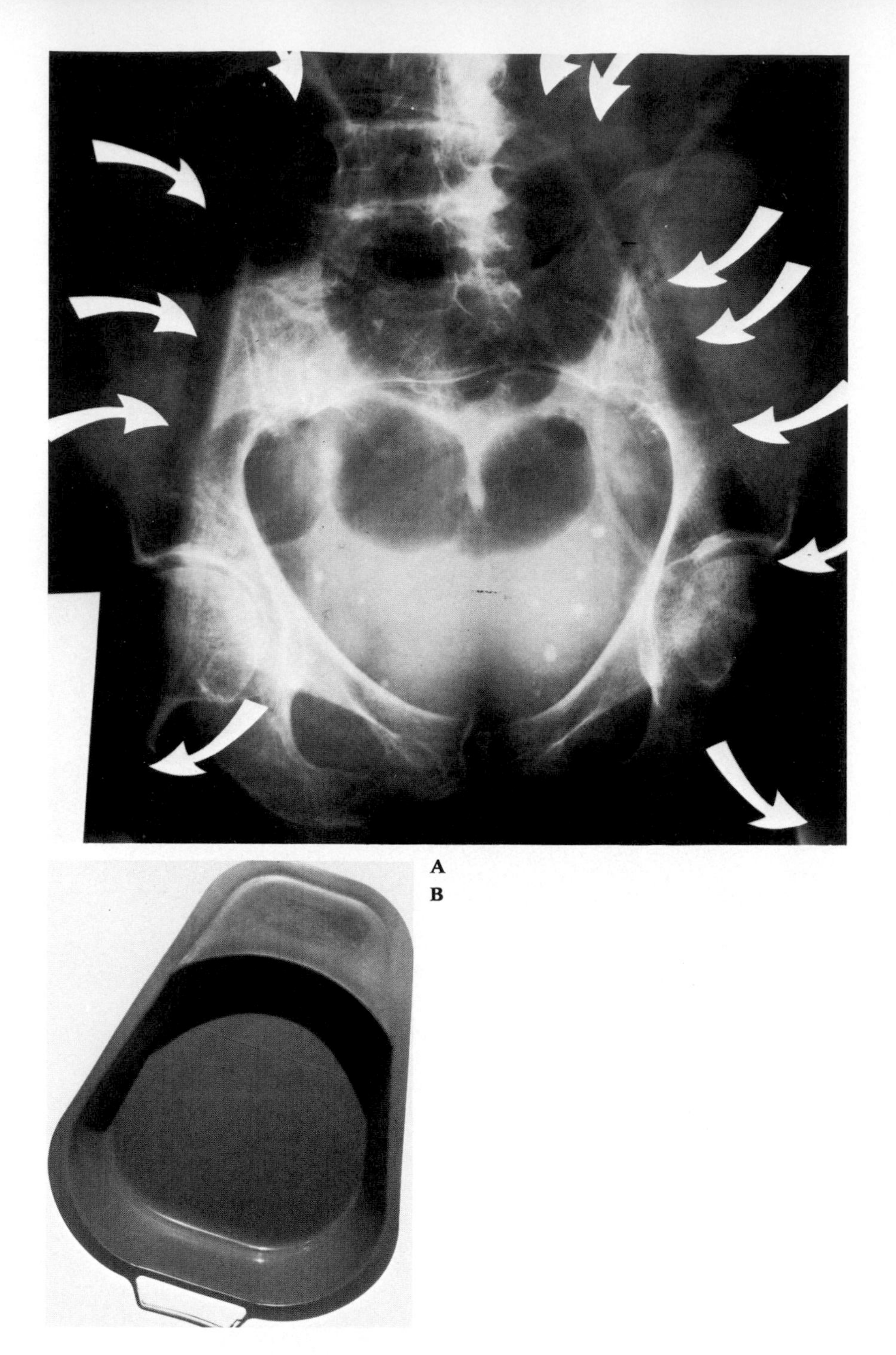

FIG. 13-30. (A) *A number of very unusual objects have appeared in the various illustrations in this book, so by now you should not be surprised at anything you find in a radiograph. The image outlined by the* arrows *in* A *is a bedpan* (B) *that was left under the patient during the radiographic examination. The height and size of this pan differ from those of the standard type. It is used when the patient has a fracture or some other type of disability. (*A *and* B*, courtesy of Ron Much, R.T. [R])*

Identify the Artifact: Solutions

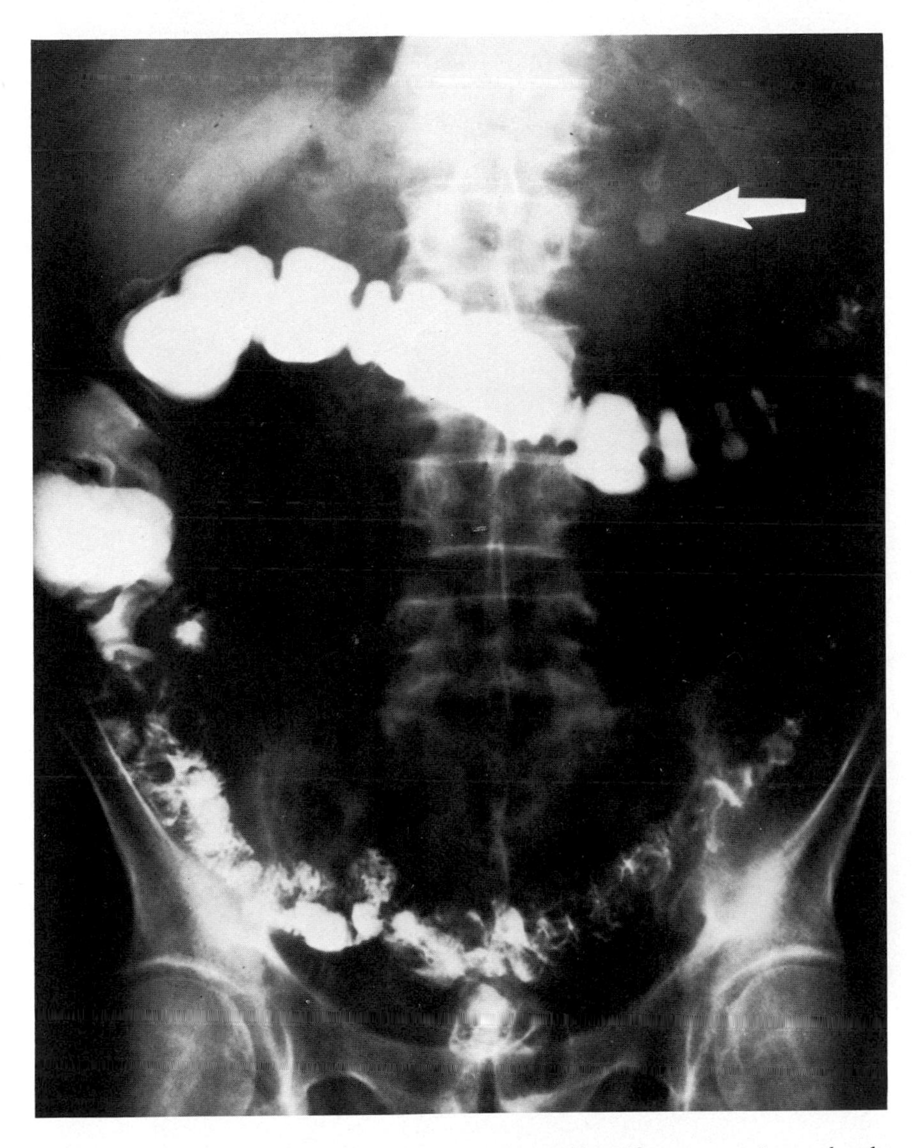

SOLUTION TO FIG. 13-1. (A) *The two small artifacts* (arrow) *are droplets of barium* (C, *close-up) located on the plastic cover under the surface of the radiographic tabletop* (B). *During certain fluoroscopic procedures, the tabletop becomes a convenient place to set a barium cup, a spoon, or a straw. This often results in spillage of barium onto the surface of the plastic cover. By exercising greater care during these procedures, as well as by periodically inspecting the cover for barium deposits, the technologist can eliminate such problems. A word of caution: the purpose of the cover is to protect various mechanical and electrical components from barium spillage, so be sure that the main circuit breaker is off before attempting any type of cleaning procedure.* (A *through* C, *courtesy of Dr. Paul O'Connor)*

(Solution to Fig. 13-1 continues.)

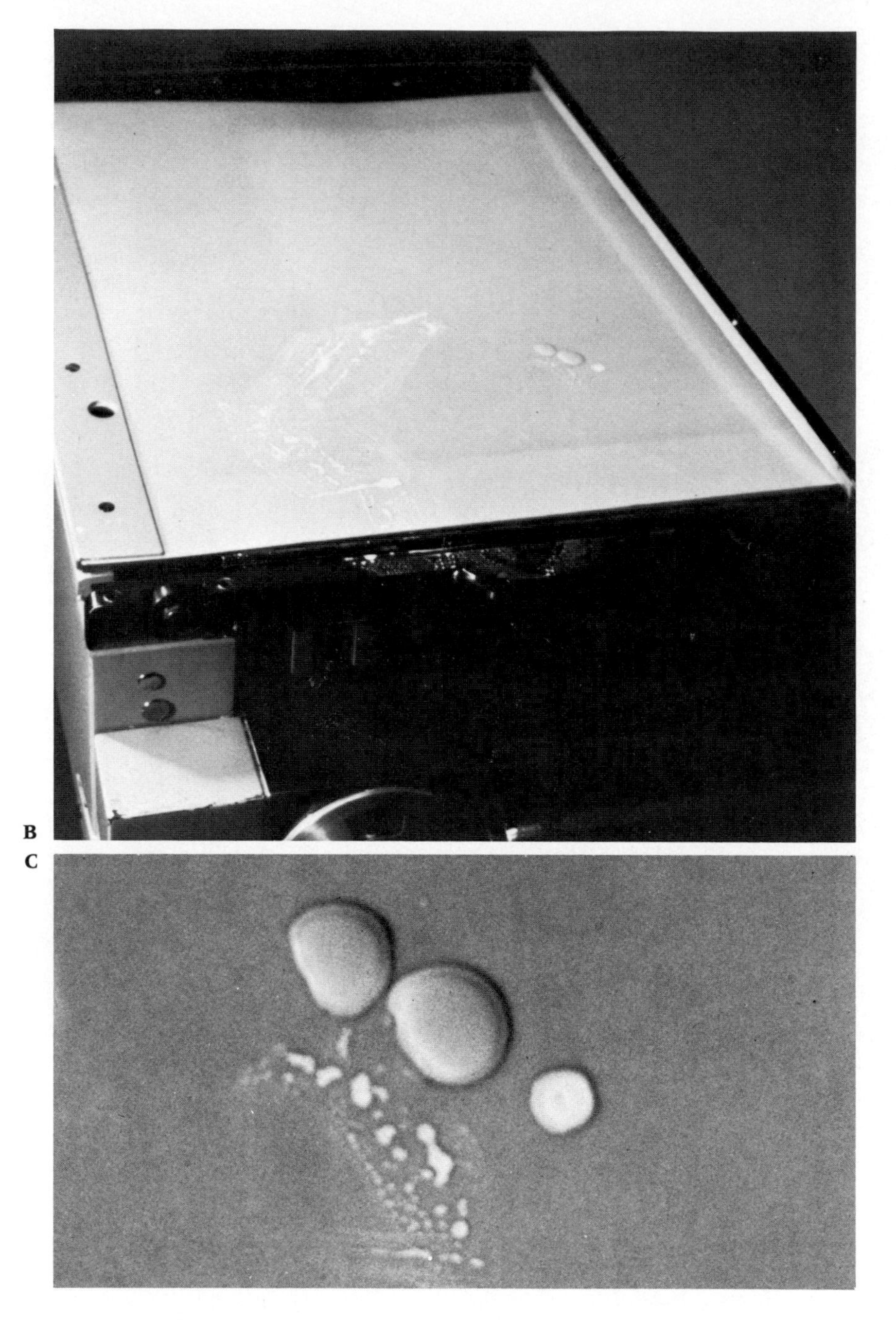
B
C

SOLUTION TO FIG. 13-1. *continued*

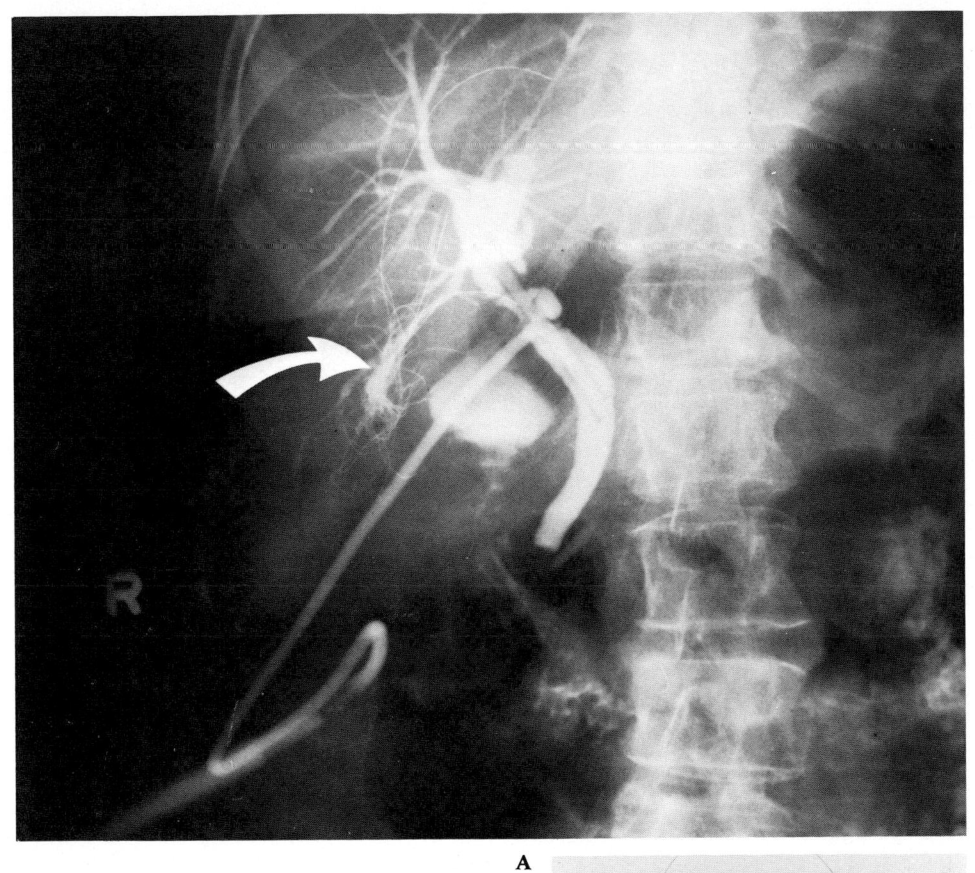

A

B

SOLUTION TO FIG. 13-2. (A) *The T-tube cholangiogram demonstrates a rather unusual image* (arrow) *in the biliary tree.* (B) *The artifact responsible for the problem is a clump of matted hair that was on the surface of the intensifying screen during the exposure.*

Suggested Reading

PHANTOM RENAL CALCULUS

Caffey J: Pediatric X-ray Diagnosis, Vol 2, pp 1582–1583. Chicago, Year Book Medical Publishers, 1972

Poznanski AK: Practical Approaches to Pediatric Radiology, a, p 12; b, p 13; c, p 18. Chicago, Year Book Medical Publishers, 1976

RETAINED SURGICAL SPONGES

Olnick HM, Weens HS, Rogers JV: Radiological diagnosis of retained surgical sponges. JAMA 159:1525–1527, 1955

Spiegel SM, Palayew MJ: Retained surgical sponges: Diagnostic dilemma and an aid to their recognition. RadioGraphics 2, No. 1:53–68, 1982

Williams RG, Bragg DG, Nelson JA: Gossipyboma—the problem of the retained sponge. Radiology 129:323–326, 1978

GEOPHAGIA (DIRT-EATING)

Gardner JE, Tevetoglu F: The roentgenographic diagnosis of geophagia (dirt-eating) in children. J Pediatr 51:667–671, 1957

Halstead JA: Geophagia in man: Its nature and nutritional effects. Am J Clin Nutr 21:1384–1393, 1968

Maravilla AM, Berk RN: The radiographic diagnosis of pica. Am J Gastroenterology 70, No. 1:94–99, 1978

Mengel CE, Carter WA: Geophagia diagnosed by roentgenograms. JAMA 187:955–956, 1964

INDEX

Numbers followed by the letter *f* represent figures.